# The Sea-Change

# The Sea-Change

## *The Worldwide Hunt for the Ancestors of Animals*

GRAHAM BELL

*Emeritus Professor, Biology Department, McGill University*

OXFORD
UNIVERSITY PRESS

OXFORD
UNIVERSITY PRESS

Oxford University Press is a department of the University of Oxford.
It furthers the University's objective of excellence in research, scholarship,
and education by publishing worldwide. Oxford is a registered trade mark of
Oxford University Press in the UK and in certain other countries.

Published in the United States of America by Oxford University Press
198 Madison Avenue, New York, NY 10016, United States of America.



CIP data is on file at the Library of Congress.

ISBN 9780197797433

ISBN 9780197797426 (hbk.)

DOI: 10.1093/9780197797464.001.0001

Paperback printed by Integrated Books International, United States of America

MIX
Paper
FSC FSC® C183721

The manufacturer's authorized representative in the EU for product safety is
Oxford University Press España S.A. of Parque Empresarial San Fernando de Henares,
Avenida de Castilla, 2 – 28830 Madrid (www.oup.es/en or product.safety@oup.com).
OUP España S.A. also acts as importer into Spain of products made by the manufacturer.

*For the faculty and staff of the Redpath Museum 1995–2023.*

*Full fathom five thy father lies;*
*Of his bones are coral made;*
*Those are pearls that were his eyes;*
*Nothing of him that doth fade*
*But that doth suffer a sea-change*
*Into something rich and strange.*

Shakespeare, *The Tempest*, Act 1 Scene II

# Contents

But I should not deceive you. This story belongs to one of the least satisfying of literary genres, a detective story without a final resolution. It is not capped by a meeting in the library of the old manor where all the suspects are assembled and the truth at last revealed (although I shall contrive something similar). In the end, it will be up to you to make up your mind about one of the most puzzling and absorbing mysteries in the whole of science.

# Introduction

## Darwin's Difficulty

The mounted skeleton of a large bipedal dinosaur called *Albertosaurus* stands just outside my office in the museum. Squinting through it, I can just make out one of the vitrines on the opposite wall of the gallery. It exhibits a group of fossils brought back to Montreal in the 1920s from a site on the Alberta–BC border by Thomas Clarke, who was a graduate student at the time but later became the director of the museum where I work, the Redpath Museum of McGill University. He was helping with the Harvard Field School in the summer of 1924, when he came across a party led by Charles Walcott, an eminent American geologist, who was searching for fossils in a bed of shale high on a hillside in the Rockies. Walcott had discovered the site—well, it is said that his wife, Helena, first noticed the fossils—several years previously, and he had returned regularly ever since to collect the extraordinary fossils preserved in the rocks there. The fauna of the Burgess Shale became well known to palaeontologists and later became familiar to a wider public through Steven Jay Gould's book *Wonderful Life*. Nowadays it is a World Heritage Site with a visitor information centre where visitors are sternly forbidden to collect any fossils at all. In the 1920s, however, there was a more casual attitude to fossils, and after Clarke had joined Walcott's party he was able to bring back several cases of specimens to McGill, where they were accessioned into the permanent collections of the museum. Some of the material that he collected nearly a century ago is now on display in our main gallery, just visible from my office through the ribcage of the intervening dinosaur. It illustrates the extraordinary radiation of animals which occurred during the Cambrian period of Earth history, a little more than 500 million years ago (Mya).*

The fossils are not much to look at. They are certainly nothing like as dramatic as the dinosaur. Most of them seem to be little more than smudges on the rock that you or I might pass by without a second glance. When they were examined more closely, however, they revealed that most of the main kinds of animal that live today had already appeared at the time of this remote epoch. I should qualify that bald statement before someone accuses me of exaggeration. Most of the main kinds of animal that live in the sea, that is, were already present in the Burgess Shale fauna: animals and plants spread onto the land only much later. And only animals that were large enough, or had hard skeletons of some sort, such as the carapace of a crab or the shell of a snail, are likely to have been preserved as fossils. With these reservations, an impressive range of different kinds of animal left their remains in the shale.

* Note: 'My' means 'million years'; 'Mya' means 'million years ago'. Thus, '521 Mya' is a date, meaning '521 million years ago'.

*The Sea-Change*. Graham Bell, Oxford University Press. © Graham Bell (2026).
DOI: 10.1093/9780197797464.003.0001

We can start with the trilobites, since these have long been the iconic fossils of the Cambrian. A typical trilobite was about the size of a matchbox, with a body divided into three longitudinal sections (hence the name): a central part that contained most of the clockwork, with lobes to right and left bearing a fringe of walking legs. The head was covered by a large shield, with a mouth below and a pair of compound eyes above. Not all trilobites were exactly like this: some were smaller, others much larger; some were blind. All were arthropods, distantly related to crabs and shrimps, and thus entirely covered—head, body, legs and all—with a tough coat of chitin, which protects them against predators, and incidentally resists decomposition, so they are very common as fossils. They first appeared in the fossil record about 520 Mya and flourished mightily for long ages—more than 10,000 species of trilobite have been described—before declining in numbers and diversity, and finally vanishing during the great extinction of marine animals at the end of the Permian period, about 250 Mya.

Trilobites were accompanied by a range of other kinds of animal. Some belonged to familiar modern groups, such as crustaceans and molluscs, and looked much like the shrimps and snails you might find at the seaside today. Brachiopods are less familiar today, although they were abundant in the early seas; on the outside they look much like a clam, with a shell of two calcified valves that fit tightly together, although their internal anatomy is quite different. The earliest echinoderms appeared at about this time, although they looked nothing like the starfish and sea urchins of modern seas; one kind looked like a rugby football tattooed with a set of spiral grooves, while others had a boxy body with a long tapering proboscis, looking something like the neck of the Guinness ostrich. Other groups that appear around this time, such as chaetognaths and priapulids, have survived down to the present day, but are familiar only to zoologists. There are also animals which flourished at the time but have since disappeared as completely as the trilobites. Among the most abundant were the hyoliths, whose narrow conical shell was propped up at the head end by a pair of long curved arms. The anomalocarids were more dramatic, gigantic swimming predators with raptorial arms able to capture a shrimp or worm and thrust it into a mouth that looked like a camera shutter. By Burgess times, then, there were as many different kinds of animal on and above the seabed as you will find today in the Vancouver aquarium.

Something like this was known when geology began to coalesce as a science, say by the end of the 18th century. Once the bands of rock could be arranged one on top of another, by observing how they were ordered at a number of localities, the sequence in which different groups of animal appeared through time could be worked out. It was soon clear that whales, mammoths and horses were found only in the uppermost beds, large reptiles lower down, the first archaic fish further down still, and below that only invertebrates such as trilobites and brachiopods. Plants followed a similar sequence, with broad-leafed trees appearing only towards the top of the sequence, conifers a little lower down, ferns and horsetails among the earliest land plants, and only seaweeds in the earliest times. This might be explained, in principle, by a series of successive divine creations, each involving the replacement of some groups—peremptorily made extinct—by those appearing for the first time. This was the interpretation supported by almost all the most eminent geologists and palaeontologists during the first half of the 19th century. There were good reasons for this: it explained, for example, why new kinds of animals, such as turtles or trilobites, seemed to appear all at once, without any series of intermediate forms linking them with another kind. Some well-respected figures, however, were beginning to doubt that species were truly immutable. I imagine that the notion of a Creator who was frequently obliged to rub out His mistakes and replace them with newer models, which subsequently turned out to be equally mistaken, might not have been fully convincing. The alternative was a succession of different but overlapping faunas, driven by some unknown natural process that was capable, over a long period of time, of transforming one species of animal into another.

The first attempt to identify such a process was made by Jean-Baptiste Lamarck, working at the Muséum in Paris during Napoleonic times. He argued that the propensity to use a structure during the life of an individual would somewhat exaggerate the development of that structure in its descendants, leading to a gradual increase in harmony and complexity over time. This constituted a simple and general mechanism that would lead naturally to the successively higher grades of organization that geologists saw in the rocks. Lamarck failed to convince his contemporaries, however, who continued to follow the creationist views of Georges Cuvier, his colleague at the Muséum. Fifty years later, Darwin was not as easily pushed aside, although cogent objections were raised by many people to his theory that random variation would necessarily lead to the spread of superior strains of a species. Fleeming Jenkin, for example, was an engineer involved in laying the first undersea telegraph cables. He pointed out that natural selection was incompatible with blending inheritance, which was then the generally accepted theory of genetics. He allowed that a rabbit might be born with the ability to run a little faster, purely through random variation, that this ability was hereditary, and even that it would outrun the fox and be more likely to rear a litter of young rabbits itself. But it would have mated with a normal, unaltered rabbit, so its offspring would be intermediate in running speed. Whatever its own edge might have been in running speed, its offspring would only have half the edge, its grandchildren only one-quarter, and so on, until the advantage of the original variant had been diluted to inconsequence. Darwin never succeeded in finding a satisfactory answer to this objection (because he shared a belief in blending inheritance) and it was not resolved until the rediscovery of Mendel's theory of particulate inheritance in the opening years of the 20th century.

A more pressing objection, though, was that the continuous series of intermediate types that must link one species with another, if his theory were correct, seemed nowhere to be found. Where were they? Perhaps they had been eliminated by competition with their more highly modified descendants. Even so, they must have existed in the past, so the fossil record should consist largely of intermediate types in the process of being modified. Yet it was common knowledge that new groups of animals usually appeared quite abruptly, without being preceded by a recognizable ancestor and a long train of gradually more highly modified types descending from it. Darwin pointed out that the time taken for a new species, once formed, to spread widely was likely to be much less than the time required for its evolution by gradual modification in some restricted locality, so that it would seem to appear suddenly in a formation. I think it is fair to say, though, that the defence mounted in this section of the *Origin*, although it has turned out to be sound, might have seemed flimsy and speculative to many of his contemporaries. Moreover, there was one firmly established generalization that even Darwin despaired of reconciling with his general theory of evolution: the abrupt appearance of animals themselves.

Trilobites and brachiopods were often abundant in the oldest fossiliferous rocks known at the time. They are clearly quite large and complex animals which must have descended, according to Darwin's theory, from a long succession of less well-organized forms. And yet there was no trace of these intermediate forms in the fossil record. Below the lowest fossil beds stretched long ages in which no fossils could be found, bar a few dubious markings that might be seaweed impressions or worm burrows. Where were the missing ancestors? Did the ultimate ancestor leave the concrete evidence of its existence as a fossil in the rocks? What did it look like? When did it live? Can we find any little-modified descendant still alive today? Darwin did not know the answer to any of these questions, and simply threw in the towel. 'To the question why we do not find richly fossiliferous records of these vast primordial periods, I can give no satisfactory answer. . . . The case at present must remain inexplicable; and may be truly urged as a valid argument against the views here entertained'.

Now that these questions could be asked, however, they demanded answers, and the search for answers has led to the most romantic quest in the history of science, a long saga of speculation, error, confusion and discovery.

# A False Start

By the time the fourth edition of the *Origin* had appeared, the missing ancestor had been found; or so it seemed at the time. To be precise, it is located two floors below my office, where the collections of the museum are stored in banks of locked and hermetically sealed steel cabinets. Cabinet number 1 is crammed with specimens of the curious fossil *Eozoon*, collected in the Laurentian Hills north of Montreal by the first Director of the museum, Sir William Dawson. The museum itself was originally built to house his extensive collections of fossils, as a bribe from Peter Redpath, of the sugar family, to persuade Dawson to stay in Montreal rather than accept an offer from Princeton. The museum collections include hundreds of thousands of invertebrate fossils collected from the Montreal region and further afield, but Dawson always regarded *Eozoon* as the most important and even wrote a book, *Life's Dawn on Earth*, to clinch its unique status as the earliest known ancestor of animals.

The specimens of *Eozoon* in Cabinet number 1 (Figure 0.1) are rather handsome striped or mottled objects with alternating bands of green and white. They were first collected by two people, an amateur geologist in Ontario and a surveyor employed by the newly-formed Geological Survey of Canada working at Grand Calumet in Quebec. Both of them sent their specimens, which they regarded as no more than interesting compound minerals, to the Director of the Survey, William Logan. To his surprise, Logan found that the Ontario and Quebec samples, although similar in appearance, were formed of quite different minerals. This suggested to him that the specimens were not minerals at all, but rather the remains of a living organism whose original structure had been replaced by different minerals in different localities. He made this alluring suggestion at the meeting of the American Association for the Advancement of Science in 1859, just in time to anticipate Darwin's difficulty and propose a solution. But few of his colleagues were convinced, and the specimens slid into obscurity.

They were rescued by Thomas Sterry Hunt, a mineralogist at McGill who wrote a paper arguing that the specimens were indeed organic and showed it to Dawson. Dawson was impressed by the idea but thought that Hunt had got the minerals the wrong way round—the green bands were the serpentine that had replaced the cytoplasm, whilst the white bands were the calcite that was the original skeleton. With this alteration, the specimens began to look like a large version of a familiar kind of organism, a foraminiferan, or foram for short. But they had been found in rocks that were indubitably ancient, far older than the earliest trilobites, and must be the earliest known ancestors of animals. With the scent of a major discovery in his nostrils, Dawson collected many other examples from the Laurentians, a range of low hills north of Montreal.

*The Sea-Change*. Graham Bell, Oxford University Press. © Graham Bell (2026).
DOI: 10.1093/9780197797464.003.0002

**Figure 0.1** Eozoon from Cabinet 1 with label 'Good canals—very finest'
Courtesy of Anthony Howell.

Large benthic forams exist today, both in warm shallow bays and in the cold dark depths of the ocean. They are not animals, but rather gigantic single-celled organisms with many nuclei, although the distinction was not made clear until much later and Dawson's contemporaries thought of animals and forams as being much the same thing at different levels of structure. The body of a foram is enclosed in a calcareous shell and in many species is divided into numerous compartments with calcite walls, connected by a complex three-dimensional system of canals. It is these canals which, according to Dawson, form the stripes preserved in *Eozoon*. Forams feed mostly on bacteria and algae, which they capture with fine threads of cytoplasm, the pseudopodia or filopodia, extending from the body either through a single opening or through separate pores scattered over the surface of the test.

Dawson and Logan got together to present a paper at the meeting of the British Association for the Advancement of Science in 1864. This was not met with the indifference shown by Logan's American audience of 5 years previously, perhaps because Charles Lyell was complimentary about the paper in his Presidential address to the meeting. Moreover, Logan met W.B. Carpenter, the great British expert on forams, later that year and convinced him that the fossils were the ancient ancestors of his favourite organisms. The fossil now had a name—*Eozoon canadense*, given to it by Dawson—and a quartet of supporters, all eminent geologists and biologists. Dawson's view, that *Eozoon* was close to the ancestry of animals and evolved into sponge-like creatures, was becoming orthodox. The final accolade was given by Darwin, who took the discovery of *Eozoon* to confirm his conclusion that "during these vast, yet quite unknown, periods of time [before the Cambrian], the world swarmed with living creatures."

Although orthodox, it was not quite unanimous. One geologist who attended the British Association meeting was not convinced. William King was lucky to be a geologist at all. He grew up in Durham, in the North of England, as a working-class lad; his mother kept a sweetshop, and his father was employed in the coal mines. He did not even have any formal qualifications as a geologist, but he was fascinated by natural history and although he managed to land a job with the Hancock Museum in Newcastle he was sacked after a few years. Surprisingly, he was then appointed to a university position, albeit at Queen's College Galway, now the University of Galway, in the Famine year of 1848. Galway is on the west coast of Ireland, almost directly across from Dublin and a long way from the power centre in London. King was never a member of the London crowd, although he knew them—Lyell, for example, supported his application to Galway. He received no honours or prizes (unlike Dawson, Logan and Lyell, for example, who were all knighted), but on the other hand he was free to express his own opinion regardless of the feelings of others. His own opinion in this case was that the original diagnosis of *Eozoon* as a compound mineral was correct and that its resemblance to a foram was illusory.

King published a string of articles denying that *Eozoon* had ever been a living creature, starting in 1865. All of them were written with a colleague at Galway, Thomas Rowney, the Professor of Chemistry. Rowney did have the PhD that King lacked, which he had earned in Liebig's lab at Giessen, so he added some academic stiffening as well as reassuring the audience that the chemistry involved in the formation of *Eozoon* was credible. The basis of King and Rowney's scepticism was very simple: *Eozoon* was found only in metamorphosed rocks, originally laid down as sediment but subsequently altered by heat and pressure, and the most perfect specimens came from highly metamorphosed rocks. They bought samples of Connemara marble, a metamorphosed limestone, and convinced themselves that the *Eozoon*-like structures that it bore were entirely mineral in nature. They extended this interpretation to *Eozoon* material from Canada (sent to them by Logan, ironically), and published their conclusions, at first briefly in a popular magazine, the *Reader*, and then at greater length in a respectable scientific journal, the *Quarterly Journal of the Geological Society of London*.

These publications sparked one of the great scientific controversies of the 19th century. At first it was quite polite, but it soon became vituperative, especially once Carpenter had become involved. I have explained previously (in *Full Fathom 5000*) how Carpenter's temperament cost him the scientific leadership of the *Challenger* expedition, and gave this example, from Carpenter's 1865 rebuttal of King and Rowney's claims in the *Reader*:

> That Professors King and Rowney, however, should take upon themselves to affirm that the structures which have been described by Dr. Dawson, Professor T. Rupert Jones, and myself, in *Eozoon Canadense* are not organic, would, I own, have greatly surprised me, if I had not had previous experience of the audacity (I use the word advisedly, and am quite prepared to justify it if called on) with which Professor King hazards denials of statements made by men who have a scientific reputation to lose, in regard to matters which they have carefully investigated. Until Professor King (I say nothing of his colleague, since it is scarcely to be supposed that a Professor of Chemistry should claim authority on Microscopic Palaeontology) shall have given some proof of his competence to estimate the value of evidence in this branch of

> scientific inquiry, I must rank him in the same category with those sagacious persons who still maintain that the flint implements were shaped out by a fortuitous succession of accidental blows, and not by human handiwork.

This is hardly the sort of calm, reasoned refutation that one might expect of an eminent scientist. King and Rowney were not sunk by this broadside, however, but persisted in their audacity until Carpenter eventually published his last word on the subject, in 1874, and refused to discuss it further. Dawson published his book-length defence of *Eozoon*, *Life's Dawn on Earth*, in the following year. It was answered in a withering review from King and Rowney in which they disputed almost all of Dawson's conclusions, from the microscopical structure of *Eozoon* to its supposed evolution into sponge-like animals. Other storm clouds were gathering; in particular, two prominent German scientists, Otto Hahn and Karl Möbius, came out in support of the mineral interpretation. (Hahn later changed his mind and decided that *Eozoon* was a seaweed; he wrote to Dawson in 1878, 'we were both wrong'. I cannot trace any reply made by Dawson.) The final blow, according to most writers, was delivered by Henry Johnston-Lavis, who found *Eozoon*-like structures in blocks of limestone ejected from Vesuvius, which would certainly be conclusive, although I have not been able to track down the article itself in which this claim is made. It probably made little difference; the climate of scientific opinion had shifted decisively towards King and Rowney's interpretation by the closing years of the 19th century, and the contents of Cabinet number 1 are no longer regarded as the witnesses of life's dawn on Earth, but merely as a collection of curiously striped pebbles.

The *Eozoon* affair has elicited a great deal of finger-wagging from different points of view. Some emphasize the clash between insiders and outsiders; others contrast metropolitan and provincial scientists, or warn us to rely on observation rather than authority. 'Lowly Irish profs trash stuck-up London toffs' might be a simplistic headline but I think conveys the tenor of most commentaries. I would prefer to underline two more technical lessons from the controversy. The first is that both sides were perfectly sincere in their beliefs. Carpenter was an expert on forams and when he looked at *Eozoon* he saw the compartments and canals of a group with which he was thoroughly familiar. Dawson was a renowned geologist who had discovered thousands of fossils and when he looked at *Eozoon* he saw the ancient equivalent of later forms with which *he* was thoroughly familiar. William King, on the other hand, saw an orderly arrangement of crystals caused by a purely physical process of segregation. With the benefit of hindsight it is obvious that King was right whereas Carpenter and Dawson were deluded. It might not have been quite so obvious at the time; and there are much more recent examples of mineral layers being mistaken for the remains of tissues—and contrariwise, examples of genuine fossils being dismissed as the outcome of purely physical processes. It is sometimes assumed that *Eozoon* is an exception to the general rule that a fossil can be immediately recognized for what it is, but this is not true. Many ancient fossils stir debate, not only about their precise identity, but even about whether they are really fossils at all.

My second point is that the articles on both sides of the controversy were illustrated by drawings, rather than by photographs. It was towards the end of the period when this was normal practice, and it has a good deal to be said for it. Details that would be difficult to illustrate with a photograph can readily be made plain by a properly designed drawing. The plastic property of drawings, however, can lead to a hypothesis leaking imperceptibly into an observation. Dawson and Carpenter's drawings depict the tissue layers and internal spaces of an organism. King and Rowney's drawings show, with equal felicity, arrangements of crystals. It is clear that, without any conscious intent to mislead, each drew what they expected to see, at least as much as what they actually did see. Photography put an end to this, at any rate for a time; now that it is so easy, and so usual, to manipulate photographs, we may see this old problem reappearing.

Anyway, *Eozoon* was just too old. The Laurentian hills were laid down hundreds of millions of years before the Cambrian, and the intervening period was (at the time) devoid of fossils. Darwin might have viewed *Eozoon* as evidence in favour of his hypothesis of natural selection, but Dawson did not see it that way at all. He knew that evolutionary change happened—he was a geologist, after all, and discovered both the lizard-like animals entombed in the lycopod stumps at Joggins and the first terrestrial plants at Cap Forillon—but he believed in original Creation and the guiding hand of God rather than the dice-throw of random variation and natural selection. For him, *Eozoon* was conclusive evidence that his point of view was true. Again, with hindsight it is easy to conclude that Darwin was right and Dawson wrong. But Darwin was right for the wrong reasons: *Eozoon* was much too early to be a credible ancestor of animals, and the problem of the enormous thickness of barren rock remained. Darwin's difficulty was as acute as ever.

# 1
# The Golden Spike

Almost 30 years before the publication of the *Origin*, Darwin had recently completed his undergraduate studies, and was accompanying Adam Sedgwick, the Professor of Geology at Cambridge, on a geological tour of North Wales in August 1831. They were looking for the Old Red Sandstone that was shown on an early geological map of the area, because it would provide an upper reference point for the older rocks of the Welsh mountains. In the event, they failed to find it (the map was wrong), and after tramping around the countryside for the best part of a month Darwin went home to Shrewsbury at the end of August. He was intending to go partridge shooting, but instead found a letter waiting for him from John Henslow, who had taught him botany at Cambridge. Henslow had declined the offer of a position as naturalist on HMS *Beagle* for a surveying voyage to South America (his wife disliked the prospect of 2 years' separation) and instead suggested that Darwin would be a suitably well-bred companion for the captain. After a little argy-bargy Darwin set sail in the *Beagle* just after Christmas, and in doing so set in motion a train of events that eventually led to his little difficulty with the abrupt appearance of animals in the fossil record at the beginning of the Cambrian period. Sedgwick, however, stayed on in the Welsh hills until he was driven out by the Welsh weather in October. It was here that he would later identify and name the Cambrian period itself, although it would be accepted by the world at large only after a long and rowdy controversy.

## The great Cambrian controversy

Sedgwick's colleague, fast friend and, eventually, bitter foe was Roderick Murchison, who came from a wealthy Scottish family. Neither was trained as a geologist; Murchison had taken up geology only after Waterloo (and the ensuing peace) had deprived him of any hope of a glorious military career, and Sedgwick seems to have been appointed to the Woodwardian chair for little more than clubbability. Both turned out to have a distinct flair for the subject, however, and worked tirelessly in the field and in the study. A collaboration that began with a field trip to Scotland in 1827 blossomed into a friendship that lasted many years and led to a regular exchange of letters between London and Cambridge. They took quite different approaches to the problem of dating the rocks. Sedgwick was a structural geologist who saw the landscape as a complex three-dimensional jigsaw puzzle made up of patches of superimposed rock, whereas Murchison used fossils to understand the succession of communities through time. This may explain why, although both were working in Wales in the summer of 1831, Sedgwick was contemplating the immense thickness of blank slate in

*The Sea-Change*. Graham Bell, Oxford University Press. © Graham Bell (2026).
DOI: 10.1093/9780197797464.003.0003

North Wales, while Murchison was exploring the sedimentary rocks of South Wales, crammed with fossils.

Their fieldwork in Wales led to a joint paper at the 1835 meeting of the British Association which separated the lower deposits from the upper. Sedgwick named the period when the rocks of North Wales were deposited after the Roman name for Wales, Cambria. Murchison, rather ironically, used Silurian for the later period in South Wales, after the local tribe who resisted the Romans after the Claudian invasion of 43 AD. The uppermost beds of the Cambrian contained fossils, although they were often sparse, and Sedgwick, emphasizing the physical structure of the rocks, gave few details; the lowermost beds were devoid of fossils. The Cambrian period, therefore, as originally defined by Sedgwick, covered the whole time when animals first appeared in Earth history, while the Silurian period, as originally defined by Murchison, saw their subsequent diversification. Order had been brought forth from chaos, and the two geologists responsible could bask in the admiration of their colleagues. There was a snag, however: Sedgwick was unable to identify a distinctively Cambrian community. Worse was to follow. When the fossils he had found in North Wales were examined by experts, they turned out to be remarkably similar to those from the Silurian. This encouraged Murchison to extend his Silurian period downwards, eventually to the very base of the fossiliferous deposits. The Cambrian had been cancelled, or at best reduced to a minor subdivision of the lower Silurian. The old friendship began to fray. The Geological Society attempted to resolve the issue by devoting a whole meeting in 1852 to the vexed separation of Cambrian from Silurian, but this only led to increasingly intemperate exchanges and the final breakdown of communication between the two principals. The general opinion of geologists, strongly influenced by the fossil evidence, shifted towards the expanded Silurian. Murchison's triumph was signalized by his appointment as director of the Geological Survey in 1855, after which he controlled the names (and the distinctive colours) of formations shown on the geological map of Britain.

At the same time, fossils were being found in Cambrian rocks outside Britain, especially in the Prague basin, where Joachim Barrande described a distinctive group of trilobites from the lowermost beds. Murchison's response was simply to shift the base of the Silurian deeper. Opinion began to change again, however, this time towards the recognition of a separate Cambrian period based on its characteristic fossils. The debate rumbled on, until it was settled in the late 1870s by a compromise offered by Charles Lapworth, another figure who began in another walk of life (he was a schoolmaster in the Scottish borders) and took up geology as a hobby. Lapworth's compromise was to set up a new period between the Cambrian and Silurian, which he named after another fierce Welsh tribe, the Ordovices. The Ordovician period saw the evolution of a characteristic shallow-water marine fauna—much of the country rock in the region where I live is an Ordovician shale which is often thickly strewn with the disarticulated remains of brachiopods, bryozoans and crinoids. The new period won general acceptance only rather slowly, but eventually ended the long Cambrian debate. Sedgwick and Murchison, who began the controversy, did not live to see its resolution; Murchison died in 1871 and Sedgwick in 1873. The world of gentleman geologists died with them, and future arguments about geological nomenclature would be conducted by professionals. This would not make them less acrimonious.

Perhaps all the fuss had been a waste of ink anyway. Time is continuous, after all, and chopping it up into periods on the basis of an opinion, however well informed, seems certain to provoke dispute. It has one obvious use: if someone refers to a deposit as being, say, lower Ordovician, then geologists around the world will be able to relate it to their local rocks. There is also a streak of egotism, however. The first person to name a period attaches their own name to some tens of millions of years of Earth history, and claims a permanent place in the annals of geology. I am sure that this was a powerful motive in the Cambrian controversy, at least for Murchison. The days of naming geological periods is not yet quite over: the latest candidate is the Anthropocene, proposed by an atmospheric chemist, Paul Crutzen, to recognize the effect of human activities on physical and biological processes, but not yet formally accepted. The older (and much longer) periods are now fixed, however, and unlikely to change. The Cambrian is the oldest period of the Palaeozoic, the age of old life, which continues with three periods named for British localities, successively the Ordovician, Silurian and Devonian periods, before branching out into the Carboniferous (when the coal beds were formed) and the Permian (after Perm, a Russian city about 1,000 km west of Moscow). It lasted about 300 million years, but was eventually brought to a close because time is not, after all, perfectly continuous but is instead punctuated by very rare events of very large magnitude. The event that closed the Palaeozoic is still uncertain, but whatever it was caused the extinction of entire groups of animals, including the last of the trilobites. It was succeeded by the Mesozoic, when the dinosaurs evolved, which lasted almost 200 million years before it was brought to an end 65 million years ago by another catastrophic event, the asteroid impact which again wiped out whole groups of animals, including the dinosaurs. We are now living in the Cenozoic, the age of recent life, the very latest phase of which we have named after ourselves.

## The base of the Cambrian

Once the Cambrian had become a fixture in the geological column it ceased to be a subject for debate. The issue was not dead, however, but only sleeping. It sprang up again like trampled heather during the 1960s, when very strange non-trilobite fossils began to be reported from lower Cambrian deposits in Australia and Siberia. Perhaps this might have been expected. After all, trilobites are large, complex and clearly highly-organised animals which must have been the outcome of millions of years of evolution from simpler ancestors. Some of these predecessors—I'll describe them later—had now been found. How much deeper into the rocks did we have to dig before all traces of animal life vanished and the base of the Cambrian had been reached?

This was not a question that could be answered by any single figure, however eminent or charismatic. Geology was by now an international endeavour, which numbered thousands of professionals spread throughout the world. Their trade union is the International Union of Geological Scientists, which has taken upon itself the authority for deciding the nomenclature and duration of geological periods. The base of the Silurian period, for example, is currently defined as the first appearance of the graptolite *Akidograptus ascensus* at Dob's Linn, Scotland (graptolites are colonial animals which floated at the surface of the sea in Palaeozoic times). Crucial horizons like this are traditionally marked with a golden spike driven into the rock as a permanent record of the boundary. I suspect that the original golden spike marked the completion of the transcontinental railroad by joining the Central Pacific and Union Pacific railroads in 1869. (The Canadian Pacific Railroad was likewise marked by a final spike in Roger's Pass, British Columbia, when it was completed in 1885; but being Canadian, it was made of iron like all the rest.)

Where should the golden spike marking the base of the Cambrian be hammered in? The first step, obviously, was to set up a broadly representative committee whose task would be to choose a particular horizon in a particular formation to mark the base of the Cambrian. This was the International Subcommission on Cambrian Stratigraphy, which deliberated for a decade or so after its formation in 1960 before giving birth to the Working Group on the PreCambrian–Cambrian Boundary. This was the committee which was to decide whether any particular deposit was laid down in the Cambrian, or whether it was older, in which case it was PreCambrian. The Working Group met in Paris in 1974 and decided that the newly-described Australian fauna (which we'll meet in Chapter 8) was PreCambrian, that any kind of early trilobite was Cambrian, and that fossiliferous sections between the two were the place to locate the boundary. This set of truisms might not seem to be very impressive progress for a decade of work, but experience has shown me that getting a score of opinionated scientists to agree on anything at all is impressive enough.

The rest of the decade was spent in organizing field trips to promising localities from the tundra to the tropics, until these were whittled down to a shortlist of three when the Working Group met in Bristol in 1983: a site on the Aldan River in eastern Siberia, a section at Meishucun in southern China, and exposures on the southern coast of Newfoundland. The Working Group decided in favour of the Siberian locality, not surprisingly in view of the variety of pre-trilobite fossils found there. Rather more surprisingly, their advice was rejected by the Subcommission itself. Ideally, the boundary between two periods should lie in a smooth transition from one to the next, since any unconformity—a break in the rock succession—might represent an indeterminate length of time. The first appearance of the pre-trilobite fossils on the Aldan River was suspected to be an unconformity, which would disqualify the locality for the golden spike accolade. Opinion swung towards the Meishucun section, but this had problems of its own, including a possible unconformity.

It was becoming clear that using body fossils, the actual remains of animals living at the time, was not as straightforward as it might seem: there were lots of different sorts, whose taxonomy was complicated and controversial, many had uncomfortably broad ranges in time, and it was often difficult to identify the same horizon in widely separated localities. Anyway, they became steadily scarcer in more ancient rocks. These difficulties shifted attention towards trace fossils, structures such as trails and burrows which reveal the activity of animals without preserving any part of their bodies. As it happens, these are common on the Newfoundland cliffs, and brought the outsider in the race to the front of the field. It was all a little like a geological Olympics, with a golden spike in place of a gold medal. The three runners had all held the lead briefly, and now raced for the tape. The finishing line was a meeting of the Working Group held shortly after the Third International Conference on the Cambrian

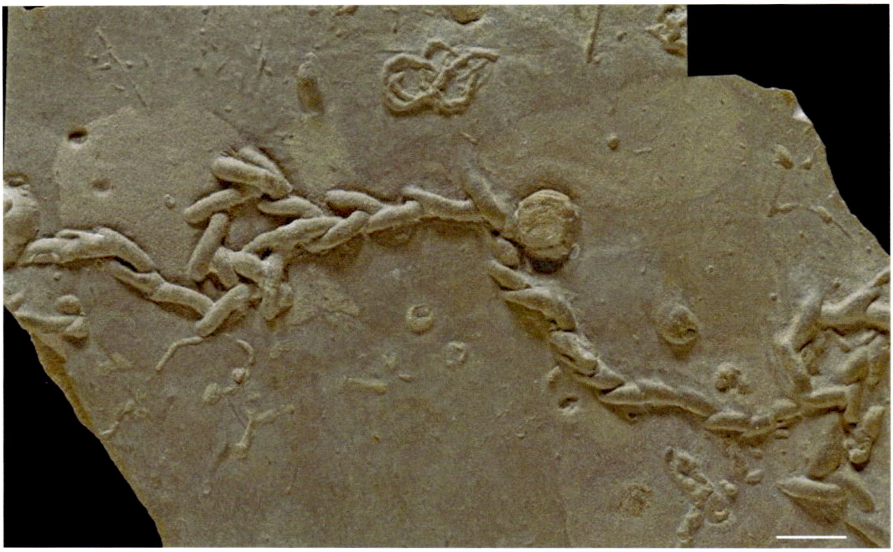

**Figure 1.1** *Treptichnus*

Reproduced from Royal Ontario Museum, © University Lyon.

System, which was held in Novosibirsk in 1990 and featured field trips to the Aldan River. There was, quite naturally, a very large Soviet delegation and a distinct emphasis on the Siberian fossils, but it would be mischievous to suggest that there were any national or political pressures being applied, any more than there are in the Olympic Games themselves. Any that were attempted were in any case ineffective, because the outsider held on to win by a nose. The Working Group voted by the barest possible majority—12 in favour, 11 against—to mark the base of the Cambrian as the earliest occurrence of the trace fossil *Treptichnus pedum*, a kind of fossil burrow, at Fortune Head, Newfoundland (Figure 1.1).

There have been fairly vociferous objections to this decision, and even plaintive calls for a stewards' inquiry to reconsider the whole business, but so far these have been unavailing. To an outsider (almost all of us) the strangest feature of the affair is that an important and long-lasting scientific problem should be decided by a *vote*, as if a town council was deciding where to put the dog run. Surely this was more than a demarcation dispute? Besides, the abrupt appearance of animals at the beginning of the Cambrian was the original reason for the debate, whereas the base of the Cambrian has now been defined as the earliest signs of animal life. Surely this is circular? And the current definition, as the earliest occurrence of a trace fossil: what if *Treptichnus* were subsequently found lower down—does the boundary then move down in step? The answer to all these questions is 'No'. The scientific problem is that animals seem to appear suddenly in the fossil record, with no sign of simpler ancestors, and the word "Cambrian" is a convenient but wholly artificial device for referring to the period of time during which animals appeared and diversified. The invention of the word sparked not one but two demarcation disputes, with Sedgwick and Murchison arguing about the upper boundary and the members of the Working Group about the lower boundary. Defining the lower boundary of a period, or any shorter length of time, as the earliest occurrence of a fossil is commonplace but risky, because a lower limit can move in only one direction, downwards. Sure enough, *Treptichnus* has since been found at lower levels, and logically the boundary should move down each time a new discovery is made. But it doesn't, because the rules (made by a committee) give stability a higher priority than logic.

Consequently, when you next find yourself on the Burin Peninsula, on the south-eastern coast of Newfoundland round the corner from St John's, you will find the official base of the Cambrian at Fortune Head. The path is difficult (it is best to take a guided tour from the Interpretation Centre) and you should be prepared to be disillusioned. In fact, there is no spike, golden or otherwise. Guy Narbonne, who led the Canadian team, persuaded the government of the province to protect the site by declaring it an Ecological Reserve, and one of the rules is that no permanent structure can be put in place, not even a spike. It seems a pity, but the guide will point to the place where it should be, the band of rock where indisputable signs of animal life first appear in the fossil record.

# 2
# Mud

The crucial horizon at Fortune Head consists of parallel layers of mudstone, showing that the rock was originally laid down as a muddy sediment in a shallow sea, close to shore but below wave base. Mud is not a glamorous substance. Most people try to avoid it, or, if unsuccessful, scrape it off. The word conjures up images of wet fields, sloppy paths and trench warfare. In my part of Canada, Spring is often called 'mud season', and this is not meant as a compliment. The few who appreciate mud, however, will point out how complex it is. Rock that is exposed on land will be split into boulders by heat, frost and plant roots; the boulders are broken up into pebbles when they are rolled about in streams, and the pebbles will eventually be reduced to sand. Even sand can be further comminuted, first to silt and eventually to the tiny flecks and flakes that clay is made of. Mud is a mixture of sand, silt and clay in any moderate proportions, made up into a paste with water. It may be cohesive enough to be squeezed into a ball (lots of clay) or so liquid that it squirts between the fingers (too much sand). Mud is not a simple compound, like salt or baking soda; a single sample may contain compounds of silicon, aluminium, iron, calcium, potassium, magnesium, sodium, sulphur and titanium, besides lesser amounts of other minerals. Its diversity and ubiquity combine to make it, in a sense, invisible. My museum has an extensive collection of minerals, and hand specimens of hundreds of them are displayed in the vitrines of the public galleries. There is no example of mud.

Mud is swept from the land by rivers and deposited on the continental shelf, where it accumulates on the seabed. The older layers are slowly buried by fresh material and become compacted into stone as the pressure increases and the water is squeezed out. They eventually become sandstone or siltstone or shale, depending on grain size, or more generically mudstone. When this is exposed on land, usually after the sea level has dropped or tectonic plates have bumped into one another, it will be split into boulders by heat and frost, and the process of weathering, erosion and deposition repeats itself. Mud is thus the juvenile stage of rock, or, more precisely, mud is the intermediate between two generations of rock. All our houses, temples, museums, walls and roads will eventually become mud, before, in the unimaginable future, they form a thin layer of new rock.

All this will take time, although it was not clear at first how much time was required. The biologists needed lots, because of Darwin's insistence that evolution by natural selection was a very slow process. August Weismann, the German biologist who clarified the process of evolution by making the crucial distinction between a mortal body and a potentially immortal germ line, wrote that 'Darwin was right in regarding transformation as taking place by minute steps, which, if useful, are augmented

*The Sea-Change*. Graham Bell, Oxford University Press. 
DOI: 10.1093/9780197797464.003.0004

in the course of innumerable generations'. I think that this extreme gradualism was forced on Darwin and his followers to counter anyone who asked for an example of natural selection in action. As late as 1949, J.B.S. Haldane, a pioneer of mathematical treatments of evolutionary change, was writing that 'Such calculations [of rates of change in gene frequency] are extremely rough, but they suggest the remarkably small magnitude of the selective 'forces' which are at work if natural selection is largely responsible for evolution, and the extreme difficulty of detecting them in action'. The physicists, however, were unwilling to grant biologists the vast stretches of time that evolution demanded. William Thompson, Lord Kelvin, calculated how long it would take the Earth to cool from an incandescent ball of gas to its current habitable state and found it no more than 20 million years. Van Helmholtz obtained a similar value for the age of the Sun. Twenty million years may sound a long time, but it does not amount to 'innumerable ages', especially when it includes the period when it would have been too hot for anything interesting to happen. As usual when biologists and physicists disagree, the biologists were eventually shown to be right. The discovery of radioactivity (by physicists, admittedly) eventually led to reliable methods of dating the rocks through the radioactive decay of one element or isotope into another. The basal Cambrian sequence at Fortune Head is not suitable for these methods, but exposures of similar age in other parts of the world are between 540 and 544 million years old. Half a billion years seems enough to satisfy the most gradualist of biologists.

## Cells

### Old cells

The Earth was already old when the sediment began to accumulate at Fortune Head. It was formed about 4.5 billion years ago from the dust cloud around the early Sun, when chance collisions created a few larger bodies that swept their orbits clear of smaller fragments, growing into the planets of the Solar System. At first it was very hot because this new material was compressed by gravity, but it gradually cooled as this heat was radiated into space, until liquid oceans could form. The oldest sedimentary rocks, which mark the beginning of the hydrological cycle of erosion and deposition, were formed about 3.8–3.7 billion years ago, and are today exposed at Isua in southwestern Greenland. The earliest structures resembling cells and filaments have been reported from rocks of about the same age in northern Quebec, but whether these are really the fossil remains of very early organisms is hotly disputed. There are somewhat more convincing structures from 3.5 billion years ago at Warrawoona in Western Australia, but these are disputed too; it was all a long time ago, and the rocks have been so altered by heat and pressure that it is difficult to be sure. By 3 billion years ago, large banded structures called stromatolites have appeared; these are formed when sheets of bacteria grow between episodes of sedimentation, and can still be found today. From this time on there is a continuous record of life, and the oceans swarmed with bacteria.

## A new kind of cell

This is a little mysterious. One of the most puzzling features of the history of life is the long, long period during which all living organisms were single cells, or at best simple filaments of cells joined end-to-end. Life began to stir soon after the hydrological cycle began, and cells appeared quite early, at least in units of a hundred million years. And yet 2 billion years elapsed before these simple organisms became assembled into multicellular individuals. This is not to say that nothing happened during this period—for one thing, the whole complex machinery of metabolism evolved—but it seems odd that keeping cells together once they had divided was such a difficult trick to master. One reason might be that bacterial cells are protected by a tough, fibrous cell wall that makes it difficult for cells to communicate directly or to cooperate. If this wall is lost, the cell is bounded only by a thin membrane, which might not be an adequate shield against physical stresses or unfriendly neighbours. On the other hand, a wall-less cell might have a broader diet. Bacteria feed by absorbing food that they have predigested by secreting digestive enzymes onto a lucky find such as a dead shrimp or a fragment of leaf. A cell without a wall can feed in a different way, by surrounding and engulfing small particles and taking them into the interior of the cell for digestion. Losing the cell wall is certainly perilous, but a wall-less cell can eat solid food—other cells, for example. This broadens its diet, at the risk of trying to eat something that declines to be digested. If this foreign body can grow and replicate within the cell, it is likely to evolve into a pathogen that manages to get into cells whether they like it or not. On rare occasions, however, some sort of evolutionary bargain might be struck, so that the two cells, originally quite separate, form together a new sort of compound organism. A partnership of this sort was eventually struck between two different kinds of cell, a familiar bacterium and a less familiar archaean. If you can't recall exactly what an archaean is, you're in good company, because most biologists regarded them as little more than bizarre curiosities before Carl Woese showed that, although both are single cells that lack nuclei, the architecture of bacterial and archaean cells is fundamentally different, to the extent that archaeans constitute a separate domain of organisms. Nowadays, archaeans are usually associated with extreme environments such as hot springs or acid bogs, although some biologists think they may be broadly distributed in ocean waters. Exactly how a bacterium and an archaean got together to form a new kind of chimeric cell has yet to be explained, but the outcome was a third kind of organism, the first eukaryote. Modern eukaryotes have a distinct cell nucleus, consisting of linear chromosomes bounded by a membrane, unlike the single loop of DNA which comprises the genome of bacteria and archaeans. All animals, plants and seaweeds, besides a host of less familiar organisms, are eukaryotes.

The original lineage of eukaryotes then engulfed another recalcitrant bacterium, which eventually became domesticated and, having given up most (but not all) of its genome to the nucleus, evolved into the mitochondrion. This new partner earns its keep by performing most of the oxidative metabolism of the cell, and because mitochondria are cheap to run, a eukaryote cell can be a thousand times larger than a bacterium or an archaean. All living eukaryotes descend from this lineage because all have mitochondria, bar a few obligate parasites (such as the agent of 'beaver fever', which you can acquire by drinking unboiled water from a stream) which have subsequently dispensed with them. Another lineage recruited a fourth partner, a cyanobacterium whose domesticated descendant is the chloroplast of green plants, which enables them to perform the complicated trick of transforming atmospheric carbon dioxide into carbohydrates. Indeed, some seaweeds and phytoplankton have recruited single-celled eukaryotes which themselves contained chloroplasts, and which subsequently evolved into complex structures arranged rather like Russian dolls, with the remains of one partner enclosing the remains of another. Animals do not have chloroplasts, although some (such as corals) maintain whole algal cells as temporary lodgers who pay rent (carbohydrate) in exchange for services (shelter and nitrate) but are liable to be evicted when times are hard.

There were plenty of single-celled eukaryotes a billion years before the first animals appeared. They are called 'acritarchs', which sounds like an ancient line of priest-kings but instead refers to the abundant but enigmatic remains of very large cells. Most acritarchs were between 50 and 150 μm in diameter, much larger than bacteria (*E. coli* is 1–2 μm) or even most modern eukaryotes (baker's yeast is about 10 μm), and exceeding most cells in animals (10–30 μm) and plants (10–100 μm). The average acritarch was about the size of a human ovum (120 μm), which is an exceptionally large cell. Acritarchs are ghosts, consisting of no more than the envelope surrounding an empty and featureless central cavity. The envelope itself must have been originally made of some exceedingly resistant material, such as the sporopollenin composing the walls of the spores and pollen grains of modern land plants, to have survived so long. Indeed, the usual way of finding acritarchs is to macerate crushed rock in strong acid, which dissolves everything except the acritarchs themselves. Nobody is sure what they were. Since they were large cells often ornamented with ridges or spines they were presumably eukaryotes, but beyond that it is difficult to say. The prevailing view is that most acritarchs were the resting stage of phytoplankton. They might be zygotes preparing for meiosis, or asexual cysts which would later divide into hundreds of small flagellated swarmers; there are plenty of suggestive living examples of either. Acritarchs would then include the ancestors of modern groups such as green algae and dinoflagellates. The name might include other kinds of organism too: persuasive claims have been made for fungal spores and animal egg cases. A conservative view might be that 'acritarch' is an artificial assemblage that encompasses a wide variety of different kinds of archaic organisms, some perhaps close to the ancestry of living groups.

## Seaweed

### Microbial mats

Without big worms, crabs and sea-cucumbers there would have been nothing to prevent cyanobacteria (plus a variety of algae and fungi, probably) from growing as more-or-less continuous sheets over extensive areas, provided that the rate of growth exceeded the rate of sedimentation. Instead of the usual soupy interface between seawater and the underlying sand or mud, continually disturbed by animal activity, there would instead have been a thick microbial mat consisting of living fibres, the sticky matrix they secrete, and embedded particles of sand or detritus. This is not just imagination: even today, these mats will form in places where animals cannot live, for example in shallow tropical bays where the sea is very salty. Nevertheless, it does take some imagination to picture the PreCambrian sea floor as being sealed over large areas by microbial mats, rather than being the naked sediment surface we are used to seeing. Adolf Seilacher had plenty of imagination, and guessed the truth even before Jim Gehling, an Australian geologist who was then a graduate student in California, made the idea respectable by showing how it gave a convincing explanation of the vague crinkles and wrinkles often found on the face of ancient rocks—'elephant skin' in the eloquent geological jargon. It did require a little imagination, though, since the microbes themselves are never preserved, and all that Seilacher and Gehling had to go on was the surface of the rock itself. These mats were not just flimsy biofilms, however. Judging from modern analogues, a microbial mat can be surprisingly cohesive and tough, easily able to resist moderate current shear and remain in place.

The mat is a habitat in its own right. Its upper surface, exposed to the water column and anything living in it, will not be simply as smooth as a sheet of clingfilm because it contains living and growing cells that will inevitably give rise to miniature hills and valleys, peaks and projections. Its lower surface, lying on top of the sediment, might be even more ragged if it contains dead and decaying cells. In between, the structure of the matrix depends on what built it. Some cyanobacteria grow as filaments, often glued together into strands by their secretions, and will form mats with more or less distinct layers, like a very large flat stromatolite. Others grow in clumps and will form mats with a more granular appearance. In either case, the matrix is unlikely to be a homogeneous solid but instead will be a labyrinth of pores, channels, caverns and blind alleys which positively invites colonization by anything small enough to wriggle inside. Where mats still grow in shallow seas (off the Bahamas, for example) they are quite densely inhabited by a wide variety of very small animals, especially miniature annelids and crustaceans. We do not know of any annelids or crustaceans from the PreCambrian, of course, but anything living at that time which could swim or crawl would have found its way into the interior of the mat.

Nowadays, the interior spaces of the mat provide a secure refuge, and many of the small, soft-bodied animals that live there may be hiding from their innumerable predators. They would not have needed to hide in the PreCambrian because there were few if any predators, but there are still some good reasons for living in the mat. There is likely to be lots of oxygen, for example, in the topmost layers of the mat. The next time you see a filmy growth of algae growing in shallow water, instead of dismissing it as pond scum take a closer look and you will probably see bubbles of gas trapped by the filaments. These are bubbles of oxygen, released as a by-product of photosynthesis. The top layers of a cyanobacterial mat might well have much higher oxygen levels than the surrounding seawater, which would have made it an attractive place to live for an organism such as a small animal using oxygen for respiration, especially if (as seems likely) the PreCambrian seas were rather poorly oxygenated by modern standards. In deep water, of course, there would be little or no sunlight and no upper layer of photosynthetic cyanobacteria. Even in shallow water, the mat would be layered like the Winogradsky column that students in elementary biology courses make out of mud and water: over the course of a term the initially homogeneous mud separates into layers of gradually decreasing levels of oxygenation, with cyanobacteria at the top and green sulphur bacteria at the bottom.

Another reason for living in the mat would be to use it as a food source. Anything that could engulf, swallow or lyse algae could browse in the mat as if they were living in a perpetual cafeteria. Among the specimens in the Redpath Museum collections is *Oldhamia*, a fan-like arrangement of hollow tubes from the lower Cambrian of Bray Head in Ireland. (It was found by Edward Forbes, the pioneer of sampling animals living on the sea floor whose early dredging trips eventually led to the *Challenger* expedition. But this is a story I have told in another book, *Full Fathom 5000.*) *Oldhamia* is usually interpreted as a trace fossil (although some think it might be the body fossil of some rod-like organism), which Dolf Seilacher interprets as an undermat miner, an elongate worm making its tunnels in the poorly oxygenated sediment at the base of the mat and feeding on the bacteria it found there.

## The first filament

Getting rid of the bacterial cell wall also makes it much easier to become multicellular, and several lineages of eukaryotes have independently evolved a multicellular body: animals, land plants and a mixed bag of photosynthetic marine organisms we call 'seaweeds'. A walk along the seashore when the tide is out will usually pass by clumps of red, green and brown seaweeds, whose handy colour coding marks separate groups which have independently evolved a multicellular body. Brown seaweeds such as wrack and kelp, which are often large and abundant on modern shores, seem to have appeared long after Cambrian times, but red and green seaweeds are ancient. Fossil seaweeds are rare and often obscure, however, being little more than a vague impression in the sediment that remained as it hardened into rock. Fortunately, there is one exception: a minute filament named *Bangiomorpha*, preserved in astonishing detail in cherts just over a billion years old which are exposed in the Canadian Arctic archipelago. The crucial innovation of *Bangiomorpha* (or its ancestors, rather) was vertical growth. Two-dimensional sheets of cyanobacteria had covered the surface of the sediment in shallow seas for long ages before, sometimes leaving traces (such as stromatolites) of their existence. *Bangiomorpha* extended its slender filaments into the third dimension, forming an underwater lawn. This feat requires a multicellular body with at least three different kinds of tissue: an anchor or holdfast to clamp it firmly in the sediment; a stalk to extend it above the sediment surface; and a frond to capture the light. *Bangiomorpha* has a complex body with all three kinds of structure: a specialized holdfast, a single filament extending upwards, and a broad frond formed by switching the plane of cell division from vertical to longitudinal. It is something like a miniature plant with root, stem and leaf. This enables it to tower above a thin sheet of cyanobacteria, capturing the light before it reached them and literally putting them in the shade. *Bangiomorpha* only towered a millimetre or so, which sounds modest but was enough to set in motion an ineluctable competition for light favouring ever-taller plants that led in time to the evolution of kelp and forest trees. By the early Cambrian this competition had been going on for half a billion years, and the shores around the site of Fortune Head would have been fringed with large seaweeds.

*Bangiomorpha* is also a convenient marker for the appearance of the characteristically biological processes of sex and death. Most living eukaryotes are sexual and the exceptions have probably lost sex quite recently, so sexuality is presumably the primitive state for eukaryotes. Perhaps it originally evolved as a way of regularizing the propensity of wall-less cells to fuse with one another, and persisted because of its side-effect of generating variability and thereby increasing the effectiveness of natural selection. The arrangement of cells in the thickened region of *Bangiomorpha* is strikingly similar to modern red seaweeds, which are certainly sexual, and there are hints that some of the fossil filaments bear sexual gametes as well as asexual spores. The reverse of the coin is that multicellularity implies death. A multicellular individual whose body is partitioned between somatic tissues (such as holdfast and stalk) and germ line (the gametes and spores) must surely die. Single-celled organisms are readily killed by any trivial accident, of course, but a complex multicellular individual is doomed to die because its somatic tissues cannot reproduce and eventually wear out. After death, the body disintegrates and its fragments are strewn by tides and currents over the nearby sediments.

## The shadow world

The mud has now become really interesting. Rather than being just a sludge of mineral grains, however complex, it now becomes loaded with organic detritus: shreds of bacterial sheets, dead acritarchs and decaying fragments of seaweed. It is no longer merely a substrate: it is, potentially at least, a food source. It can only be used, however, by organisms able to penetrate the surface and live underground. Bacteria will get there eventually—they get everywhere—and fungal hyphae can penetrate soil and sediment, but rapid and efficient harvesting requires two features that have not yet evolved in these pre-Cambrian seas: motility and a mouth. How these are deployed depends on how large the animal is. We are very large animals and take our size for granted. We find it difficult to understand how the way of life of other animals is governed by scale: how they move, and what they eat, depends on how large they are. Very large and very small animals must move in different ways using different structures, because they live under different dynamic regimes. They might just as well live in different worlds.

### Scale

The smallest object that we can see or feel is about the size of a grain of sand. It has to be brought close to the eye, of course; we can't see a grain of sand across the street. There is a limit, however: the closest that an object can be brought into focus is about

10 cm from the eye. Whether or not we can distinguish separate objects at any distance depends on the spacing of cells in the retina, which limits the resolution of our eye to about 1 arc-minute. Putting these two constraints together, the smallest object we can see distinctly (if we are young and have good vision) is about 30 μm in length. We can see a grain of sand; we struggle to see a particle of silt, even in a good light against a white background; we cannot see a tiny flake of clay at all. The components of mud, then, straddle our field of visual perception.

We can extend the range of size to the left, towards smaller and smaller objects, down to a hydrogen atom, or to the right, towards larger objects, up to our own body size. To fit everything in we use a logarithmic scale, on which each interval, instead of marking a fixed increment in size, marks instead a fixed factor (Figure 2.1). For example, a factor of 10 separates a guppy (about 1 cm) from a sprat (about 10 cm), or a sprat from a pike (about 1 m). On this scale, about half the objects in the world are invisible. We can't see anything much smaller than the thickness of a hair, which means that atoms, molecules, viruses, bacteria and most single-celled eukaryotes are invisible to us. Consequently, the early naturalists, from Aristotle to Conrad Gesner, knew nothing of the microbial life that surrounded them. Just as bad, animals smaller than about 1 cm could be seen, but their parts could not be distinguished and their anatomy could not be described. Adjacent to the invisible world, therefore, there is a shadow world of small or minute animals which could be listed, at least in broad categories, but could not be dissected or even clearly seen (Figure 2.1).

All this changed with the invention of the microscope, based on the magnifying property of the convex glass lens. The Romans made glass bowls and must have noticed that objects on the far side looked larger, but if so they made no use of the observation. Reading glasses with a magnification of about 2x seem to have been invented in northern Italy around 1300 (there is a portrait of Cardinal Hugh of Saint-Cher wearing glasses dated 1352), but it was not until about 300 years later that the first microscopes appeared, in circumstances that are still mysterious. By the

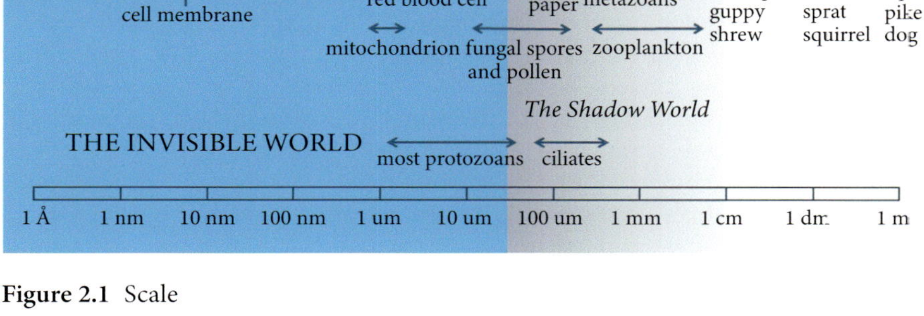


**Figure 2.1** Scale

middle of the 17th century the first book dedicated to microscopy (Robert Hooke's *Micrographia*) appeared, and Antoni van Leeuwenhoek drew bacteria scraped from the inside of his mouth. Linnaeus did possess a single-lens microscope, but almost completely ignored microbes in his classification, merely describing the Microcosmos as consisting of organisms with a variety of body coverings.

Every organism inhabits a landscape in which it finds food and shelter. How it perceives the features of this landscape depends on their size relative to its own. A mountainside or a square mile of forest is about a thousand times bigger than we are tall. We perceive it not as a single unit but rather as a patchwork of threats and opportunities, and we probably share it with several other people. A grain of sand is about a thousand times smaller than us. It cannot be shared and in any case it is much too small to be worth exploiting. Between these two extremes are objects about our own size, such as a garden or a tree, which can be defended and exploited by a single (rather antisocial) individual. This is how the world appears from our point of view, but a much smaller organism will perceive it quite differently. We treat mud as a fluid, because its individual particles are all much smaller than us. A large worm that makes its burrow in the mud, such as the lugworm dug up by anglers on muddy beaches, perceives it in much the same way. The minute animals which actually live in the mud, however, are similar in size to the particles they encounter. A bacterium, for example, is about the same size as a tiny flake of clay, a micron or two across, and from its point of view a grain of sand is as large as a mountain or a forest is to us. Single-celled eukaryotes such as flagellates and amoebas span the range of silt particles, while the smallest metazoans are about the size of sand grains, from 200 μm to 2 mm or so. These tiny animals move between sand grains as we would move in a jumble of large boulders, twisting and turning to examine each grain individually. They are members of the meiobenthos, an abundant and diverse community of minute animals, unsuspected before we harnessed the power of the convex glass lens, which swim in the thin film of water between sand grains and scraps of detritus on the seabed, in the sediment of lakes and marshes, or even on land where the soil is damp enough.

## The interstitial animals

You can visit the shadow world and watch its inhabitants if you have access to a low-power binocular microscope. The standard optical models can be pricey, but a digital camera with a computer screen can be bought for a couple of hundred dollars or so. A spoonful of pond mud will usually provide lots of minute animals. These rather portly earnest-looking citizens bustling about in midwater are rotifers, propelled by a pair of continuously beating rings of cilia which pull them through the water. The stream of water they generate is searched for algal cells, which are first swallowed then pounded to a paste with a peculiar structure called a mastax, which operates like a set of internal jaws and looks much like a beating heart in the middle of the animal's body. The slim animal that is gliding smoothly as a ship up a fragment of dead leaf is a gastrotrich, using the field of cilia on its underside for locomotion as it hunts for bacteria and protozoans (Figure 2.2). The long, thin animals ploughing

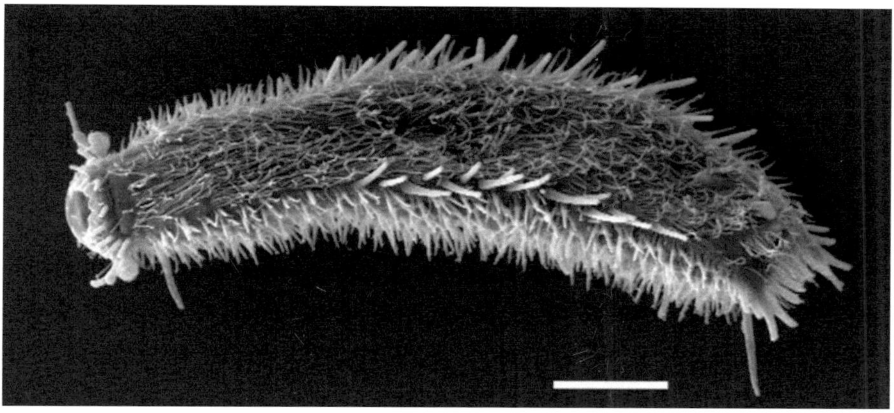

**Figure 2.2** Interstitial animal. Scale bar is 20 μm



through the sediment are worms that usually reproduce by transverse fission, so that just before the two halves separate they look like a short railway train. Some are flatworms called catenulids ('catena' is Latin for 'chain') which have a mouth, a blind digestive cavity, a rudimentary kidney, gonads and not much else, except sometimes a statocyst. Their taxonomy is too difficult for all but a few dedicated souls because their bodies are so simple that there is not much to go on. Other worms you might see have much more complicated bodies, arranged in segments and studded with bristles. These are annelids, distantly related to earthworms (although much smaller) and bearing the rather attractive name of naiads, although any resemblance to water nymphs is strictly in the eye of the microscopist. I rather like *Aelosoma*, a mysterious worm—nobody seems sure which group it belongs to—with a round head that sucks up algae from the sediment. The white lance-shaped animal that is forever thrashing about is a nematode much like *Caenorhabditis elegans*, the darling of laboratory biologists. The animal shaped like a miniature shrimp is a harpacticoid copepod, scrabbling around among the detritus. It is clearly an arthropod, because it has jointed limbs and segments armoured with plates of chitin. It is a member of the community of animals that consumes the shreds and tatters of decaying vegetation and grinds them up smaller to prepare them for fungi, part of the long slow cycle of growth and decay.

If you have instead sampled a sandy beach, there are two other kinds of animal you might find, although I doubt you've heard of either of them before unless you're a zoologist. One of them is a priapulid worm. Some priapulids are quite large, up to several inches in length, but others are only a millimetre or so and crawl among the sand grains. The second is a kinorhynch, which looks like a miniature shrimp with a segmented body and a badly drawn head. Both are misleading because the segments are no more than rings of calcium carbonate and the head looks badly drawn because instead of bearing eyes, antennae and so forth it has an invertible spiny proboscis, just like a priapulid, which it likewise uses for capturing its prey as it clambers among the sand grains.

## Burrows

Preston Cloud—what a splendid name!—was the scourge of pseudofossils and the dupes who took them for real. In his view, the literature of the time (about 50 years ago) was full of dubious reports of PreCambrian fossils that could easily have been produced by purely physical processes. Fragments of arthropod carapace more than a billion years old were just curled and cracked flakes of dried mud. Jellyfish medusae were radiating clusters of pyrite crystals growing between sandstone beds. He was particularly hard on worm trails, which could be produced quite casually by a variety of processes. The long sinuous trails found on the muddy shore of Great Salt Lake looked as though they had been made by worms crawling over the surface, but in fact were clearly scratched by fragments of the surface crust washed about by a recent flood. Harry Clemmey, a geologist at Leeds University, claimed to have discovered worm tracks about a billion years old from a copper mine in Zambia; Cloud showed that they were termite burrows only a few years old.

Ironically, he fell into the trap himself when he described worm burrows from a bedding plain 620 My old from North Carolina. He named the find as *Vermiforma* and claimed that it established the United States rather than Australia or Russia as the site of the earliest traces of animal life. His nemesis was Dolf Seilacher (of whom more later), who noticed that the sinuous outline of a *Vermiforma* burrow, in itself a rather convincing trace fossil, was almost exactly repeated by another burrow a short distance away on the same slab. This could scarcely be a coincidence. Seilacher showed that a similar pattern could be produced by bits sticking out from the overlying layer when it was shifted by tectonic forces relative to the bedding plane where the fossils were found. You can imitate tectonic shear at home if you have two small sheets of window glass, or glass from a picture frame. Coat one side of each with a thin film of margarine, like preparing a baking tray. Cut a few small pieces of wine cork and trim them to about the same size. Now lay the pieces of cork between the margarined sides of the glass sheets, and gently rotate the top sheets. The cork will make swirls in the margarine film that would look like fossil worm trails if they were made in sediment that subsequently hardened into rock. What gives the game away, of course, is that the patterns of swirls look much like one another instead of the random pattern and orientation that real worm trails would show. *Vermiforma* is a pseudofossil.

## Worms

Worm burrows, however mundane they might appear, remind us that big worms, more than about a few millimetres thick, have big problems. In particular, if the body of a big worm is constructed like a small worm it will find it difficult either to breathe or to swim. It can't breathe because diffusion, which works well enough in a small worm, is too slow in a large worm to sustain a reasonable rate of gas exchange in its core. It can't swim because the cilia that are perfectly capable of propelling a small worm are too weak to accelerate the bulky body of a large worm.

Large worms have a breathing problem because the time taken for diffusion is proportional to the *square* of the diffusion path. For pure diffusion, tissue thickness must be less than 100 μm at 1% of the current oxygen concentration, which is a reasonable guess for the early Cambrian atmosphere. Gases would take several hours to diffuse from a site 1 cm inside the body of a large worm, which is too long for even the most sluggish metabolism. Large sea anemones and jellyfish don't have this problem because seawater is constantly sloshing around in their digestive cavity and can be renewed at will. Large worms do have a problem because their gut is relatively narrow or, worse, has been isolated by piercing an exit, the anus, for the undigested remains of their food. A through-gut is much more efficient in processing food, because specialized digestive tasks can be allocated to different regions of the gut, but it creates dead space between the gut wall and the body wall which needs ventilation. This is provided by a secondary body cavity around which fluid—modified seawater—can be circulated. This secondary body cavity, the coelom, is not a badge of superior design for the convenience of comparative biologists; it evolves as a respiratory organ that compensates for one of the main disadvantages of getting bigger.

Burrowing in the mud brings back the problem of friction in another form. A very small animal will perceive the particles of the mud as individual entities which have to be searched or circumvented, as we would pick our way through a field of large boulders. A big worm perceives mud as a bulk substance because it is much larger than its component particles, so it has to force its way through a viscous medium. This requires muscle power, because cilia are much too weak to do more than scratch a surface trail. Muscles alone are not enough, however: contracting a muscle will shorten the body but when the muscle is relaxed the body lengthens again without moving in any particular direction. Muscles will produce movement from place to place only if they are coupled with a skeleton able to translate contraction into locomotion.

The skeleton might be no more than a sheet of elastic tissue, provided that the body is the right shape. This is how jellyfish move. Their bowl-shaped body has a layer of muscle fibres running around it which force a jet of water out of the base of the bowl when they contract, propelling the animal by reaction. Underlying the muscle layer is an elastic tissue which passively re-extends the muscle fibres when they relax, allowing a further episode of jet propulsion.

Although you can swim like this, you can't burrow, at least not fast enough to make it worthwhile. Worms with a coelom can burrow rapidly through coherent substrates by using two layers of muscle, one running from end to end (the longitudinal muscles) and the other from side to side around the body wall (the circulars). Contracting the longitudinals makes the body shorter and fatter; contracting the circulars makes it longer and thinner. In itself, this will not move you forwards. But the secondary body cavity constitutes a sort of skeleton, because the coelomic fluid is incompressible and acts as a hydrostatic skeleton (think of the brake system of your car; if the hydraulic fluid were compressible, treading on the brake pedal would not have much effect). Now, contract your longitudinals (you are a worm, remember) to make your body short and fat, then relax them in the front part of your body and contract the circulars.

This re-extends the longitudinals, because the fluid inside the body is incompressible, and pushes the front part of the body forwards through the mud against the anchor provided by the expanded rear part. Now do the reverse: contract the longitudinals at the front, relax the circulars at the rear, and draw up the rear section of the body. You have now moved a little distance through the mud, and by repeating the process you can make a burrow.

All this is rather onerous: it requires two layers of muscle, a hydrostatic skeleton so they can antagonize one another, and a nervous system to coordinate contraction and relaxation. Burrowing may seem a humble sort of achievement, but it requires preparation. A good deal of evolution must have taken place before the body plan of an active burrowing worm became possible. Once it had appeared, worm burrows disrupted the surface of the sediment, broke up the old microbial mat and altered the ecology of the seafloor forever. *Treptichnus* is not just a convenient geological marker; it really does signify the advent of a new kind of animal and a new kind of world.

## Reynolds number

The wide difference between an interstitial animal like a rotifer and a burrowing worm is due to scale, because small and large animals live in different dynamic regimes and must use different means of propulsion. Locomotion through a fluid must overcome two sources of resistance, inertia (which resists acceleration) and friction (which causes deceleration). The balance of these two forces is summarized by a quantity known as the Reynolds number, Re for short. Osborne Reynolds was a 19th-century scientist who was interested in the flow of fluids through pipes and perhaps did not appreciate how his principle could be applied to the locomotion of swimming animals. His number expresses the difference between two dynamic regimes: a low-Re regime dominated by friction, and a high-Re regime dominated by inertia. For animals swimming through water, Re is governed by velocity and size, or in practice mainly by size, since most motile animals cruise at about one body length per second. (Time yourself the next time you are heading towards a traffic light with the pedestrian clock counting down; you should find that one stride takes about one second at a brisk walk.) Locomotion in low-Re and high-Re regimes requires different power sources, and to accommodate them small and large animals must have different kinds of body plan.

There are two (and only two) families of contractile proteins which are used by animals to move themselves or parts of themselves: dynein/tubulin, which powers flagella and cilia, and actin/myosin, which powers muscles. Which of the two is used for locomotion depends on size. The very small animals of the interstitial fauna live in a low-Re regime where they move slowly and are hindered mainly by frictional drag. They overcome this by a continually acting low-power motor provided by a field or ring of cilia. Small benthic animals that creep slowly over the mud also live at low Re, so they can deploy a ventral field of cilia beating in mucus. Larger swimming animals, such as big jellyfish, squid or sharks, are hindered mainly by inertia and use muscles to execute a power stroke followed by a glide, as we do when swimming in a lake. Small animals are not just miniature versions of large animals. They evolve different kinds of body for straightforward physical reasons, although from the standpoint of a large animal we may find it difficult to appreciate how their size affects their morphology.

## *Treptichnus*

If you can picture a small twig of cedar made of sand and stuck to the base of a layer of sandstone you have a reasonable idea of what the *Treptichnus* that marks the base of the Cambrian at Fortune Head looks like. It has been found at many other sites of similar age throughout the world, from shallow water to the deep sea, and varies considerably in appearance, from a cedar twig to a more or less linear string of lobes or even a series of unconnected lobes; sometimes it is curled and looks to the eye of faith like a bishop's crozier or pedum, hence the rather presumptuous binomial *Treptichnus pedum*. It is the remains of the burrow, afterwards filled with sand, made by an animal probing underneath the microbial mat or emerging at the surface. There is no trace of the animal itself beyond the burrow that it made. Nevertheless, we can confidently infer the leading feature of the body plan of the *Treptichnus* animal: in order to burrow forcefully through sediment, even unconsolidated sediment, it must have been a worm with a secondary body cavity and a full suite of circular and longitudinal muscle bands. It was by no means the only animal that made burrows around this time. *Skolithos*, for example, made larger and simpler burrows that are sometimes so numerous that the rock is named after them as 'pipe-rock'. Vertical burrows are unknown in the PreCambrian and become common in the Cambrian, implying that the worms that made them had already evolved circular and horizontal muscle layers, plus a coelom to connect them.

Present-day animals living in the sediment span the whole range from the shadow world to the dissectable. We can be confident that the same range of body size was present on the sea floor in Cambrian times. The *Treptichnus* animal has vanished without trace, leaving only its burrow behind. The interstitial animals that lived at the same time have also vanished, without even making a burrow. In seeking the ancestor of all animals, one place to start might be the least romantic of places, in the mud.

# 3
# Animals and Individuals

## Animals

A zoologist might be mischievously defined as someone who is not quite sure what an animal is. Most people know an animal when they see one; even confirmed city-dwellers will be familiar with dogs, squirrels, pigeons and crows, and would immediately recognize a worm or a spider or the delicate long-legged house-centipede as animals, although without necessarily appreciating their interest and beauty. But what is this intuition based on? A rather long-winded definition might be something like this: animals are motile organisms able to move themselves (or parts of themselves at least) by muscles activated by nerves, with bodies consisting of many different kinds of cell arranged in tissues and organs, which have a mouth to swallow solid food and a gut where they digest their food internally. These rules can be used rather confidently to decide whether any familiar object belongs to the animal club: spiders and pigeons pass the test; mushrooms, seaweed and trees do not (they are alive, but do not move around or swallow their food); rocks and stars are not even alive.

Less familiar organisms, however, might not be as easy to classify. For example, consider a mat of filaments that catches small worms with a sort of lasso, throttles them and then digests its prey. Is it an animal? (No, it is the carnivorous fungus *Arthrobotrys*.) Or how about an undifferentiated bulbous mass of tissue that grows on crabs and sends root-like extensions into their body? (Yes, it is the parasitic barnacle *Sacculina*.) Or an actively swimming sphere that splits apart to give birth to small but fully-formed offspring? (No, it is the green alga *Volvox*.) All of these cases are now understood, but in the early years of scientific zoology they were small parts of a greater puzzle. There were whole swathes of organisms which were undoubtedly living, but appeared to be neither animals nor plants, the only two divisions of living things then recognized.

Animals are everywhere, and you don't need to buy expensive equipment nor travel to exotic locations to study them. You can find enough diversity for a lifetime of study by sampling any meadow or woodland or beach or rocky shore. If all you have is a muddy pond in the corner of a field, a fine-mesh net will usually be squirming with shrimps, insect larvae, snails and a newt or two after a few passes after through the weeds. Tipping the contents of the net into a shallow enamel tray will show much more, especially under a lens: copepods, ostracods, water fleas, tubifex worms and, if you are lucky, a flatworm called *Mesostoma ehrenbergii*. It is named after Christian Gottfried Ehrenberg, a distinguished 19th-century German zoologist who pioneered the study of minute organisms in water, soil and sediment. It is not itself very distinguished at all, except for one feature: it is as clear as glass. It is the ideal introduction to

*The Sea-Change*. Graham Bell, Oxford University Press. © Graham Bell (2026).
DOI: 10.1093/9780197797464.003.0005

**Figure 3.1** Mesostoma

Reproduced from www.lebendkulturen.de.

the anatomy of a simple animal because every feature of its body can be seen as clearly as if it were a model, without the need for sectioning or dissection (Figure 3.1).

## *Mesostoma*

To begin with, it is quite large (about a centimetre long) and very flat. Its body basically consists of two sheets of cells, together with a mass of cells in between. The outermost sheet is an epithelium which covers the entire surface of the body, above and below. Like any epithelium, it is a two-dimensional layer of cells connected to one another, arranged with their longitudinal axes parallel, and glued underneath by a thin, pliable basement membrane (our skin is a complicated version of the same kind of structure). Each cell bears a short hair-like projection, the cilium, which can be waved actively to and fro. The combined beat of the cilia on the underside of the animal moves it smoothly forwards, almost as though it were being towed by a cord. The inner sheet of cells is also an epithelium, which forms the lining of the gut tube. The entrance to the gut is a mouth, situated not at the front of the animal but about in the middle of its underside, so that the gut itself extends forwards and backwards along the midline of the body. *Mesostoma* is a carnivore which leaps upon any small animal it encounters, folds its body around it, everts its pharynx through its mouth, and then subdues the unfortunate animal by shoving the pharynx into its body and liquefying its tissues. It digests its food externally, through the secretion of digestive enzymes, and also internally, by the cells of the gut engulfing particles produced by the break-up of the prey. Any indigestible remains must be voided through the mouth because the gut is blind and there is no anus.

The main structures of the body lie between the outer (skin) and inner (gut) epithelia. Immediately beneath the outer epithelium there are layers of muscles that can throw the body into fantastic convulsions, for example to escape a predator (as you will see if you gently prod the animal with a blunt needle). There is a nervous system, with several pairs of nerve cords running the length of the body connected to a ganglion, a distinct group of neurons at the forward end of the body that might by courtesy be called a brain. This is hooked up to some sense organs, especially a rather basic kind of eye which can tell whether the sun is out or not, and a statocyst that can tell up from down. The only complicated part of the body is the reproductive system. *Mesostoma* is a hermaphrodite with pair mating, and has both male and female gonads with the full complement of accessory structures, including a bursa copulatrix and seminal receptacle (to store sperm), an intromittent organ (to transmit it), a shell gland and all the tubing to link them together. That's about it: there is no circulatory system because it doesn't need one; being very flat, it can exchange gases with the water by diffusion. There is also no body cavity, apart from the gut; the space between body wall and gut wall is occupied by a mass of mesenchyme, consisting of more or less mobile amoeboid cells able to roam around in a semi-fluid matrix.

## *Hydra*

*Mesostoma* is obviously a simple kind of animal: it moves, uses its mouth to eat solid food, and has internal structures, including muscle layers and a nervous system. If you continue to poke around in the pond, you are likely to find *Hydra*, the only hydroid that has become abundant and widely distributed in freshwater. It is a small solitary polyp, a column of tissue attached to the substrate (usually the leaf of a pondweed) at one end and trailing long tentacles into the water at the other. Its body is even simpler than *Mesostoma*, consisting of the two sheets of cells alone, one pushed inside the other. The ectoderm is the epithelium that covers the outside of the body; the endoderm is the layer that is folded into the ectoderm and lines the interior gastric cavity. They are not separated by tissues, as in *Mesostoma*, but merely by a thin layer of gelatinous material that contains collagen and probably originated as an expanded basement membrane. It has muscle strands and a diffuse nervous system (although nothing resembling a brain), and will contract if prodded. Apparently it can move, by performing a sort of clumsy somersault, although I have never seen it in motion and it is usually fixed in place, like a plant. It normally reproduces by budding, again like a plant, and may even be bright green (because it can farm green algae inside its cells). Marine hydroids, which grow as rather inconspicuous felts on pebbles and shells, are even more plant-like, forming colonies in which shoots arise from a basal creeping rhizome, much like reeds or grasses. On the other hand, *Hydra* is a voracious carnivore, which captures live prey using the stinging cells of its tentacles and can cram animals larger than itself into its distended gut. Here the food is processed, again by a mixture of external and internal digestion, and the remains expelled through the mouth. This is a creature, then, with a mixture of features: more or less fixed in place, reproducing by buds and lacking internal organs, like a plant, but eating live prey with a mouth, like an animal.

## *Spongilla*

Another creature you might find in the pond is even more difficult to place. It will not turn up in the net, but you might notice an inconspicuous grey or green mound growing on rocks or on twigs trailing in the water. It is a sponge, *Spongilla*, again the sole freshwater representative of a group that is much more abundant and diverse in the sea. It has no mouth or digestive cavity, and on closer examination it has no epithelia, muscle or nerve either. It cannot move and does not react to a prod. On the other hand, if you look at it under a lens you will see that a stream of water passes continually through its body. Sponges are able to filter bacteria from this stream and eat by engulfing and then digesting them inside cells. It is difficult to say whether *Spongilla* is an animal or a plant or something halfway between the two.

This is a very old dilemma. It has been known since the time of Aristotle, who lived and worked on Lesbos, by the shores of the Mediterranean, when he was not teaching at Athens or tutoring Alexander. Anything which could move was clearly an animal, but sessile creatures, more or less fixed to the substrate, were more difficult to place. He used two main criteria for deciding whether an object was an animal or a plant: whether it responds to touch (and therefore has nerve and muscle) and whether it has a mouth and stomach (to eat solid food). Movement, sensitivity and feeding all correspond closely with our contemporary common-sense notion of what an animal should be. What did he decide? His 'tethya' are clearly tunicates, which draw in a stream of water through one siphon, strain out the edible particles with slits in the pharynx, and expel it through a second siphon. Aristotle thought that the incurrent siphon was a mouth and the excurrent an anus, an example of a mistake that led to a correct conclusion, that tethya were animals. The 'acalephe' were sea anemones and sea cucumbers, which likewise have mouths and are therefore animals. This left the sponges, which he considered to be animals because they seemed to be capable of sensation: sponge fishers reported that the sponges were easier to detach from the rocks if the approach were made stealthily. In this way, reasoned (if sometimes mistaken) arguments led Aristotle to admit tunicates, sea anemones and sponges as animals, despite being firmly fixed in place.

Two thousand years later, as scientific biology began to gather steam again, the same difficulties presented themselves. If we come across an organism with a complex body that includes muscle and nerve, and is motile, and has a mouth and gut with external digestion, then we can be sure that it is an animal. Another organism may lack some or all of these features, however, and yet still be classified (however reluctantly) as an animal. In particular, there are many kinds of animal that are fixed in place and unable to move. How can we recognize a sessile creature as an animal rather than a clump of seaweed?

## Zoophytes

Anyone who strolls along a rocky shore will come across many organisms which might be animal or vegetable or somewhere in between. Here is a bluntly branched grey bush; there is a purple-red tuft, gritty to the fingers; what are they? John Ellis wrote a book about them in 1755 which described a confusing mixture of soft corals, hydroids, bryozoans and calcareous red seaweed. Ellis was aware that many of these organisms were colonial, which sounds like a plant; and yet the minute zooids which formed their compound body were each furnished with a mouth and stomach. His Eschara, for example, is a bryozoan (probably *Flustra*) that looks like a seaweed. It has broad straplike branches, grey or brown in colour, which are made up of little boxes of calcium carbonate fused into a single structure, each inhabited by a tiny zooid only about half a millimetre long. The whole structure is the result of vegetative budding by the zooids, starting with a single founder. Each zooid has a tuft of tentacles which capture floating cells and transports them to a mouth which opens into a complete through-gut, so there is no doubt they are animals. His Corallina is also calcified, and seems to consist of short hard segments separated by a thin flexible membrane that allows its fronds to bend in response to waves or current. It has no zooids, however, because it is a plant, a sort of armoured red seaweed. Eschara, Corallina and other puzzling objects were classified as Zoophyta, a rather foot-shuffling group from which each different kind was gradually extracted as being either an animal or a plant. I think that sponges were the last to leave, recognized as a distinct kind of animal in 1826 after a long and unsuccessful search for the polyps that many naturalists felt sure must be there.

It was at this point, around the middle of the 19th century, that Zoophyta was retired, and three distinct kinds of animal were recognized. The simplest were the sponges, viewed as large aggregates of cells tacked together by mineral rods. Jellyfish and hydroids had a more elaborate body consisting of sheets of tissue that incorporated muscle and nerve but lacked many organs (such as brain, gut and heart) and had only simplified versions of others (such as the reproductive system). Everything else fell into a third category that included all the unequivocal animals that had been recognized as such since ancient times: everything from worms to whales, from snails to starfish, mostly having more complicated bodies with a full range of organs. This is an advance on Aristotle, because he didn't have a microscope, but has led to a similar conclusion.

A zoologist might make a more convincing list of animals by expanding the number of features that delimit the category. They would point to the characteristic ovum and sperm of animals, and the expulsion of polar bodies during the formation of the ovum; the motile larva of animals like tunicates or bryozoans that are sessile as adults; the universal presence of collagen in connective tissue; the specialized junctions which connect animal cells and permit signals to pass between them; and the particular molecules that are involved in cell-to-cell signalling. A palaeontologist might not be impressed. These features may well characterize living animals, at least after careful study, but none of them are likely to be preserved in fossils.

## Individuals

We now have a list that (in principle) includes all the animals and none of the non-animals. It has a rather arbitrary feel, though, because we began with a list of attributes that all animals are supposed to have, and ended by discarding most of them. We were forced to do so because we began with a word, 'animal', that was founded on our everyday experience of fish and dogs, and then found that there are organisms, especially sessile organisms, that do not fit comfortably into the category that we have created. The fault is with the category, not with the organisms. There is a similar difficulty with the word 'individual'. Sessile animals are often colonies consisting of thousands of small repeated units, the zooids, formed by vegetative budding. When we are looking at a hydroid or a bryozoan, is it the zooid or the entire colony which constitutes the individual? The zooid has a full complement of somatic structures, and in principle could live autonomously, so it has a reasonable claim to be considered an individual. The colony has a definite form, often specific to a particular species, and its units can pool resources because they are all connected together by a continuous strand of cytoplasm, so it might also be considered an individual. In some colonial animals reproduction is carried out only by a particular type of zooid, and others take this further by allocating somatic tasks such as protection to a caste of non-reproductive zooid. In some bryozoans, for example, there are zooids which become transformed into a fearsome structure like the beak of an eagle, which will seize and hold any creature trespassing on the surface of the colony. Siphonophores are motile hydrozoan colonies which have reached the limit of this tendency by modifying zooids for flotation, propulsion, feeding, protection and reproduction. Perhaps the extremes need cause little concern. A simple boulder coral, basking in the warm sunlit waters of a shallow tropical sea, can be viewed as an assemblage of more or less autonomous zooids which happen to be able to share food, as social animals often do. A siphonophore, fishing with its long stinging tentacle in the dark cold water a kilometre deep, is unmistakeably an individual, albeit one whose structure is based on zooids rather than organs. Intermediate degrees of delegation imply intermediate degrees of individuality, with neither the zooid nor the colony itself being unequivocally an individual. We have again been misled by a word which was invented from our commonplace experience of fish and dogs—and ourselves—and causes no difficulty in the day-to-day world. It is only when naturalists begin looking closely at zoophytes that the hidden ambiguity of the word is revealed.

### Eggs and individuals

We are thoroughly familiar with the concept of individuality because we live as individuals ourselves. We are by no means alone: frogs, sea urchins and snails (to take three examples at random) all live as individuals like us, each sharply distinct from all others and having its own unique fate. An individual begins life as a fertilized egg, goes through a regular course of development, and finally matures as an adult with a complex and highly integrated body. Its body is complex because it consists of many different kinds of somatic cell: muscle cells, liver cells, skin cells and so forth. It is highly integrated because the concerted action of all these different kinds of cell, arranged in tissues and organs, enables the individual to move, to feed, to survive stress and to reproduce. A simplified body might work to a point, and an imperfectly coordinated body might work to a point, but only to a point. Frogs, sea urchins, snails and people are very different kinds of animal, but in each case any reduction in complexity or integration would be almost certain to impair the normal activities of the individual.

This might seem too obvious to be worth stating, but it conceals the requirement that the lineages of cells descending from the fertilized egg must each follow a particular pathway during development in order to produce a normal individual. Since all these lineages have the same genome, it is not easy to understand how this could be possible. The alternatives identified by early developmental biologists were preformation and epigenesis. On the one hand, all the adult structures of the individual might already be present and appropriately arranged in the egg, so that development is no more than unfurling what has been preformed. This is a temptingly simple theory, but leads to an infinite regress, because if the adult woman is preformed in the egg, then her eggs must likewise be preformed in her ovary, and in each egg . . . you see the problem. The alternative is that development proceeds by an epigenetic process in which one cell induces its neighbours to differentiate in a particular way. This is a more satisfying theory that contains its own motor, but runs up against the problem of the first mover: it does not explain how the initial formation of the embryo can arise from a homogeneous and featureless cytoplasm.

Towards the end of the 19th century, better microscopes and microdissection made it possible to remove or destroy one of the cells produced by the earliest cleavage divisions of the embryo. If all adult structures are preformed, this experiment should result in severely defective adults which entirely lack some body parts, whereas epigenetic development might produce a morphologically normal adult despite the initial injury. The results were mixed: experiments with frogs produced embryos with half their structures missing, which supports preformation, but similar experiments with sea urchins produced small but perfectly formed larvae, suggesting epigenesis. It was not until a century later that it became possible to understand how both points of view were correct: the information which specifies an individual is indeed preformed, as genes in the nucleus of the fertilized egg, but the instructions that this information encodes are executed sequentially, causing cells to be induced by their neighbours along one pathway or another. The instructions are proteins, encoded by genes, which act as signals to turn genes in other cells on or off and thereby differentiate as a particular cell type.

These instructions might be altered, however, by a somatic mutation—a mutation that occurs in the ordinary cells of the body during development. Like mutations that occur in the germ cells and are transmitted to offspring, somatic mutations are usually deleterious because any random change in the fate of a cell lineage is likely to impede the normal function of an organ or tissue. It would be a disaster, for example, if a mutation caused some cells in the growing kidney to develop into muscle, producing an individual with a defective excretory system. Any such individual would be unlikely to be able to reproduce as successfully as a normal individual, if at all, so the accumulation of mutations of this kind, occurring in somatic cells during the course of development, will be prevented by natural selection. Complex and highly integrated bodies can evolve and persist only because the population is continually screened by purifying natural selection, eliminating types with abnormal development.

The effectiveness of purifying selection, and thus the stable propagation of individuals, depends on reproduction from eggs, rather than from buds or cuttings. Although reproduction by eggs is commonplace, it cannot be simply taken for granted. If you pass a fine-mesh net through pondweed and examine the catch under a lens, or better a low-power microscope, you will usually find a few minute worms, the catenulids and naiads that we have met already, that look like short railway trains with two or three carriages arranged end-to-end but moving as a unit. These are preparing to reproduce: the foremost is forming a new tail, and the hindmost a new head, before they separate into two complete new worms. This is called vegetative reproduction, the formation of a new individual from a mass of parental tissue. It is very common among animals and plants: hydras and corals reproduce by budding, and most plants can be propagated as cuttings. It has the great advantage that the newly formed offspring can immediately move, eat and defend itself. Eggs can do neither, so restricting reproduction to single cells (as we do) requires some special explanation.

Although vegetative reproduction is common, every animal, plant and seaweed has a life cycle which sooner or later is reduced to a single cell. This generalization was first clearly stated by William Harvey, a 17th-century English physician who is best known for demonstrating the circulation of the blood, but who also coined the phrase 'ex ovo omnia': every living thing comes from an egg. I do not think that this can ever be proven, because this would involve scrutinizing millions of species, many of them with cryptic lives, but I don't know of any clear counter-example. Why should reproduction by eggs be (apparently) universal? Imagine that a somatic mutation occurs early in the development of an individual which reproduces vegetatively. The tissue that forms the bud is likely to include cells bearing the mutation, which will therefore be transmitted to all or most vegetative offspring. Purifying selection is ruled out, or at the most might favour budded offspring which received fewer mutant cells by eliminating those with more, which would be unlikely to prevent a general trend towards deterioration. On the other hand, suppose that the individual reproduces by eggs. Each egg, being a single cell, will either bear the mutant allele or not. The individuals which develop from the mutant eggs will be at an unequivocal disadvantage with respect to normal individuals, and purifying selection will act with full force to protect the normal course of development.

## Selfish cell lineages

An even more insidious threat to the individual is posed by mutations that re-direct cells from a somatic lineage into the germ line. A multicellular individual (such as yourself) consists of a large number of somatic cells, its body, which act in concert to facilitate the production and deployment of a small number of germ cells. The somatic cells and the germ line of an individual have identical genomes, of course, since both descend from the same fertilized egg. The development of a distinct soma will be favoured by natural selection if a multicellular individual produces more surviving offspring than the equivalent bulk of isolated germ cells, by virtue of the ability of the body of an individual to move, feed, avoid danger and resist stress. The somatic cells can only do this by making the supreme evolutionary sacrifice. It is only the germ cells, the gametes, which can contribute to the next generation; the somatic cells have foregone any hope of an evolutionary future, and die when the individual dies. Nevertheless, the genome of the soma is propagated through the germ line it nurtures, so its sacrifice is worthwhile. Now, suppose that mutation in a somatic cell causes it to rescind this bargain and differentiate as a germ cell. This will almost certainly disturb the course of development and result in a defective adult which is less likely to survive and reproduce than the normal type. The mutation has some chance of being passed on to the next generation, however, while the unmutated gene has no chance at all. The mutation spreads because it drives the mutant cell lineage into the germ line, despite crippling the individual.

In this way, the evolution of multicellular individuals introduces an extra level of selection. In a population of single-celled organisms, there is a single level of selection: any lineage with a higher rate of increase will tend to spread, eventually replacing all others. Once multicellular individuals have evolved, there are two levels of selection: among lineages of individuals, and among the lineages of cells that develop into an individual. Selection acts on variation among individuals, created by mutations in germ cells that are transmitted to offspring, and on variation among cells, created by mutations in somatic cells that are transmitted to tissues and organs. These two sources of selection may pull in opposite directions; in particular, somatic mutations which redirect cells into the germ line have an advantage in competition between cell lineages within an individual, despite being at a disadvantage in competition between individuals. Suppose that a mutation of this sort has spread through the population. Now a second mutation occurs which drives another cell lineage into the germ line. By the same logic, this too will spread through the population; and so forth, until the complex and highly integrated individual with which we began has been reduced to little more than a formless heap of squabbling cells.

Somatic mutations like this are not just a theoretical possibility. They have been described and studied in simple organisms such as slime moulds, which have a life cycle that includes both unicellular and multicellular stages. They normally forage in the soil, eating bacteria, as single amoeboid cells. When the food supply runs low, however, all the amoebas in the vicinity aggregate to form a small slug-like organism that can move slowly (I'm not sure how) over the surface of the soil. At some point it stops and sends up a stalk. At the tip of the stalk, the cells differentiate into spores within a sac which are eventually released when the sac bursts, and hatch into the next generation of feeding amoebas wherever they land. The point of the stalk is to raise the spore sac above the ground surface, so that the spores are dispersed as widely as possible. The cells which form the stalk, however, have given up their own chance to reproduce in order to help the cells which become spores. Somatic mutations which cause stalk cells to climb into the sac and become spores are quite common in laboratory cultures of slime moulds. The result is, of course, a shorter stalk and less efficient dispersal of spores, but the mutation spreads nonetheless (especially if it is encouraged by the experimenter). A succession of such mutations can result in a stalk-less spore sac which can only dump the spores back where they came from in the first place.

The price of multicellularity, therefore, is eternal vigilance against somatic mutations which disrupt normal development by trying to muscle into the germ line. Individuality is not something to be taken for granted, because a complex and highly integrated body will be handed down from parent to offspring only if it is equipped with powerful devices to prevent the spread of selfish cell lineages. There is an obvious difficulty here: such devices are pointless unless individuals have already evolved, but individuals can only evolve once they are in place. It is resolved because the differentiated body and the systems which protect its integrity evolve together, gradually making possible more stable development and a greater degree of differentiation. We should not be surprised that individuals took so long to evolve, for they are highly improbable objects that persist from generation to generation only if they are continually defended. Leo Buss, the curator of invertebrates at the Peabody Museum of Yale University, was among the first to appreciate that individuals cannot simply be taken for granted, and published an extraordinary book called *The Evolution of Individuality* to define the powerful defences that must be in place before the regular development of individuals is possible. He emphasized two of them in particular which either prevent selfish cell lineages from altering the course of early development or deny them access to the germ line.

## Defending development

The first line of defence is the maternal control of early development. Exactly how the mother guides the development of the egg depends on the kind of animal, but the rules are particularly clear in simple animals such as hydroids. These are very common in shallow seas, where they form inconspicuous colonies of polyps—basically hollow cylinders of cells, attached to rock or seaweed at one end, with a mouth and stinging tentacles at the other. The polyps produce medusae, small jellyfish, by budding. These jellyfish swim and feed in the open water and reproduce sexually by eggs and sperm. The egg starts out as a normal body cell with the full complement of chromosomes. When it enters the sexual cycle it goes through a special kind of cell division, meiosis, in which the number of chromosomes is halved. Meiosis consists of two divisions which result in four nuclei, each with half the full complement of chromosomes. Three of these are discarded as the so-called polar bodies, leaving only a single reduced nucleus in the oocyte. The oocyte may seem to consist of a featureless ball of cytoplasm, but it already incorporates a source of asymmetry, the site at which the polar bodies have been extruded. Moreover, this is the site at which the sperm enters the egg (in hydroids, although not in other animals). This initial asymmetry is used to arrange substances which direct the course of development into particular parts of the fertilized egg, setting up a distinction between the animal pole (usually the livelier region, with more rapid cell division) on one side of the egg and the vegetal pole on the opposite side. The cleavage furrow which initiates the first division of the egg is formed at the animal pole. After cleavage has resulted in the formation of a blastula, cells first move inside at the animal pole. Gastrulation eventually results in a minute ciliated larva which settles on the seabed and grows into the founding polyp of a new colony, closing the life cycle. This polyp is again attached to a solid surface at one end and bears a mouth at the other, so its body is organized along an axis with an aboral (attached) pole and an oral (mouth-bearing) pole. Now, the oral pole forms at the site of the animal pole of the oocyte. Thus, the original asymmetry of the oocyte, used to partition maternally derived substances into different regions of the cytoplasm, is carried all the way through development to organize the body regions of the offspring. This effectively safeguards the embryo against nefarious somatic mutations, which would be powerless to override the original maternal instructions.

The second line of defence is to segregate the gametes and their precursors into distinct gonads provided with a duct to carry them to the exterior. This sets up a physical barrier which is difficult for any selfish cell lineage to breach. Ideally, the cell lineage that forms the gonads and germ cells is specified as early as possible in development, giving it a priority that it would be difficult for any mutation that occurred subsequently to overcome. If early development proceeds under maternal control and the germ line is sequestered immediately after this period comes to an end, somatic mutations cannot be transmitted to offspring, no matter when they occur.

It is feasible to sequester the germ line only if there is no roving population of totipotent stem cells that remain able to differentiate into any cell type, including gametes, during growth and development. There is no general rule, however; in some animals the germ line is sequestered very early in development, whereas in others it remains open, at least in principle, until much later. Some animals have evolved a complete ban on somatic stem cells. For example, the nematode (roundworm) *Caenorhabditis elegans* is common in soil and has been used to study development in hundreds of laboratories throughout the world (the name is longer than the worm, which is barely 1 mm from snout to tail). The only stem cells in its body belong to the germ line. Consequently, somatic cells at any stage of development must be committed to a particular fate, and any tissue must differentiate through a succession of intermediate somatic stages. The muscle cells of its body wall, for example, arise from one of the two cells resulting from the first cleavage division (which also go on to contribute to the pharynx and other structures), or from two of the four lineages proceeding from the other cell, (both of which also contribute to other tissues of the body) or from a lineage arising from a third cell. Selfish somatic mutations can make no headway in an animal like this. Other animals (such as hydroids) maintain populations of stem cells throughout life, but these are not necessarily capable of unfettered differentiation. In particular, they may not be able to enter the germ line, even when there seems to be no obvious barrier. Coral colonies, for example, may consist of hundreds of thousands of polyps, and somatic mutations can occur in any of them. Although they can be identified in somatic tissue, however, they cannot be found in gametes. Even in animals which have stem cells and an open germ line there may be cryptic ways of preventing somatic mutations from being passed on to offspring.

In land plants and seaweeds the situation is quite different because growth is accomplished solely by a group of totipotent cells, the meristem, which gives rise to all cell types and tissues: roots or holdfast, stems or stipes, leaves or fronds, reproductive structures and the gametes themselves. There seems to be no barrier that prevents somatic mutations from entering the germ line, at least early in development, although the rigid cell wall that plants have re-evolved prevents cells from moving around once they have been formed. Perhaps this is why seaweeds are simpler than animals.

The differentiation of cell types within the body is an example of the cooperative division of labour, with each specialized type contributing to the well-being of the whole. Adam Smith pointed out long ago that the division of labour within an organization has at least three advantages: saving the loss of time involved in turning from one task to another; increasing the dexterity with which a repetitive operation can be performed; and encouraging mechanical innovation to replace human labour. The third of these has no biological parallel, but the first and second can be translated into the cost of re-differentiation (it takes time and material to transform one cell type into another) and the advantage of specialization (two specialized types can perform a complex task better than a generalist). Smith concluded that the degree of cooperative division of labour will depend on the size of the market, in this case the number of employees in an organization, and this can be expressed in biological terms as the number of cells in the body of an individual. There is indeed an overall trend for larger organisms to have more kinds of cell. At the same size, however, animals are more complex than seaweeds: a salmon and a kelp are about the same size, for example, but the salmon has a hundred or so distinguishable types of cell, whereas the kelp can barely muster ten.

The rise and spread of selfish somatic cell lineages is a grave threat to the integrity of the individual. It is difficult to appreciate in modern animals and plants because over the last few hundred million years their ancestors evolved effective means of blocking somatic mutations from the germ line. It must have been a very different situation in the early days of multicellularity, when these means were slowly being put into operation. There must have been a long period when nothing more complex than a slime mould was possible, followed by the evolution of increasingly complex bodies once the main difficulties of defending a regular pathway of development had been overcome.

# 4
# Ancestors and Embryos

I live in the Canadian province of Quebec, where the first European settlers started to arrive from France some 400 years ago. This flow was abruptly halted in September 1759, when the British general James Wolfe took Quebec City, which is rammed like a cork in the throat of the St. Lawrence and controls access to the interior of the country. The unintended consequence of his victory was that French Canadians have the best-documented genealogy of any large group anywhere in the world. About 8,500 people had already arrived from France, and all were Catholics who were baptized, married and buried by the Church, from the earliest times almost until the present day. Piecing together these documentary records (a very painstaking project, to be sure) has produced an almost complete genealogy of the Québecois, so that most (at least those who are *de souche*) can trace their ancestry back ten generations, all the way back to the 17th century. Few of us can say as much. My sister has managed to trace our family (in England) back to the middle of the 19th century, but then the documents and photographs peter out, like an old trail in the woods.

If there is documentary evidence, or a reliable oral tradition, we can identify a particular individual who lived in the past as standing in the direct line of descent leading to a particular individual living in the present. The two are literally ancestor and descendant; one will have transmitted some fraction of their genome (diminished by the number of generations that have elapsed) to the other. We could not know this for sure, however, from information about the genomes alone, no matter how detailed or extensive. Genetic evidence may tell us that the two belong to the same lineage, but the earlier individual might have been the brother or sister of a direct ancestor of the later, rather than the direct ancestor itself. Bones are even less informative; without a headstone and a death certificate, it cannot even be established that a skeleton dug up from a graveyard belongs to the same lineage as a particular person alive today. Old fragments of bone are worse still. Few fields of science are as rancorous as human palaeontology, partly because the most ancient scraps of bone are so rare that many finds are controversially named as new species, but partly because the notion of ancestry becomes increasingly ambiguous as we move backwards in time.

## Ancestors

It is sometimes possible to place a puzzling organism because it is clearly related to a more familiar form and must be a member of the same group. Take *Sacculina*, for example, the blob that grows on the abdomen of crabs. How do we know that this is a barnacle? It certainly doesn't resemble the crust of hard white shells that live on the

*The Sea-Change*. Graham Bell, Oxford University Press. © Graham Bell (2026).
DOI: 10.1093/9780197797464.003.0006

rocks of the intertidal. It was a mystery until its life cycle was worked out (by Fritz Müller in the 1860s) and found to include a particular kind of larva, the nauplius, which closely resembled the larva of barnacles. *Sacculina* is therefore a barnacle itself, even though the adult had become modified out of all recognition for its parasitic way of life. The nauplius is a minute animal with three pairs of legs which is easily recognized under the microscope because it swims in a series of short bursts, resting in between. Barnacles themselves were originally classified as molluscs because of their hard outer shell, although their jointed limbs (which they protrude from the shell when covered by the tide and kick to capture small plankton) caused some hesitation, because jointed limbs are characteristic of arthropods. When John Thompson showed in 1830 that they shared a nauplius larva with shrimps and crabs he assigned them to the crustaceans, a major group of aquatic arthropods, where they have remained ever since. Hence, *Sacculina* is a kind of barnacle; barnacles are a kind of crustacean; and crustaceans are a kind of arthropod. In this way, using morphology and development, a few generations of zoologists were able to build a complete classification of animals, so that almost any newly discovered species could immediately be placed in the correct group.

The classification itself survived the development of evolutionary biology in the last few decades of the 19th century, but its interpretation was radically changed. Rather than (at its grandest) revealing the divine plan or (at its most mundane) serving as a useful simplification of the diversity of living beings, the purpose of classification was instead to trace the course of descent through modification. Indeed, Müller used *Sacculina* as evidence of Darwinian modification for a particular way of life. Evolution shifted classification from using similarity as the basis for placing organisms into the same group to using common descent. This introduces a difficulty. A system of groups based on similarity can be arranged simply as a list, with no necessary connection between the categories. A system based on descent, however, implies that any group is a member of some more inclusive group. This is inconvenient to represent as a list, but it can easily be expressed as a branching diagram showing related groups diverging within larger groups. This branching diagram is a phylogenetic tree.

## The phylogenetic tree

The phylogenetic tree is the iconic diagram of evolution, with much the same status as the geographical map or the geological column. It is a versatile plant which can take a variety of forms, from a purely abstract diagram to a realistic picture of a tree, but it is easy to understand because of its close resemblance to a family tree. The main difference is that the tips of its branches and all the nodes—the interior branching points—are species (or sometimes groups of species) rather than individuals. The branches never fuse, as they do through marriage in a family tree, but instead continually diverge as a group radiates over time into a variety of different types. I suppose it has become so popular because it is such a compact and visually striking way of summarizing the relationships among the members of a group of organisms.

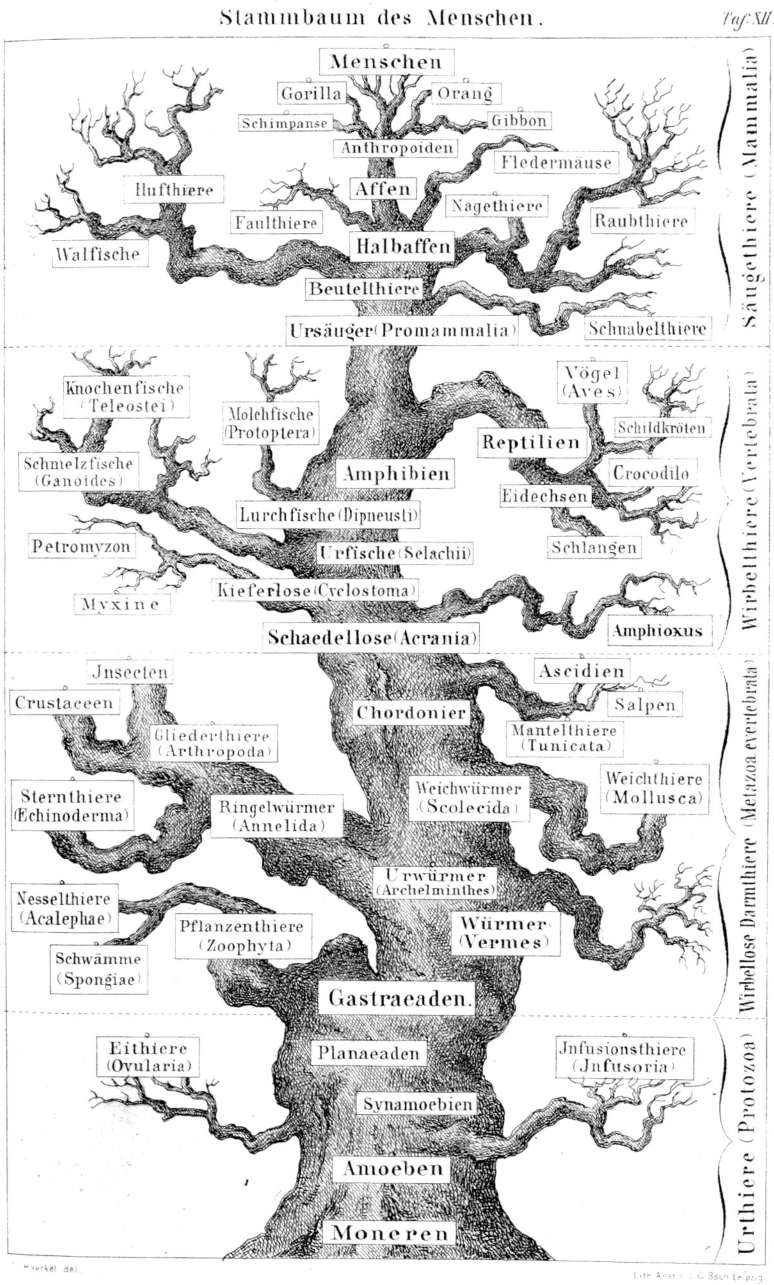


**Figure 4.1** Haeckel's phylogenetic tree

Reproduced from Haeckel, E., 1874. *Anthropogenie. Entwickelungsgeschichte des Menschen*. Wilhelm Engelmann, Leipzig. Taf. XII following p. 496.

The first tree-like figures displaying how a group of organisms is classified are actually pre-Darwinian, but after the publication of the *Origin* (with a sort of phylogenetic herb as its only figure, in the chapter on natural selection) they became almost universal. The trees published by Ernst Haeckel, in particular, made Darwin's hypothesis more concrete by illustrating the lines of descent thought to connect specific groups of living organisms. I still use Haeckel's classic oak tree (Figure 4.1), showing the evolutionary history of fifty or so groups of modern animals, to introduce my lectures on phylogenetics (for its time, it is surprisingly accurate). But how do we trace the course of descent? The only method available to classical zoologists was morphological similarity, whether of the adult or any other stage of the life cycle. The groups themselves are made up of species judged to be similar by some competent anatomist, such as molluscs (with a shell and foot) and arthropods (with segments and jointed limbs). Subordinate groups were then arranged into larger groups to give the familiar Linnean hierarchy of species, genus, family, order, class and phylum. Despite the radical change in its guiding principle, the day-to-day practice of classification went on much as before, appealing to similarity to assign species to groups. All such groups are artificial: the choice of characters used to define each group in this hierarchy is wholly subjective, and the arrangement preferred by one taxonomist is likely to be disputed by equally qualified colleagues who chose to use a different set of characters. Such disputes cannot be resolved by any objective procedure, which is why taxonomy has traditionally been such a rancorous discipline. The ambiguity of similarity is the crippling flaw of this approach to systematics.

Haeckel's oak tree, and the hundreds of similar diagrams, are pictures of classifications: the system of groups within groups has been decided previously, on the basis of similarity, and the tree then drawn to represent it. Different classifications would yield different trees. Imagine that we did the reverse: the tree is known in advance, and we use it to classify species into groups. The tree that we start with depicts the channels along which a group has evolved as a branching diagram extended in time. Each branch represents a lineage that connects a set of ancestors which lived at some time in the past with their descendants at some later time (usually the present). The lineage is typically a species, although in trees depicting higher-order relationships it will be a group of related species. Any branch may split in two at any time, when two new species arise from a single ancestral species, or it may end, when a species becomes extinct. The tree therefore depicts not only modern species, but also, implicitly or explicitly, a complete series of their ancestors all the way back to the oldest species at the root of the tree.

In this way, a phylogenetic tree expresses both descent and ancestry. Descent is the lineage forwards-in-time from any interior node of the tree. It is the outcome of diversification, the tendency of a lineage to split in two, whose outcome is the branching structure of the tree. Ancestry is the lineage backwards-in-time from any tip or interior node of the tree. It is the outcome of modification, and traces a unique sequence of ancestors that goes all the way back to the root of the tree. Diversification and modification are not contradictory, nor even in conflict, although they are sometimes supposed to be in debates about 'tree-thinking' versus 'linear thinking'. Both are inevitable consequences of Darwinian evolution, understood as modification through descent.

## The new systematics

This is all very well, but it can be put into practice only if we have some method for constructing a phylogenetic tree that does not rely on mere similarity. The answer was not found until long after Haeckel's day, not because it was particularly abstruse, but rather because the need for it was not widely recognized. The problem is to distinguish resemblance attributable to ancestry from resemblance due to parallel evolution. The best solution so far was found by a German prisoner-of-war, who scribbled it in pencil into a cheap notebook while in captivity. His name was Willi Hennig.

Like so many of the characters in this story, Hennig found his way into the local museum, in his case the Dresden State Museum of Natural History, while still at high school. People who work on museum collections are often meticulous to the point of obsession, but by Hennig's time museums had ceded the leading place they had occupied during the 19th century to the new experimental sciences of transmission genetics and embryology, located in the laboratories of university departments. Hennig went on to revise a rather obscure family of flies for his doctorate. It is difficult to get further from the leading edge of biology than the taxonomy of a rather obscure family of flies, but this was the period when he was beginning to develop his ideas about classification. They were interrupted, like much else, by the outbreak of war, but they led to a post in an antimalarial unit, one of the very few military applications of insect taxonomy. He was captured by the British right at the end of the war in northern Italy, but he was soon moved from the POW camp to the corresponding Allied antimalarial unit; there are some enemies that everyone has in common. It was here that he began to write the manuscript of *Grundzüger einer Theorie der Phylogenetischen Systematik*, which was eventually published as a book in 1950. It made little impact, partly because only a few copies were printed—the postwar paper shortage—and partly because it was written in German, which had ceased to be part of the routine education of English-speaking biologists after the First World War. It was not until 1966 that an English translation was published as *Phylogenetic Systematics*, which, together with a paper published the previous year in English, finally woke up the British and (especially) American systematists, like an ants' nest poked with a stick.

Hennig distinguished between primitive characters, whose state is the same in all members of a group, and derived characters, which have been modified, usually by natural selection, so that they have a range of different states among the members of the group. Primitive characters are useful in defining the group itself, but cannot help us to arrange species into subordinate groups. A derived character may be unique to a particular species, in which case it can be used to identify that species, but likewise cannot be used to classify species within the group. On the other hand, a derived character state which is shared by several species can be used to assign those species to a group distinct from another group, which comprises species in which the same character takes a different state. Hennig's insight was that only shared derived characters should be used to sort animals into groups which correspond to branches of the phylogenetic tree.

Let us take a point in the interior of a tree that has been constructed using shared derived characters. This interior node represents a given species in the past which is about to give rise to two new species. Now trace the complete lineage forwards in time from this given species by following each separate lineage—each line on the diagram—through all their intermediate branching points, all the way to the present (or ending when a lineage becomes extinct). The result is somewhat like an eccentrically cut slice of pizza, starting at some arbitrary point and ending at the rim. This slice of pizza cut out of the phylogenetic tree constitutes a monophyletic group, meaning one whose members all descend from a common ancestor. This is often called a 'clade' (though not by Hennig). I shall use the term 'natural group'. More formally: a natural group comprises a defined ancestor and all its descendants, living or extinct. The group may even be given a name by some ambitious taxonomist, if they think their colleagues can be persuaded to adopt it. This is how taxonomy, the artificial procedure of naming groups, becomes aligned with evolution, the natural process by which these groups arise, as a logically consistent system of classification—systematics—describing the natural relationships among groups, that is, the pattern of relationship that follows from common descent.

In principle, the branching structure of the phylogenetic tree can be used to identify and name all the natural groups that make it up. In practice, this is hopelessly unwieldy. One of the main objectives of classification is to partition the incomprehensible diversity of life into a manageable set of categories. A few minutes with pencil and paper will convince you that the number of natural groups (that is, the number of interior nodes) in a tree with N tips is N-1; so a tree which represented the descent of 100 species would require 99 names for all the natural groups. This is a trivial economy; you might just as well try to remember all the species. In practice, the system of nomenclature pioneered by Linnaeus in the mid-18th century has survived down to the present day, and a comparative anatomist of 200 years ago would be perfectly comfortable with a modern taxonomic treatise. The only difference is that a taxonomic category, of whatever rank, must correspond with a natural group; so that the most prominent natural groups do, after all, receive a name.

## Are animals a natural group?

With this background, we can ask a sharp question: do animals form a natural group? If not, then 'animal' can be used for any organism with a suite of characters such as motility and a mouth, or, to accommodate sessile creatures, at least some of them. But any of these characters could have evolved independently in unrelated lineages, as wings have evolved independently in birds, bats and butterflies. We could then abandon the search for the ancestor of all animals because there would be several or many ancestors, each leading to an organism of animal grade. However, one of the few points on which the old comparative anatomists were agreed—once Zoophyta had been cleared out of the way—is that animals do indeed constitute a natural group, stemming from a single ancestor. This group certainly deserves a name. 'Animalia' is obvious and available, and had even been used by Linnaeus as his Regnum Animale. He was a little ambiguous about sponges; he grouped them first with seaweeds, but later changed his mind and put them with the other animals. For once, however, Linnaeus was not followed—perhaps Animalia didn't seem technical enough. Instead, the term in general use was coined by Ernst Haeckel, who defined 'Metazoa' on embryological grounds, consisting of everything with ectoderm and endoderm. Even sponges had these two primary layers, according to Haeckel, who was inclined to be rather bold in his interpretation of sponge anatomy. If you need a term that refers to real animals only you can always use Eumetazoa, which excludes sponges. I don't recommend it, but sometimes it can't be avoided.

If we accept Metazoa as a natural group (we can always change our mind later) then we can go a little further. Let us go a little deeper into the tree—back further in time—to the next branching point. From here, one lineage leads to Metazoa, while the other leads to another natural group. These two are sister groups. Any natural group has a sister group, which is also a natural group. Sister groups are more closely related to one another than either is to any other group on the tree. For example, the sister group of humans among living animals is the chimpanzee (common chimp and bonobo). Sister groups can be living or extinct, however, and if extinct species are included then the sister species of *Homo sapiens* is *Homo erectus*, an early human which lived in Africa and southern Asia, and was able to control fire and make stone tools.

Sister groups can be defined in some absolute sense, implying complete knowledge of phylogeny, but it is often more useful to define them from the list of groups represented on a particular phylogenetic tree. Any phylogenetic tree is drawn up from a fixed list of taxa, and the tree as it stands refers to these taxa alone. Adding species to the list does not change the branching structure of the original tree and therefore does not change the nesting structure of sister taxa, although it will change their composition. For example, suppose that we draw the tree of modern humans and apes, but omit chimpanzees. The sister taxon of the gorilla will be humans. If we add back chimpanzees, the sister taxon of gorilla will still include humans, but with the addition of chimpanzees, that is, {humans + chimpanzees}.

What is the sister group of Metazoa, given a wide sample of unicellular groups? The old comparative anatomists came up with the rather surprising nomination of choanoflagellates. These live in both freshwater and the sea (although they are much easier to find in freshwater, from my experience) and have a distinctive structure of an oval cell body provided with a flagellum which beats inside a sort of collar or skirt. The flagellum draws in a stream of water which passes through closely-spaced fibres in the collar, filtering out bacteria which are then engulfed by the cell. Choanoflagellates often form colonies, consisting of several cells joined together to form a star-shaped or tree-shaped structure. They would be little more than a curiosity, except that remarkably similar cells, the choanocytes, are found in sponges. The sponge body is in effect a very efficient device to filter bacteria. The beating of the flagella of choanocytes massed in the interior of the sponge draws in a stream of water through pores in its coat, from which the bacteria are filtered, just as they are by choanoflagellates. Moreover, similar collar cells capable of capturing and digesting food particles are widespread among animals, in the gastric cavity of corals, for example. The resemblance of choanoflagellates to the choanocytes of sponges, the broad distribution of similar kinds of cells among animals, and the absence of collar cells in other kinds of eukaryote combine to suggest that choanoflagellates are the sister group of Metazoa.

The same line of argument seems to identify the ancestor of Metazoa as a colonial choanoflagellate which, along the lineage leading to Metazoa, evolved a multicellular body with an external layer of protective cells while its flagellated cells were relocated to the interior. The earliest fossils in the animal lineage should, then, be sponges, or at any rate proto-sponges. Before we accept or abandon the appealing notion of a primeval colony of bacteria-filtering cells, though, we should take a closer look at what we mean by 'ancestor'.

## The elusive ancestor

The first problem is the sheer improbability of finding a remote ancestor. Take any population at some time in the past, and for the time being assume that it is asexual and constant in size. Each individual represents a lineage from the previous generation. Some of the individuals will reproduce, giving rise to the same number of individuals in the next generation, whereas others will die before reproducing. There are always fewer lineages in the next generation than in the current generation, and by extending this argument the number of lineages must continually fall over time. It follows that a single individual living in the remote past eventually gave rise to all the individuals alive today; all the other individuals in the population, no matter how many, left no descendants at all. This is a familiar outcome of experimental evolution, where we might, for example, expose a population of billions of bacteria to a high concentration of an antibiotic. A mutation that occurs in a single cell confers resistance to the antibiotic, giving the lineage stemming from this cell a competitive advantage that will eventually result in it replacing all others. What is true on a small scale is equally true on a larger scale. During the early history of cells the oceans must have swarmed with primitive versions of metabolism and replication. One such version, initially arising in a single cell, flourished and eventually gave rise to modern bacteria and archaeans; others may have flourished for a time, but all eventually became extinct without leaving any direct descendant. Any contemporary population therefore coalesces to a single individual in the remote past.

Coalescence is easy to understand in an asexual population, but sexual populations likewise lose lineages over time and eventually coalesce to a single mated pair, although the details are more complicated. This mated pair was the ancestor of all the current members of the population, or the species, or, come to that, of all Metazoa. The coalescent is a special kind of ancestor: it is the most recent common ancestor of the group it belongs to. A single individual, or a single mated pair, is not, however, the usual meaning of 'the most recent common ancestor'. The phrase usually refers to an entire species, which in palaeontology usually means a set of individuals from a given time period whose character states fall within a given range. Leaving aside the problems of this definition, there must have been a great many species in the 20 or 30 million years before the Cambrian, and the probability of finding any one of them is surely very small. Even given a rather relaxed definition, we might never find the most recent common ancestor of Metazoa.

## Stem and crown

Even if we were to find it, how would we recognize it? The answer seems straightforward: it would have all the shared derived characters of the group which descends from it. It is still most unlikely that we shall stumble across the veritable ancestor, a member of the sole lineage from which all subsequent members of the focal group descend. What we might find, if we are lucky, is a species which, judging from its shared derived characters, resembles the true ancestor and is related to it. I have called this species the sibla taxon (for shortest inferred branch length from ancestor). The term has not caught on, but nevertheless the term 'ancestor', when used by a palaeontologist, invariably means 'sibla taxon'.

Now let us move backwards in time along the lineage leading to the most recent common ancestor, until we reach the next branching point, at which one lineage leads to Metazoa and the other to its sister group, the choanoflagellates. The attributes which define the Metazoa would almost certainly have been accumulated sequentially, one by one, rather than all arising simultaneously. The lineage connecting the series of ancestors would show a stepwise incorporation of the specific attributes of Metazoa over time, with each species in the series retaining the attributes of its predecessor and adding one of its own. We do not know, of course, whether any particular fossil is a member of this series, but it might be a sibla taxon, a member of some extinct group fairly close to the ancestral lineage itself. This succession of ancestors, together with the extinct groups that descend from them, has been called the *stem group*. The stem consists of all the ancestors and their descendants of a living group, except its common ancestor and all its descendants. All members of the stem are now extinct.

The species descending from the common ancestor itself, living or extinct, collectively constitute the *crown group*. The crown group includes the lineage leading from the ultimate common ancestor to the most recent common ancestor, so the total group is neatly partitioned into stem and crown.

All this sounds rather forbiddingly taxonomic, but in practice the distinction between stem and crown can be very enlightening. An example may make it plain. Birds have feathers, wings and a horny toothless beak, which seem likely to have evolved in that order. The stem group includes feathered dinosaurs (such as *Ornithomimus*, which had feathers alone) and proto-birds (such as *Archaeopteryx*, which had feathers and wings), while the crown group comprises the common ancestor of modern birds (an unknown animal thought to have lived about 90 million years ago) and all its descendants (which have feathers, wings and beak).

If we follow the stem group backwards in time, almost but not quite to the next branch point, we should find a species which lies on the direct line of descent of the crown group, but has none of its attributes, because even the first has not yet evolved. This is the most remote common ancestor of the group. But how could we recognize it as such? We might suspect that it belongs to the ancestral lineage because it resembles early members of the stem group in other respects but lived earlier still, for example a small featherless bipedal dinosaur. Given a complete fossil record, we should be able to reconstruct the ancestry of the crown group all the way back to its most remote common ancestor. The fossil record, however, is famously incomplete, and grows more fragmentary as we probe further back in time.

## Embryos

Charles Darwin and Ernst Haeckel had similar backgrounds, to a point. Both were middle class, of course, so it was easy for them to begin studying for a medical degree and possible for them to give this up in favour of their natural inclination towards a poorly paid or unpaid career as a naturalist. Both travelled when young (Darwin much more extensively, to be sure) and afterwards settled down and never moved again: Darwin at Down House, in the countryside south-east of London, Haeckel in a large house in Jena, on the corner of Berggasse and what is now Ernst-Haeckel-Strasse. Both wrote very long taxonomic accounts of obscure groups, Darwin on barnacles and Haeckel on radiolarians. Less expectedly, both married a first cousin. But their lives then diverged. Haeckel's wife, Anna Sethe, died very young, whereas Darwin's wife, Emma Wedgwood, outlived him. Haeckel held down a senior academic position at Jena throughout a long career, whereas Darwin cannily invested the Wedgwood money in railway stock and never held a salaried position. Above all, their personalities were radically different. Darwin was famously a reclusive figure who avoided publicity and controversy, even shielding his wife from the implications of his ideas, whereas Haeckel took centre stage, published several books on evolution for a general audience and played a major role in public life. Everyone seems to like Darwin, as a rather meek and unassuming Victorian liberal scarcely aware of his transformative position in history. Haeckel is stronger meat, attracting virulent criticism during his lifetime and since, down to the present day, because of his uncompromising promotion of Darwinian principles.

### Haeckel's universal Darwinism

In the summer of 1860 Haeckel had just returned to Berlin from travels in Italy and was working on the radiolarians he had collected when he read Georg Bronn's German translation of the *Origin of Species*. He reacted with Pauline vehemence. He reasoned that evolution driven by natural selection was by no means restricted to biology, but was instead a general theory of everything, capable of explaining the whole of the natural world. He called this general theory 'monism', probably borrowing the term from August Schleicher, a linguist at Jena who, before the *Origin of Species*, drew family trees of languages which earned him the derision of his colleagues. Monism is the belief that mind and body are aspects of the same phenomenon, in contrast to Cartesian dualism, which regards them as fundamentally different. Human beings are animals like any other, part of the natural world, and human society, civilization and culture are the outcome of natural laws and owe nothing to supernatural causes such as divine creation. These laws include evolution by natural selection, which implies that all organisms descend from a single common ancestor, and in particular that humans descend from an ape-like predecessor. Haeckel wrote several popular books that explained his worldview, including *The History of Creation* (*Natürliche Schöpfungsgechichte*) in 1868, *The Evolution of Man* (*Anthropogenie, oder Entwicklungsgechichte des Menschen*) in 1895 and his monist manifesto *The Riddle of the Universe* (*Die Welträtsel*) in 1899 (I have a battered paperback translation issued by the Rationalist Press Association). It has been said that Haeckel introduced more people to evolution than Darwin, and this may be true: his books sold hundreds of thousands of copies in many languages. Haeckel disregarded the boundaries (notably human consciousness) limiting the application of Darwinism set by other biologists, such as Emil du Bois-Reymond in Germany and Alfred Russel Wallace in England, declared that the concept of a mind independent of physical processes was the product of poetic imagination, and attacked Christianity as a misleading superstition, one of those 'mystic views which would subdue reason under the yoke of an alleged revelation' (Joseph McCabe's translation). These revolutionary views naturally drew vitriolic attacks from conservative academics and the Church, making Haeckel a prominent and very controversial figure.

Haeckel founded the German Monist League (Monistenbund) in 1906 as a platform for his philosophical views, and for a while it promised to be a serious political movement; but it fizzled out after the First World War and was eventually closed down by the Nazis in 1933. The contribution that was made by monism in general, and by Haeckel in particular, to the rise of fascism in Germany and communism in Russia has been tirelessly debated by historians and philosophers, but I doubt that it was ever consistent or decisive. The references to higher and lower races in *The Riddle of the Universe* certainly jar on modern sensibilities, but its insistence on secular ethics and the equality of men and women, outspoken at the time, now sound unexceptionable, at least in Western countries.

Philosophers continue to earnestly debate the mind-body problem and have come up with a bewildering variety of interpretations. Scientists who study the brain, either in humans or in other animals, do not seem to debate the issue at all. I doubt that editors or reviewers of neurobiology journals would look kindly on manuscripts which referred mental phenomena to immaterial, vitalistic or supernatural causes, and my neurobiology colleagues take it for granted that any particular behaviour is caused by specific neural pathways in the brain, driven by chemical reactions. This is the monist position, but one could try to finesse the issue like Stephen Jay Gould, a very erudite evolutionary biologist, who argued that science and religion constitute two 'non-overlapping magisteria', separated by a fenced-off demilitarized zone that prevents any conflict. If so, it works as well as any fence between irreconcilable opponents who itch to conduct raiding parties into enemy territory.

The unconscious monism of most scientists, and of evolutionary biologists in particular, has a disturbing and widely feared implication. If thought has a physical basis, then in principle, and perhaps in practice, it can be controlled by the appropriate physical interventions. Technocratic organizations, such as the state, large corporations and political movements, might be able to direct our thoughts, or seek out and punish us for subversive thoughts. Orwell's thought police, and their more sophisticated modern equivalents, are the ultimate nightmare of the rationalist. We read that the truth shall make us free (John 8:32), but what if it instead leads to an enslavement more absolute than any previously known? Haeckel's exposition of Darwinism casts a long shadow.

## Recapitulation

Haeckel's success as a popularizer was aided by his skill as a draughtsman. Darwin made little use of illustrations; the only figure in the *Origin of Species* is that phylogenetic herb, a rather fussy branching diagram illustrating identity by descent. Haeckel, on the other hand, believed that a picture was worth a thousand words, and (in *The Riddle of the Universe*) included drawing as an important part of the ideal education: 'Every pupil must be taught to draw well, and from nature . . . it gives the pupil his first lesson in seeing correctly and understanding what he has seen'. He was a gifted artist who had the good fortune to be assisted by highly skilled lithographers. His plates of the glassy architecture of radiolarians put me in mind of the intricate tessellations of M.C. Escher or the geodesic dome that Walter Bauersfeld constructed at Jena, and his free-flowing pictures of medusae and siphonophores would grace any gallery (where I would prefer them to dull landscapes or the stiffly posed portraits of long-dead royalty). I can't think of any other scientist whose artwork was published—and has recently been republished—as separate stand-alone books. Photographs may be a more faithful depiction of reality than drawings (or were, before digital technology made it possible to manipulate them at will), but drawings have the advantage that features which rarely occur together in a single organism can be combined, and particularly important features emphasized or exaggerated. This can make them more useful than photographs as a teaching aid, and Haeckel would have made good use of them during his 40 years at Jena. But this advantage conceals a trap: the artistically facile scientist may be led insensibly to add features which are not really there, or to omit features which are, in order to correspond with some preconceived view. Haeckel fell into this trap, and caused a scandal a century after his death.

Haeckel's mission was to contrive a Darwinian interpretation of the leading facts of comparative morphology. To begin with, there was the old problem of whether the process of development itself was rolled out by preformation or by epigenesis. Haeckel cut the knot by arguing that preformation was essentially a creationist theory because it required that the whole of living nature had been packaged by God at the beginning. Epigenesis, on the other hand, was consistent with Darwinism because every individual evolves (so to speak) from simple beginnings to a more complex adult form. Besides, the most careful observations of embryogenesis seemed to rule out preformation. Caspar Wolff had studied chick development (because hens' eggs are large and easy to obtain) a hundred years earlier and found no trace of a long chain of smaller and smaller beaks and feathers, but only the orderly differentiation of sheets, tubes and organs.

The next topic was the sequence of stages during development. Earlier in the 19th century, Karl von Baer had traced the origin of the ducts and organs of the adult to the membranes of the early embryo, which he called the Keimblätter, usually translated as 'germ layers'. Thomas Huxley called them 'foundation membranes', and wrote that 'a Medusa consists essentially of two membranes'. Perhaps the idea can be traced back to Cuvier, who described the polyp as being formed from a double membrane whose two layers correspond to ectoderm and endoderm.

Von Baer described how the early embryos of different kinds of animal usually resemble one another more closely than the adults do, and concluded that more general forms, earlier in development, change into more specific forms later on. This seems reasonable. If development is epigenetic, with each stage contingent on the preceding stage, then the pathways taken by different kinds of animal will tend to diverge from egg to adult. Darwin concurred; after all, he had recognized barnacles as crustaceans because of the characteristic body plan of their nauplius larva, which they share with other crustaceans such as crabs, shrimps and copepods. He took this as evidence for evolution, since, in his words, 'community in embryonic structure reveals community of descent', although he is careful to point out that this will apply only when embryos continue to follow the ancestral way of life.

Haeckel took the idea a little further. It had long been known that the course of development often depended on the way of life of an animal (terrestrial versus aquatic would be an extreme example), so it is first necessary to specify the original way of life that was followed early in animal evolution. This was imagined (on good grounds) to be external fertilization, followed by a rather stereotyped succession of stages living in the open sea, leading to a swimming larva which eventually becomes the characteristic adult of a particular group. This focused attention on marine invertebrates with this sort of life cycle: hydroids and jellyfish, flatworms, polychaete annelids (such as ragworms and lugworms), molluscs, echinoderms and several smaller groups. The cleavage of the fertilized egg produces a stack of cells, twisted in some groups, which eventually forms a hollow ball, flagellated on the outside, called the blastula. Cells move into the interior, either by migrating inwards from the surface or (and originally, according to Haeckel) by invaginating the surface of the blastula, like a finger poked into a hollow rubber ball, forming an external opening, the blastopore, leading to a blind tube. This stage is called the gastrula, in which the distinction between ectoderm (the sheet of cells on the outside) and endoderm (the sheet of cells forming the wall of the tube) becomes established. Gastrulation creates the two germ layers of ectoderm and endoderm, which in the simplest animals, such as sea anemones and jellyfish, go on to form all the tissues of the adult. In bilaterians there is a third germ layer, the mesoderm, which gives rise to many of the internal organs of the body. The similarity of development, Haeckel argued, is the consequence of ancestry and shows that animals form a natural group because they all descend from a common ancestor.

Haeckel bolstered his argument with drawings of vertebrate embryos which purported to show that groups in which the adults were very different—salamander, bird, dog—were almost identical as embryos. His drawings were reproduced in many textbooks as important evidence for evolution. A century later, Michael Richardson of St. George's Hospital Medical School in London, together with colleagues in France, Australia, Canada and the United States, was sceptical, and checked Haeckel's conclusion by conducting a much broader survey of vertebrate embryos. They found that there was much more variation among vertebrates than Haeckel had allowed, and concluded that their survey 'seriously undermines the credibility of Haeckel's drawings, which depict not a conserved stage for vertebrates, but a stylized amniote embryo'. The article was picked up by *Science*, one of the most respected of international science journals, where Elizabeth Pennisi wrote an article headlined 'Haeckel's Embryos: Fraud Rediscovered'. Creationists, who still flourish in backward areas of the world, were delighted. They had said all along that the evidence for evolution was lies, and this had now been confirmed by the scientists themselves! Richardson and his colleagues were horrified, and quickly published a clarification in *Science*. Haeckel was fundamentally correct, they admitted: all vertebrates develop a similar body plan which reflects their shared evolutionary history. 'Fraudulent' was downgraded to 'overzealous'. It was too late. The choice of a sensational headline had encouraged folly and ignorance, blackened the reputation of a leading early advocate of evolution, and brought into lasting doubt the editorial judgment of *Science* magazine.

## The Gastraea

Haeckel coined the term 'monophyletic' (*monophyletische*) to denote a group whose members all descend from a common ancestor, and like many of his coinages ('ecology' and 'phylogeny' are a couple of others) it has stuck. His conclusion that the Metazoa (another Haeckel coinage) is a monophyletic group was accepted by zoologists in the classical period and has survived the rise of gene-based systematics in recent decades. One might be tempted to go further: if common ancestry leads to resemblance, then perhaps resemblance can be used to infer the ancestral state. Darwin came to this conclusion in the first edition of the *Origin*: 'the embryo is the animal in its less modified state; and in so far it reveals the structure of its progenitor'. A few years later he was more explicit: 'it is probable that with many groups of animals the earlier larval stages do show us, more or less completely, the form of the ancient and adult progenitor of the whole group'. Haeckel picked up the ball and ran with it. Many marine invertebrates pass through a characteristic series of stages marking the course of their development from a fertilized egg, which cleaves to give a more or less solid clump of cells, then divides further to form a hollow blastula, which invaginates to produce a gastrula. Since the germ layers fundamental to the animals are formed during gastrulation, the common ancestor of metazoans must have been a minute cup-shaped animal that lived in the plankton and looked like a gastrula (Figure 4.2). This was his Gastraea theory (which sounds much more impressive in German as *die Gastraea-Theorie*). It was a brilliant stroke that captured the imagination of zoologists everywhere. In place of the bland statement that animals must all descend from a common ancestor there was now a vivid concrete image of the ancestor itself.

Haeckel developed these ideas while he was studying calcareous sponges, an obscure group of animals that are undramatic even by sponge standards. He was interested in them because they are simple animals with a simple mode of development. He saw the fertilized egg develop into a hollow egg-shaped larva consisting of two cell layers, a ciliated surface ectoderm and an internal digestive endoderm, with an opening at one end. This is the stage for which he coined the term 'gastrula' and which was the

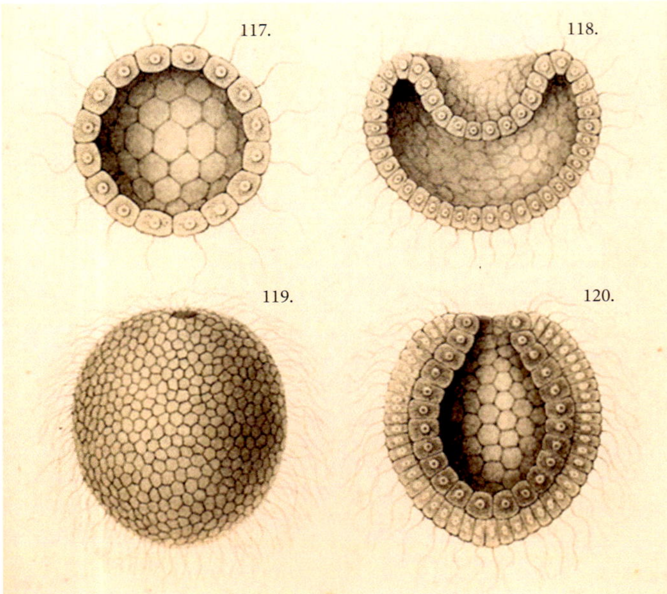


**Figure 4.2** The Gastraea

Reproduced from Haeckel, E., 1875. Die Gastrula und die Eifurchung der Thiere. *Jenaische Zeitschrift für Naturwissenschaft*, 9, 402–508.

source of his Gastraea theory. The larva soon settles on a hard surface, with its opening upward, and metamorphoses into a young sponge in which the ectoderm secretes spicules and the endoderm changes from a gut precursor into the lining of the internal choanocyte chambers. Not many people study the embryology of calcareous sponges, but those who did at the time mostly disagreed with Haeckel about the details of this process. He cannot be severely faulted, since arguments about sponge development are still going on, but he could certainly have chosen an easier group to study. No matter; his grandiose vision extended to all animals, and technical arguments about sponges were soon forgotten, at least outside the narrow circle of sponge specialists. It was this vision, with its tempting promise that the nature of the common ancestor had been revealed, that engrossed the attention of zoologists, and continues to be discussed down to the present day.

## The ciliation constraint

The Gastraea theory is built on the observation that animals developing from eggs in the open sea go through a characteristic series of stages in which gastrulation is the central event. This begs the question of why gastrulation should occur in the first place. After all, plants and seaweeds have evolved multicellular bodies that develop from an egg without forming a gastrula at any stage. Gastrulation is unique to animals because of a unique feature of animal cells that was pointed out by Leo Buss in his revolutionary book *The Evolution of Individuality*: no metazoan cell can divide while ciliated. This constraint does not apply to other multicellular organisms. The giant colonies of *Volvox*, for example, are built by ciliated cells that are able to divide. Choanoflagellates, though, can't do this, and must absorb and depolymerize the whole of the flagellar apparatus before replicating the basal bodies of the mitotic spindle and proceeding to divide. I don't know how this constraint arose, but it was inherited by the first animals from their choanoflagellate predecessors, and blocked development altogether. The problem is that the blastula depends on its uniformly flagellated cells to keep it in suspension, and if these withdraw their flagella in order to divide and differentiate it will sink. Sponges have found several ingenious solutions, for example the foremost cells can retain flagella while those at the rear divide, so the larva as a whole can swim and divide at the same time. But these solutions work only in very small animals and allow only a limited degree of differentiation.

Buss pointed out that gastrulation is a more satisfactory resolution of the conflict between swimming and differentiating. The cells that are relocated to the inside of the embryo are free to withdraw their flagella in order to divide and differentiate, while those remaining on the outside can retain their flagella to propel the gastrula. Gastrulation is not merely an idiosyncratic feature of development that sets animals apart from plants and seaweeds, but is rather a process that makes functional sense, given the constraint imposed on animal development in general.

## The Plakula

Haeckel had done more than propose a novel solution to one of the problems that Darwin's theory had raised. He had founded a whole new field of speculative zoology, in which generations of theoretically minded biologists have used their imagination to devise plausible candidates for the common ancestor of animals. Otto Bütschli, for example, was another grand old man of biology, professor at Heidelberg for 40 years, who has nowadays been almost completely forgotten—even the site of his grave was uncertain until recently—although he probably did more than anyone to further our understanding of single-celled organisms. He was sceptical that a primordial gut could evolve gradually in a more or less spherical gastrula, because it would be of little or no value before it was able to admit, retain and digest food particles. He accepted the concept of ectoderm and endoderm as the two main germ layers, but stacked them on top of one another to make a plate-shaped animal (the 'Plakula' of later authors) that crawled on the sea bed (Figure 4.3). The lower layer was a digestive endoderm, the upper layer a protective ectoderm. Curving the body would create a space beneath where food particles could be trapped and digested. Increasing curvature would enable larger particles to be captured, eventually culminating in a narrow mouth-like opening to a gut cavity whose digestive function had already been established. All this seems quite reasonable.

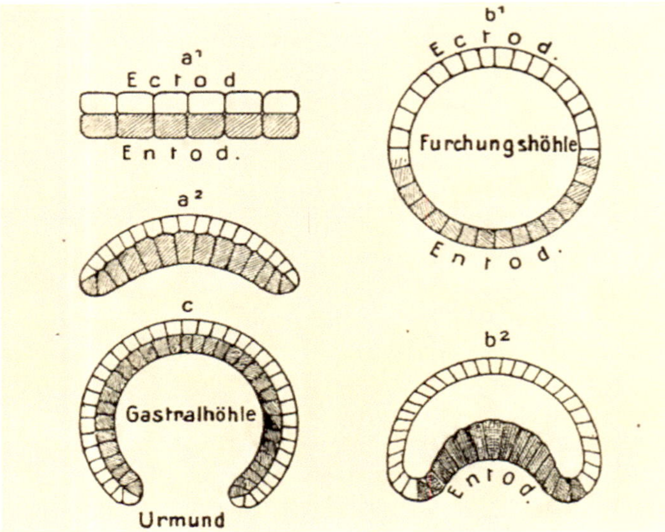


**Figure 4.3** The Plakula

Reproduced from Bütschli, O., 1921. *Vorlesungen über Vergleichende Anatomie*. Julius Springer, Berlin, p. 9.

Other reasonable ideas were put forward. Perhaps the original metazoan was a minute solid-bodied animal resembling the planula larva of a sponge or hydroid. Once it had evolved a mouth it would look something like a simple flatworm, with a ciliated epidermis outside for locomotion and a mass of cytoplasm inside for digestion. This is the planuloid-acoeloid theory of Élie Metchnikoff and Ludwig von Graff. Metchnikoff was a biologist who moved from Odessa to Paris in 1882 (forced out, I think, after taking the wrong side after the assassination of Alexander II) and is best known for his discovery of white blood cells that engulf bacteria, the basis of the immune response. As a side-project he proposed that the ancestral animal was not Haeckel's Gastraea, arising from the invagination of the blastula wall, but rather a solid-bodied Parenchymella, similar to the planula larva of hydroids, whose interior is filled, not by invagination, but rather by cells moving from the outside to the inside of the blastula (Figure 4.4). Von Graff was a less charismatic zoologist at the University of Graz in Austria at about the same time, in the closing years of the 19th century. He was an authority on flatworms, which are fascinating to zoologists although repulsive to everyone else, and even wrote a book about the very simplest kind of flatworm, the acoels. These are small and exceedingly simple animals which have a mouth and a nervous system but not much else. Acoels have neither brain nor eyes, have no circulatory or excretory systems, are not segmented, and lack a coelom or any other kind of body cavity; even their blind gut is a solid mass of cells. If the Parenchymella were the ultimate ancestor, then it might easily have evolved into an acoel flatworm from which all more complex Bilateria were derived.

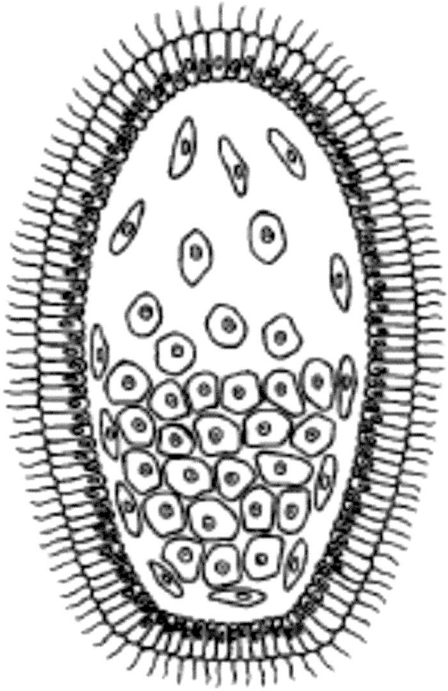

**Figure 4.4** The Parenchymella

Reproduced with permission from Ereskovsky A.V. and Dondua, A.K., 2006. The problem of germ layers in sponges (Porifera) and some issues concerning early metazoan evolution. *Zoologischer Anzeiger, 245*, 65–76, fig. 5C.

Some ideas were not so reasonable. For example, the notion that metazoans originated when large protozoans such as ciliates were divided up into cells won many adherents for a time, but was discredited as soon as it became clear that ciliates were not closely related to animals and could not make credible ancestors. Others were frankly bizarre, such as the suggestion that metazoans evolved from bacterial slime moulds which somehow acquired a nucleus. All that was needed was a fertile imagination and a good knowledge of general zoology, and even the latter was not absolutely essential.

# 5
# The Grand Design

## Morphologists

Haeckel wrote in German, of course, and most of us know him only through translations of his main works. Before 1914, indeed, the literature of biology was written mainly in German, and much of it has never been translated. Those who wrote it did not always think very highly of their English-speaking colleagues, even the most eminent. Here, for example, is part of a review by 'F.M.B' of the German translation by J.W. Spengel of T.H. Huxley's *Manual of the Anatomy of Invertebrated Animals*:

> In this country we have been accustomed in our simplemindedness to think that Prof. Huxley possesses a singular talent for exposition, while his reputation amongst us as an anatomist is based on our belief that his knowledge of anatomical facts is as wide and extensive and as well kept up as his critical judgment is acute, and his treatment of morphological problems broad and original. We have for some time past been under the idea that Prof. Huxley has had a good deal to do with the progress of animal morphology during the last twenty or thirty years. But we live to learn, and we feel very grateful that a man of Dr. Spengel' s standing should show us how imperfect and unequal (*liickenhaft und ungleichmassig*) is Prof. Huxley's treatment of the subject to which he has devoted his life.

The worm turned, of course. At the time of Haeckel's death, German society, already shocked by military defeat, was about to be buffeted further by hyperinflation and a worldwide recession. Military defeat was followed by scientific eclipse. The approach through morphology and development that had been the backbone of evolutionary biology in its first half-century seemed increasingly tired and old-fashioned next to the new Mendelian genetics. Particulate inheritance made it possible to reconceptualize evolution by natural selection as the replacement of one allelic form of a gene by another, and to analyse this process with mathematical rigour. This was mainly the work of British and American scientists, and English began to supplant German as the primary language of biology. This trend was firmly cemented in place by the imposition of ideological theories of human racial variation in Germany and of transmission genetics in the Soviet Union during the 1930s. These ensured that no important work in biology was published in German or Russian for a generation.

It was during the next war that Julian Huxley published the manifesto of neo-Darwinism as *Evolution: the Modern Synthesis*. The contrast with the earlier literature is quite stark. About half the book is devoted to species and speciation, with just two

*The Sea-Change*. Graham Bell, Oxford University Press. © Graham Bell (2026).
DOI: 10.1093/9780197797464.003.0007

paragraphs devoted to palaeontology, while the entire theme of development is relegated to a short and rather eccentric section late in the book. Haeckel has disappeared: his name does not appear in the index (not from any lack of historical perspective, since Darwin has twenty entries), none of his books are listed in the bibliography, and, beyond a brief mention of 'recapitulation', there is no mention of the biogenetic law or the Gastraea theory. The old morphological school had been consigned to the dustbin of history.

The problems that the old morphologists had grappled with, however, had only been shelved; they had not been solved, nor had they gone away.

Libbie Hyman wrote two laboratory manuals in her early thirties, one on general zoology and one on vertebrate anatomy, and intended to write a third on her real love, invertebrates. Her colleagues persuaded her to write instead a comprehensive treatise on all the non-vertebrate animals, and, supported by the royalties from her two successful manuals, she took an apartment near the American Museum of Natural History in New York in 1932 and settled down to work. She had a doctorate in zoology (from Chicago) but no paid academic position, then or later. The project soon expanded beyond a single volume to become a set of three volumes, but each of these became divided into two or more parts, until it became clear that a whole shelf of books would be required to cover the subject in sufficient detail. She had underestimated the amount of work required, too, and the first volume (*Protozoa through Ctenophora*) took 6 years to write and was not published until 1940. The second volume (*Platyhelminthes and Rhynchocoela*) did not appear until a decade later, although a third volume (*Acanthocephala, Aschelminthes and Ectoprocta*) was published in the same year; the fourth volume (*Echinodermata*) was published in 1955, and the fifth (*Smaller Coelomate Groups*) in 1959. Time was running out by now. Her health collapsed in 1961, and she struggled to complete a sixth volume (*Mollusca I*) in 1967. She died in 1969 at the age of 80 with her magnum opus nearly 5,000 pages long but still seriously incomplete—no annelids, cephalopods or arthropods—and out of date. It is nevertheless my first port of call when I come across an unfamiliar group of animals and seldom fails to enlighten. It was perhaps the last time that a single mind attempted to understand animals in their entirety (except vertebrates, which, despite the success of her manual, she never cared for much) and in detail. Libbie Hyman's *The Invertebrates* is to zoology what Samuel Johnson's *Dictionary* is to English literature.

She stood at the end of a long tradition of comparative zoology. Until the 1960s, university zoology departments were staffed with specialists in groups such as echinoderms, crustaceans or insects, and the curriculum was based on a detailed survey of animal groups. After that time, new hiring increasingly favoured ecology and molecular biology, until the old professors retired and the traditional curriculum became unsustainable and was abandoned, or relegated to a superficial treatment in an introductory course. Most biology students nowadays have never heard of a sipunculid (one of the *Smaller Coelomate Groups*) or a ctenophore, and would be hard pressed to recall the basic body plan of a starfish or a snail. A century of comparative zoology, however, culminating in Hyman's treatise, had led to a clear and compelling account of how animal body plans have evolved.

## The choice of characters

The first problem was to choose features that help us to reconstruct animal evolution. There is a very long list of candidates, but the choice fell on features that occur early in development. The reason is straightforward: features that are expressed late in development (such as the colour of feathers in the head crest of a bird, or the thickness of a snail shell) are readily modified to produce a new variety or species, whereas altering a feature of the early embryo (such as the pattern of cleavage or the structure formed by a cell lineage) is likely to lead to an inviable or severely disabled adult. The body plan of the adult is circumscribed by events early in embryonic development, each of which switches development onto a pathway leading to a particular kind of adult. This is why the main features of early development are very similar in most animals, at any rate in most free-living marine animals.

Once the germ layers have been formed during gastrulation, the course of development follows a series of switch-points that define a number of different routes. The members of any given group of animals follow a characteristic route so long as (Darwin cautions) their way of life has not been modified. The interior tube formed during gastrulation or shortly afterwards remains blind in some groups but develops a secondary opening to the exterior to form a continuous gut in others. The original opening, the blastopore, becomes the mouth in some or the anus in others. The gastrula develops into a larva with bands of cilia arranged in a characteristic fashion. These propel it through the water, before it settles on the seabed and undergoes a more or less rapid series of changes, or metamorphosis, which transform it into the adult. The space between the gut wall and the body wall may be a simple gelatinous layer, or may be occupied by a solid mass of cells, or may develop into a secondary body cavity (the lumen of the blastula being the primary cavity).

A secondary cavity may or may not be lined with tissue arising from the third germ layer, the mesoderm, which subsequently forms most of the internal structures of the body. A secondary body cavity lined with mesoderm is called a coelom, a word unknown in common speech but capable of exciting the strongest passions in zoologists. The coelom may be a simple open space, or it may be partitioned by sheets of mesoderm into two or three separate compartments, or it may be serially partitioned into a large number of segments. The body of the adult also contains a nervous system and sense organs (formed principally from the ectoderm), a variety of internal structures such as circulatory and excretory systems (formed principally from mesoderm), and of course the gut tube (formed principally from endoderm).

The two main lineages of bilaterians are protostomes and deuterostomes, distinguished by their different arrangements for filter-feeding. Any elementary zoology text will list other features which distinguish between them. There are usually three of these. In early development, deuterostomes have radial cleavage so that the second quartet of cells sits directly on top of the first quartet, whereas protostomes have spiral cleavage, with the plane of division twisted to one side or the other, so that the second quartet sits in the furrows between the cells of the first quartet, like a pile of oranges arranged by a greengrocer. Subsequently, the developing embryo forms a hollow ball, the blastula, which is filled in by cells originating at a particular point, the blastopore, on its surface. The blastopore forms the mouth in protostomes but the anus in deuterostomes. Later on, a secondary body cavity, the coelom, develops from splits in the internal tissue in protostomes ('schizocoely') but by outgrowths from the gut in deuterostomes ('enterocoely'). Protostomes are thus schizocoelous animals with spiral cleavage and the mouth developing from the blastopore, whereas deuterostomes are enterocoelous animals with radial cleavage whose anus develops from the blastopore. These clear textbook criteria, however, are rules of thumb rather than rigorously consistent distinctions. Mark Martindale, the renowned invertebrate zoologist who runs the Whitney Marine Laboratory at the University of Florida, has emphasized the variability of development: some rules (such as the development of the anus from the blastopore in deuterostomes) seem to followed to the letter, others (such as spiral cleavage in protostomes) can be modified or even annulled, and then there are some (such as the manner of coelom development) which are sometimes irrelevant (if the coelom has been lost) and often difficult to ascertain. Nevertheless, the distinction between protostomes and deuterostomes remains a useful way of recognizing the two major groups of animals that have evolved from polyps.

## The phylum

This is a crudely simplified account of development, but it will serve to identify some of the main signposts that have been used by zoologists. Most of them are rather abstract, or at least far removed from everyday experience. The next step, however, is to sort animals into distinctive groups, many of which are familiar to everyone—flatworms, molluscs, arthropods, annelids, echinoderms and so forth. Some are less familiar (bryozoans and rotifers, for example) and a few are known only to zoologists (such as sipunculids and phoronids). Each such group is a phylum. All are defined by some unique morphological feature, such as the stinging cells of cnidarians or the five-fold symmetry of echinoderms, which has no obvious homology in other kinds of animal. The word 'phylum' was coined by Haeckel (of course), who used it for the six main branches of his classification. It came to be used for animals having the same basic body plan, or 'general anatomical construction' in Hyman's phrase. I think of it as the highest taxonomic level for which membership is unequivocal. If you hand a strange animal to a zoologist, they might or might not be able to identify it, but they could almost certainly tell you which phylum it belonged to. There are exceptions, of course, such as *Sacculina*, but it makes a good rule of thumb.

Hyman herself recognized twenty-two phyla, although this includes six classes of Aschelminthes which are all usually regarded as separate phyla. From time to time an animal is discovered which cannot be placed in any existing phylum, so one has to be created for it. Phyla are occasionally demoted when they are found to branch within some other phylum, which they are necessarily a part of. These novelties and demotions tend to cancel out, so the number of phyla recognized by the people who write the textbooks is more or less constant at about thirty.

There is, of course, no Hennigian justification for giving special treatment to phyla. They are merely different levels of a phylogenetic tree, and the reasons for treating them as something special are purely subjective. Nevertheless, the list of phyla is much more stable than that of any other taxonomic group, and this may be telling us something important about animal evolution. It certainly provides a convenient dichotomy separating the relationships of species within phyla from the relationships among the phyla themselves. We are concerned with the latter problem, and in particular with the first steps in the radiation of animals. Having decided on the characters that we are going to use, the next job is to put them in order of appearance.

## The great debate

It is difficult for us to imagine nowadays how exciting zoology must have been in the closing decades of the 19th century. Darwin had supplied the key, Haeckel had turned it in the lock, and all that was needed was a gentle push for the door to open and reveal the whole prospect of animal evolution. All the questions that had puzzled zoologists for centuries would be resolved at last. In particular, the origin of germ layers, the coelom and segmentation would be revealed.

The germ layers were easy. The existence of two germ layers, ectoderm and endoderm, in cnidarians, and the addition of a third, mesoderm, in bilaterians has never been challenged. Consequently, the cnidarians are the sister group of bilaterians.

The coelom was much more difficult. The term was coined by Haeckel (in his monograph on the calcareous sponges), although he was mainly concerned to distinguish between the gastric cavity of cnidarians and the space between gut wall and body wall in bilaterians. His term was immediately accepted, but led to a prolonged argument about what he really meant. The notion that the coelom was a body cavity lined with an epithelium derived from the mesoderm was put forward by Oscar and Richard Hertwig in 1881 and soon became canonical (it still is). In many animals, however, the coelom, defined in this way, is only a minor or even vestigial compartment, being almost entirely replaced by another kind of cavity. In molluscs, for example, the only trace of a coelom corresponding to the Hertwigs' definition is the space surrounding the heart; the fluid-filled space in which the gill and other organs are slung is the mantle cavity. The coelom of arthropods has been reduced to a space around the gonads, while most of the body is occupied by a cavity filled with blood, essentially a large blood vessel. In the development of amphioxus and tunicates, the coelom shrinks as it becomes largely replaced by the atrium. Nevertheless, any cavity lined with mesoderm is, by the Hertwigs' definition, a coelom, and how it develops is a crucial clue to how it evolved.

Many animals, such as nematodes, have a spacious body cavity which is not lined by a mesodermal epithelium. Consequently, it cannot be a coelom and came to be called a 'pseudocoel', a spurious or sham coelom. All the animals whose body cavity was a pseudocoel were grouped together as the Aschelminthes, an intermediate stage in the evolution of a true coelom from ancestors which lacked a secondary body cavity altogether.

In 1933, long after the argument had died down, it was summarized in a remarkable doctoral thesis by Ernestine Sarvaas, who begins (in my translation from the original French): "In the whole field of zoology one can scarcely find a term so vaguely defined and yet so universally employed as the word 'coelom'." She then fights her way through a thicket of highly technical articles towards a clear view of what a coelom is, how it originated and how it subsequently evolved. The view had been clouded by conflating these three questions. Animals form their body cavities in different ways, and every biologist was swayed by how their favourite organism developed. Someone studying echinoderms would find that their coelom developed as pockets from the embryonic gut, whereas someone studying annelids would find instead that their coelom arose as widening gaps in blocks of mesenchyme. Both observations were correct, but which represented the original course of development? The association of the gonads with the coelom suggested to some biologists that the coelom originated as the hollow interior of a gonad. Others were more impressed by the association of the excretory system with the coelom, and thought that the coelom might instead have originated as the hollow interior of a nephridium.

Sarvaas explains all these theories and reviews the structure and development of the coelom in animals, but points out that the problem of the evolution of the coelom hinges on its original function, which most previous authors had overlooked. She proposed that the coelom acts as a hydrostatic skeleton which allows the circular and longitudinal muscles of the body wall to oppose one another. This would be especially important in burrowing animals which need to force their bodies through the sediment. In this way she transformed a field which was mainly concerned with embryology and its phylogenetic interpretation into one focused on function. The way seemed clear to a brilliant career, but she did not follow up her thesis with a series of papers and never held an academic position. Perhaps this was because she married in the following year, but I cannot even trace her date of death, and perhaps—this was the mid-1930s after all—there might be a more sinister reason for her complete disappearance.

The origin of segmentation is also a classical problem in zoology. Some zoologists have suggested that segmentation is merely the extension of the trend in flatworms towards a repetition of body parts such as gonads and gut diverticula along the length of the body. This led eventually to a fully segmented coelomate worm, each segment having its own musculature, innovation, blood supply and excretory system, as in annelids. This sounds reasonable, but entails the unexpected and unwelcome corollary that the first segmented worm must already have had a coelom, so that all unsegmented coelomates evolved by losing segmentation. An alternative view is that segmentation evolved from the vegetative fission characteristic of catenulids and naiad oligochaetes, as a frozen chain of zooids before they have developed fully adult structures such as a brain. This theory allows a segmented coelomate to evolve from an unsegmented precursor, but vegetative reproduction is quite different from segmentation in many respects (the fission zone, for example, forms in the middle of the body rather than at one end), and it has attracted few proponents since the end of the 19th century. Perhaps segmentation is merely the inevitable consequence of evolving an elongate body; but many flatworms are elongate, and indeed the record is currently held by the nemertean *Lineus longissimus*, which is said to reach a length of more than 50 m despite being completely unsegmented.

Such arguments about the evolution of the coelom and segmentation dominated comparative biology during the half-century after Darwin's death. They were based almost exclusively on morphological and embryological features of extant organisms to imagine how the coelom and segmentation evolved, with scant regard to the function of either. It was only towards the middle of the 20th century that several authors (including Sarvaas and Hyman) came to the conclusion that segmentation evolved to permit undulatory swimming. This is probably incorrect, since unsegmented animals such as marine flatworms (and even the ribbon-shaped ctenophore *Cestus*) can swim by undulation, but it serves to underline the importance of attributing some clear advantage to the origin of a body plan.

The utility of a segmented coelom was eventually explained by R.B. Clark, who published a remarkable book, *Dynamics in Metazoan Evolution*, in 1964. Clark was head of the zoology department at Newcastle and a specialist in polychaete annelids, but he had trained as a physicist and was well aware that 'the principles of dynamics do not change', as he puts it, and operated just as strictly in the Cambrian as they do at present. He argued persuasively that the hydrostatic isolation of body compartments made peristalsic burrowing much more efficient and powerful. This provides a clear advantage to having a segmented body. Whether the body cavity is a coelom or a pseudocoel is beside the point, because the dynamics would be the same for both. It also raises the possibility of parallel evolution. If the coelom and segmentation have a clear functional advantage, then this body plan might have evolved independently in several lineages. It might also have been lost on several occasions, whenever a lineage adopted a different way of life. The fundamental features of the body plan might reflect the shifting imperatives of adaptation rather than being a reliable guide to ancestry.

## The progressive theory of animal evolution

The guiding principle that has been used by comparative morphologists is that bodies evolve from simpler to more complex layouts. This has the merit that it must be true on the whole, given that the sister group of the multicellular Metazoa is unicellular. Some of the attributes I have mentioned serve to define major groups, although they cannot be arranged from simpler to more complex in any straightforward fashion: the pattern of early cleavage, for example, the fate of the blastopore or the manner in which the coelom is formed. Three others can be polarized. First, there is the transition from no epithelium to two epithelia (ectoderm and endoderm) to three (with mesoderm). Secondly, there is the route from no internal digestive cavity to a simple blind tube (mouth but no anus) to a through-gut (both mouth and anus). Thirdly, the body wall and gut wall may be separated by no more than a gelatinous layer, by a solid mass of cells, by a secondary body cavity, and finally by a coelom lined with mesoderm. Putting all these characters together gives us a general scheme for the evolution of animal body plans.

The first step is the origin of multicellular animals from unicellular precursors. Flagellates, amoebas and ciliates all had their champions who argued the case for a particular group while denying that any other was possible. Or perhaps each of these candidates gave rise to a different group of animals?—as (among others) Steven Jay Gould believed. In that case it would not make sense to look for a single ancestor of all animals. The similarity of choanoflagellates to the choanocytes of sponges made them the favourites, however. This was the view supported by Hyman, so I take it to be the orthodox morphological position, although the point continued to be disputed for 30 years afterwards. It implies that animals constitute a single coherent group, with a single ancestor, whose closest relatives are the choanoflagellates. The choanoflagellates are credible ancestors because they readily evolve coloniality and have the beginnings of an extracellular matrix, in which the cells of a multicellular animal could be embedded, and cell connections that would allow cell-to-cell communication. All that remains is to evolve the simplest kind of differentiation into somatic cells responsible for feeding and a germ line responsible for reproduction. This is not implausible, either, since it has happened several times in eukaryotes. Indeed, by Haeckel's time it was known that all stages from strictly unicellular types like *Chlamydomonas*, through an incipient differentiation in types like *Euglena*, to a complete separation of soma and germ in *Volvox*, can be found in a single family of green algae, and if you are lucky in a single glass of pondwater. This transition is even reflected in gametes of different gender, which are the same size in *Chlamydomonas* but clearly separated into small motile male gametes and large immotile female gametes in *Volvox*. The ancestor of all animals was then a colonial choanoflagellate with a twofold differentiation, between specialized somatic and germ cells and between eggs and sperm. Sponges are the surviving representatives of this stage.

The second stage was the evolution of epithelia, with the two germ layers of ectoderm and endoderm as the basis of development. Ectoderm and endoderm were the outcome of gastrulation, which led to the formation of a mouth and a blind gastric cavity. Muscle and nerve evolved at the same time, although at first they were poorly coordinated and there was no central nervous system. This animal was a two-layered polyp, something like a sea anemone. Its lineage radiated into the diverse forms of cnidarians, such as hydroids, jellyfish, corals and siphonophores. It also gave rise to ctenophores, which share a two-layered construction and are grouped with cnidarians as the Coelenterata.

The next step is a solid-bodied flatworm with directed locomotion, the original bilaterian. This retained a blind gut, but the space between body wall and gut was filled with tissue arising from a third germ layer, the mesoderm. Flatworms have a simple central nervous system at the leading end of their body, usually hooked up to sense organs that give information about the environment ahead.

A cavity between body wall and gut next evolved in animals like nematodes and rotifers. These were grouped together as Aschelminthes, on the grounds that the cavity is not lined by mesoderm but rather develops directly from the primary body cavity, the original inner cavity of the blastula.

Lining the secondary body cavity with mesoderm produces a coelom and gives rise to the coelomate bilaterians, which include all the remaining animals: annelids, arthropods, starfish, vertebrates and a host of smaller groups. The two main lines of descent are defined on morphological criteria as the protostomes and the deuterostomes. Among the protostomes, annelids and arthropods are closely related because both have conspicuously segmented bodies.

This scheme provides an intelligible and more or less complete arrangement of the main kinds of animals (Figure 5.1). It is by no means perfect. The formation of the mouth does not necessarily mirror the fate of the blastopore; several groups have idiosyncratic patterns of cleavage that are not clearly either radial or spiral; the

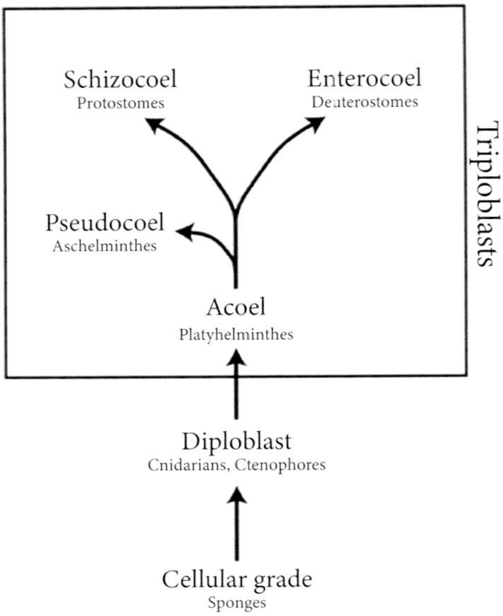


**Figure 5.1** The progressive theory of animal evolution

Reproduced with permission from Halanych, K.M. and Passamaneck, Y., 2001. A brief review of metazoan phylogeny and future prospects in Hox-research. *American Zoologist*, *41*(3), pp.629–639.

nature and formation of the secondary body cavity are often obscure and contentious; and the Aschelminthes look uncomfortably like a convenient dump for animals that do not fit neatly into any predefined category. But it brings order from the chaos of biodiversity, and ranks as one of the great intellectual achievements of scientific biology.

It is more than a scheme of classification, however, because it is an explicitly evolutionary scheme that defines the sequence in which different groups of animals evolved: first sponges, then polyps, followed by flatworms and roundworms, with coelomate bilaterians appearing only after these earlier stages had passed. This predicts the order in which fossils should be found, starting with sponges and ending with arthropods and vertebrates. In particular, it answers our main question: all animals descend from sponges, or rather from a sponge-like most recent common ancestor of Metazoa. The earliest fossils will therefore resemble sponges. Whether or not they did was quite unknown when the grand design of comparative morphology was reaching its climax in the 1950s with the publication of Libbie Hyman's treatise.

## Relics

The zoologist's prayer is that they may find a living animal so simple that it could serve as a sort of type or model from which any other sort of animal could have evolved. It would not be itself an ancestor, but rather a relic: the almost unmodified descendant of an ancestor, retaining its characteristics and recognized by its simplicity. Why the ancestral body plan should have persisted over geological time, like a flint-knapper still chipping away in downtown Montreal, might be hard to explain. (Actually, there are still flint-knappers in Montreal, or at least a McGill Flint Knappers Club. I doubt they do it for profit, though.) It does not seem likely that the very first version of an animal should be so resilient that it would survive in competition with its modified and improved descendants. It is much more likely that a distant ancestor would have been driven to extinction long ago. This obvious objection has not dampened the enthusiasm of zoologists searching for relics, however, and several candidates have been nominated over the years. Here are a few.

## The obvious ancestor

If choanoflagellates are the sister group of Metazoa, then one obvious relic would be a multicellular choanoflagellate. Fortunately, there are lots of them, because choanoflagellates seem predisposed to form colonies, which consist of several hundred cells in some species. The crucial link between choanoflagellates and sponges was found in 1880 by the British microscopist William Saville Kent. He took water samples from the lake at Kew Gardens (which I imagine many of us have passed by, on our way to the greenhouses) and cultured them for several days in shallow glass dishes. When he inspected them he found films of tissue on the glass or the surface of the water which contained two kinds of cell: typical choanoflagellate cells on the outside, and amoeboid cells on the inside. The amoeboid cells could reproduce, either by binary fission, or by division into a large number of spores which subsequently produced a typical colony. There was no fixed difference between the flagellated and amoeboid cells: the flagellated cells became amoeboid when mildly stimulated, but the distinction between soma and germ is clear. This interesting organism received a formal description and a name: *Proterospongia haeckeli* (Figure 5.2). Saville Kent realized that *Proterospongia* would have to be turned inside-out to make a sponge, with the collar cells lining chambers inside the animal, not outside, but nevertheless claimed that his discovery showed how the transition between choanoflagellates and sponges could be initiated. There was just one snag: nobody has found *Proterospongia* since. It has been suggested that the amoeboid cells were simply collar cells viewed from the base, so that their collars were hidden, but Saville Kent was a very accomplished microscopist and this seems unlikely. It has even been hinted that he made the whole thing up to fit in with his view that sponges were compound protozoans, rather than true metazoans, but this seems unlikely too. Saville Kent's drawing of *Proterospongia* has been reproduced dozens of times (by Hyman amongst others) but the animal itself has been strangely elusive.

The original description declares that 'this interesting type is herewith dedicated to, and specially commended to the notice of, the illustrious evolutionist of Jena'. This is a very barbed dedication. Saville Kent disagreed strongly and volubly with Haeckel about the nature of sponges. He concluded that they were compound protozoans, whereas Haeckel thought they were true metazoans, complete with

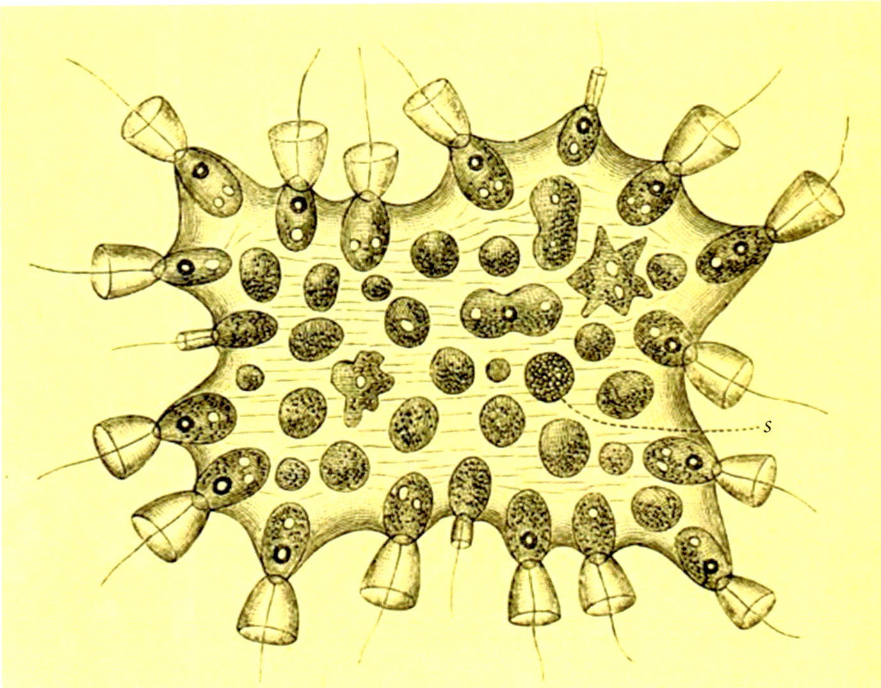


**Figure 5.2** Proterospongia

Reproduced from Saville Kent, W., 1882. *A Manual of the Infusoria*. Vol III, Plate X, Fig 20.

ectoderm, endoderm, gastrulation and the beginnings of a gastric cavity. Nonsense, wrote Saville Kent: 'not only has it been shown that errors do exist [in Haeckel's account of sponge development], but that these are of such a radical and fundamental nature that the inference is most reluctantly arrived at that Haeckel, carried away in his ardent pursuit of the Metazoic archetype, has lost for the time his power of discrimination between matters of fact and hypothesis, and so evolved from his own inner consciousness those details that are wanting to complete and perfect his theory'. In other words, Haeckel made it up, so 'it is clearly evident that the so-called "sponge-gastrulæ," described and figured over and over again in Professor Haeckel's 'Monograph,' are an entire myth, and that the superstructure of the gastræa theory, so far as it rests upon this basis, is entirely worthless.' Saville Kent was quite right about Haeckel's optimistic interpretation of sponge development, as Haeckel himself probably realized, since he later partially retracted it. Saville Kent had made a powerful enemy, though, and shown himself a little too intemperate in debate. He had long had the ambition of heading a national research laboratory for marine biology, but when the Marine Biological Association was founded in 1884 it was Ray Lankester who was appointed Hon Secretary; T.H. Huxley, a long-time colleague and friend of Haeckel, was its first President. Saville Kent was instead appointed Inspector of Fisheries in Tasmania, and left England for good in the same year.

## The magician's ball

Haeckel had his own candidate, of course. He never did find a living Gastraea (neither has anyone else), but he did find the next best thing, the Blastaea, an organism still existing which retained the form of a blastula. He was taking samples off the coast of Norway in 1869 when he noticed a minute ball of cells which at first he took for one of the annelid or mollusc larvae which swarm in these waters. Back in the lab, however, he noticed that the ball could break up into individual ciliated cells that eventually sank to the bottom of the culture vessel, where the cilia became transformed into the ordinary cytoplasmic extensions or 'pseudopodia' of amoebas. The amoeboid cells presumably then divided to reconstitute the pelagic colony, but although Haeckel inferred that this stage must occur he did not observe it directly. He named the organism *Magosphaera*, the magician's ball, from its uncanny ability to perpetuate a blastula-grade organism as one stage in a complex life cycle. It does not seem to me that *Magosphaera* makes a very convincing blastula, since it is a solid ball of cells each bearing a dozen or so cilia, whereas the classical blastula is a hollow ball of monociliated cells, but Haeckel was a much better zoologist than I am, and besides he actually watched the animal in life. Unfortunately, nobody else ever has. *Magosphaera* was never seen again, and all that remains of it is Haeckel's characteristically detailed drawings. Perhaps he was simply carried away by Darwinian enthusiasm and read more than he should have into a mixed collection of plankton.

## Simplified bodies

There are many multicellular animals that are not so much simple as simplified. They are not credible as ancestors because they are clearly derived from a more complex kind of creature, but they nevertheless serve to show that a certain kind of simple body is viable.

Parasites may be highly simplified, because the host does most of their work for them, and I suspect that the enigmatic mesozoans are an example. They were studied by Edouard Van Beneden, a Belgian zoologist who is still remembered for the first clear demonstration that the number of chromosomes is halved during the maturation of the egg; after a long and complex controversy this led to the modern interpretation of meiosis. His work on certain minute worms found in the kidneys of squid has understandably been forgotten, although they are most remarkable animals. They consist only of a few dozen ciliated cells wrapped around a single axial cell, which is about as simple as an animal can get. Indeed, Van Beneden coined the term 'Mesozoa' to indicate that they seemed to be intermediate between single-celled protozoans and Metazoa. They have no internal organs, no distinction of ectoderm and endoderm, and the axial cell is reproductive rather than digestive. It is most unlikely, however, that their simplicity is the retention of the ancestral state. They are obligate parasites, after all, and parasites are often highly simplified because most of their needs are provided by the body fluids of their host—think of *Sacculina*, for example. Mesozoans are most likely to be highly modified members of some familiar group, perhaps flatworms or annelids, although their true relationships are still uncertain, nearly 150 years after Van Beneden gave them a name.

The ultimate simplification would be the reduction of the complex multicellular body of a metazoan to a single cell. This has evolved in myxozoans, a mysterious group of organisms which are obligate parasites of fish and a range of invertebrates, such as annelids and bryozoans. Although few people have heard of myxozoans, there are lots of them (about two thousand species) and some cause severe or lethal infections, such as the whirling disease of trout and salmon. Most are single-celled and were naturally classified with protozoans, although there were suspicions that they might have a more distinguished parentage. They even have a structure which looks rather like a cnidocyte, hinting that they might be related to hydroids and sea anemones. This implausible hypothesis is now known (from molecular studies) to be correct: myxozoans are cnidarians which have become highly simplified after adopting a parasitic lifestyle. A few are multicellular worm-like creatures that have even retained traces of epithelia and muscles. There is no suggestion, of course, that evolution has proceeded from simple myxozoans to more complex corals and jellyfish; myxozoans are a clear example of simplification.

## The animated tube

The very simplest multicellular animal might be a tube with a mouth and not much else. We know that a body plan as simple as this would work because examples of it can be found today. If you fish around with an aquarium net in the upper few inches of sediment in brackish localities such as estuaries, for example, you might find a minute brown or red thread which winces when you touch it. This is the hydroid *Protohydra*, a simple two-layered tube with no tentacles (but plenty of cnidocytes). It has the hydroid body plan reduced to its simplest elements, no more than ectoderm and endoderm, with a mouth at one end and a holdfast at the other. Its cnidocytes enable it to paralyse quite large animals, such as copepods, and it can somehow manoeuvre them into its mouth and engulf them, so that it ends up looking like a miniature python that has swallowed a goat. *Protohydra* itself has presumably evolved from tentaculate precursors through living in unconsolidated sediment where tentacles would not be useful. What it shows is that a minute tube with a mouth is a perfectly viable predator (although the cnidocytes help, of course).

## The salt animal

Johannes Frenzel found an animal that seemed like the perfect ancestor, or so he said. Frenzel was trained in zoology at Berlin and Göttingen during the early 1880s, and worked at the Stazione Zoologica in Naples after receiving his doctorate. He later returned to Germany, but when he heard of a post at the University of Cordoba, in Argentina, the prospect of exploring a new fauna was irresistible. He might not normally have had much success—Cordoba tended to favour local naturalists—but on this occasion the selection committee judged that the prestige of a German zoologist outweighed the obvious difficulties of hiring a foreigner, and offered him the job. Once the Minister in Buenos Aires had given permission, Frenzel set off almost immediately, accompanied by his wife (herself the daughter of a well-known German zoologist), and took up his new position in September 1887. For the next few years he concentrated on the microscopic organisms of the region, the protozoa, and published two very long and very dull articles which would have been completely forgotten by now except that one of them described a new and extraordinary creature that has puzzled zoologists ever since.

*Salinella salve* appeared when Frenzel moistened a pellet of soil, collected by a colleague from a nearby salt flat, with saline solution and then forgot about it for several weeks. When he eventually inspected the sample under a microscope he found it wriggling with minute creatures that seemed to him to bridge the gap between unicellular organisms—the protozoa he was investigating—and multicellular animals. They consisted of little more than a single layer of cells, forming a sort of hollow cylinder pointed at both ends, enclosing a central cavity (Figure 5.3). This cavity constituted a rudimentary gut, into which food particles entered through a mouth and the remains evacuated through an anus. The cells themselves bore cilia, short hair-like processes that wave backwards and forwards to move the animal through the mud (those on the outside) or to move food through the gut (those on the inside). It was almost a caricature of the simplest conceivable animal, but to make sure there was no mistake Frenzel prepared very detailed drawings of its anatomy. Zoologists everywhere were elated, and wove *Salinella* into the narrative of early animal evolution. There was only one snag: the animal itself has never been seen again, and to this day Frenzel's description and painstaking drawings remain the only evidence for its existence.

Frenzel himself seems to have been discontented at Cordoba, and returned to Germany in 1891, the same year that he reported the discovery of *Salinella*. Once back home he branched out in a new direction and founded the Biologische- und Fischerei-Versuchsstation (Biological and Fisheries Research Station) on the Müggelsee. Before he could accomplish much, however, he fell while working on a footbridge and died

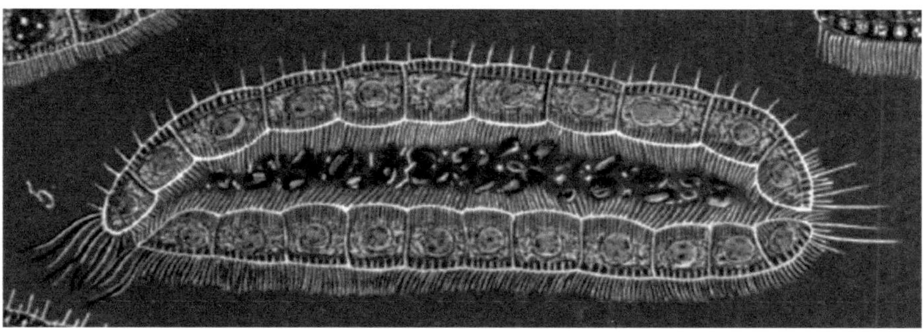

**Figure 5.3** Salinella

Reproduced from Franzel J., 1891. *Untersuchungen über die Mikroscopische Fauna Argentiniens.* Nicolaische, Berlin.

of his injuries, just before his thirty-ninth birthday. He is only remembered today as the discoverer of *Salinella*.

As the years have passed with no new specimens turning up, interest in *Salinella* has waned, and nowadays it usually appears, if at all, only as a rather puzzled footnote to discussions of animal ancestry. What should we make of it? Well, a glance at Frenzel's careful drawings shows that the animal is zoologically impossible: the single layer of cells is ciliated on both surfaces, inside and outside, a condition never seen in animals and quite incredible in any form taken to represent their ancestral condition. Frenzel was a trained zoologist who must have known that animals do not have cells with a cilium at either end, and in any case an epithelium is usually bounded on its inner side by a basement membrane that precludes any possibility of ciliation. These drawings accompany his long and detailed description of *Salinella*, which was written just before he left Cordoba for ever. The name he gave to the species, *salve*, may be connected with 'salveo', to be in good health, but another meaning of 'salve' is 'good-day', or even 'farewell.' The article itself is dated (by him) 1 April 1891.

## A ciliated sack

An almost equally simple animal whose authenticity is not in question (because it can be collected routinely) was found a few years later off the coast of southern Sweden. In the summer of 1915, Karl Sixten Bock was cruising in very shallow water, scarcely 100 m deep, and bringing up trawls with the usual mixture of mud and animals. He sorted the animals from the mud and stored them, preserved with spirits, in jars that found their way to the shelves of the Swedish Museum of Natural History in Stockholm. The remained there through two World Wars until one of the animals brought up by the trawl was examined by a puzzled Einar Westblad, who had never seen anything like it before. He thought it was probably a flatworm, but hedged his bets by naming it *Xenoturbella*, the strange flatworm. It is a pale, rather corpulent animal about an inch long, which in life glides steadily across the seabed, propelled by cilia on its underside. Its mouth is underneath and leads to a capacious but blind gut. There are layers of muscle beneath its body wall, and a rudimentary kind of nerve net. That's about all. It has no circulatory, respiratory or excretory system, and no secondary body cavity. It has no central nervous system, not even ganglia or nerve cords, and no sense organs except for a statocyst. It does not even have gonads, let alone any other sexual structures. What is it? Its featureless body gave few clues to anatomists, but once it became possible to identify organisms from gene sequences it was duly ground up and found, surprisingly, to be a bivalve, strangely modified to be sure. Better still, it nursed oocytes in its gastric cavity which shared a distinctive feature of the nucleus only with a particular group of bivalves. This conclusion was firmly based on two independent lines of evidence, but at the same time it was so surprising that some sceptical zoologists looked more closely at *Xenoturbella*. They found lots of eggs and embryonic stages, and these certainly came from molluscs, but this was not because *Xenoturbella* was a mollusc; instead, it was because it ate mollusc larvae. Even when this mildly entertaining mistake had been cleared up, however, it was still not clear where it belongs. Most recent (molecular) phylogenetic work puts it back in the flatworms, although in a special group, the Acoela, which has been hived off from the turbellarians and which we shall meet again later. I think its current placement is unstable, though, and I would not be surprised to see it moved again. Its extreme simplification would not be surprising in an obligate parasite, but it is certainly freeliving, and how its way of life has reduced it to a mere ciliated bag remains a mystery. I expect that the mystery will be solved one day, since *Xenoturbella* seems to be quite widespread and is sometimes locally abundant, but for the time being we must leave this strange animal to glide over the seabed, blindly hoovering up the bivalve larvae on which it feeds.

## The sticky hairy plate

Only one man has ever had the zoologist's prayer answered, and even he lived to see his discovery discredited and his reputation slighted. Franz Eilhard Schulze was particularly interested in sponges and other simple metazoans, and even in xenophyophores (there is no common name, but we shall come across them later), which live in the deep sea and look a little like sponges, although they are now known to be colossal protists. He was chosen to write the report on the glass sponges collected by the *Challenger* expedition in the 1870s, which meant he had a European reputation, and he served as President of the German Zoological Society for several years in the 1890s, which meant that he was regarded as the head of his profession by his peers. In short, he was an eminent and authoritative figure whose opinion was not to be taken lightly. In 1873 he moved from Rostock, where he had risen from graduate student to full professor, to the Institute of Zoology at Graz, in Austria, where he was a colleague of Ludwig von Graff, of flatworm fame. Here he made a most surprising discovery. The Institute had an aquarium, filled with seawater from the Adriatic, which Schulze used to maintain his sponges and other marine animals. At some time in 1882 he began to notice small, pale grey spots on the glass walls of the aquarium. Most people would have dismissed them as uninteresting crusts of some kind, but Schulze watched them and found that they moved slowly, continually changing shape. They could not be scraped off with a scalpel, which just reduced them to shreds, but a strong jet of water from a pipette could be used to detach them intact into a tube, and then onto a microscope slide for investigation. They turned out to be multicellular, eliminating the possibility that they were simply gigantic amoebas, but with a simple and peculiar organization. Their body was a thin disc which consisted of little more than a lower sheet of cells, on which they crawled using cilia, an upper sheet, and a sparse middle layer of fibrous cells. They were not in the least like a sponge—Schulze, of all people, would know—but they had no mouth or digestive cavity, so they were not coelenterates or worms either. Schulze concluded that they were a distinctive kind of animal of very simple structure, and because at the time he was busy with the *Challenger* glass sponges simply dashed off a short note which described their structure and gave them a name: *Trichoplax adhaerens*, the hairy (that is, ciliated) plate which sticks tightly (Figure 5.4).

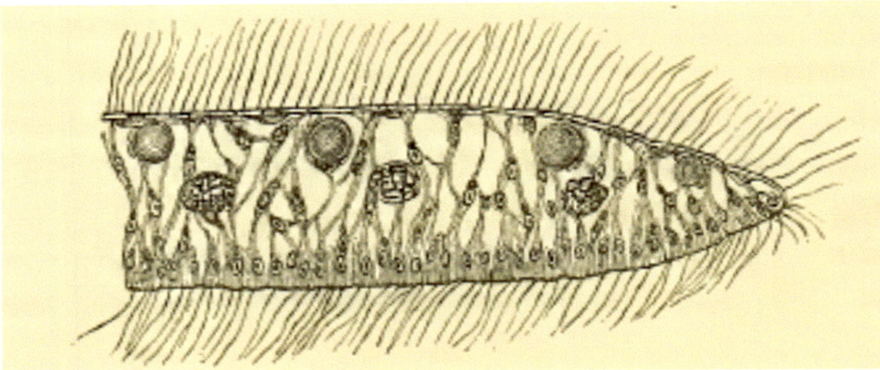

**Figure 5.4** Trichoplax

Reproduced from Schultze, F.E., 1883. Trichoplax adhaerens, nov. gen., nov. spec. *Zoologischer Anzeiger*, 6, pp.92–97.

*Trichoplax* immediately attracted the attention of zoologists because it was a good candidate for the long-sought ur-metazoan, displaying the metazoan body plan in its simplest form. If the upper and lower layers could be interpreted as ectoderm and endoderm (as Schulze himself proposed), then the middle layer might be mesoderm and the whole animal a very simple kind of flatworm. Or had it been secondarily simplified for its particular way of life? It seemed that the problem had been put to rest in 1907 when a graduate student at the Zoological Institute of Breslau presented his thesis. The rather portentously named Thilo Krumbach had been working for several years on a hydroid called *Eleutheria*, which he collected in the Gulf of Trieste and maintained in small seawater aquaria at Breslau. As in most hydroids, the life cycle of *Eleutheria* includes both a vegetative polyp phase and a sexual medusa phase; the two are linked by the fertilized egg of the medusa, which develops into a solid-bodied ciliated larva, the planula, which after a period of swimming around settles on a substrate and metamorphoses into the polyp. Krumbach's thesis is a rather brief and conventional account of the structure of the tentacle and other features of *Eleutheria*. There is nothing the least controversial about it. His aquaria, however, were provided with sea lettuce, *Ulva*, a crinkly green seaweed that provided a home for a variety of animals—including *Trichoplax*. Krumbach noticed that when the *Trichoplax* had disappeared from the sea-lettuce, following a period of dim light, they were replaced by polyps. He leapt to the conclusion that the polyps, medusae and *Trichoplax* were all part of the same life cycle, with *Trichoplax* as the planula larva. A graduate student was calling into question the conclusion of an eminent and authoritative professor: *Trichoplax* was not, after all, an entirely different sort of animal, still less the ur-metazoan incarnate, but merely the rather odd larva of a common hydroid. Surprisingly, Krumbach's intervention was completely successful. His interpretation, although based on purely circumstantial evidence, was widely accepted—even Libbie Hyman uses it to demote *Trichoplax* to a single sentence in an obscure paragraph—and nothing further was published, so far as I can find out, for half a century afterwards.

Krumbach may have been very persuasive, but he was also quite wrong. Schulze's original interpretation was vindicated in the 1970s by Karl Grell, a protozoologist at the Max-Planck-Institut at Tubingen who knew how to operate one of the new electron microscopes and used its novel powers to describe the ultrastructure of *Trichoplax*. Grell's work confirmed the three cell layers that Schulte had originally seen, added a host of further details, and coined a name for the group: Placozoa. This made *Trichoplax* respectable again, and in the next half-century a slowly growing number of zoologists around the world began to investigate its anatomy, life cycle, ecology and genomics. It seems that placozoans are far from being merely an aquarium curiosity, but instead are much more abundant, widespread and diverse than anyone previously thought. They remain a bit of a puzzle, though: a very simple kind of animal, not clearly related to any other. To the eye of faith, they might appear to be Bütschli's Plakula made flesh; at all accounts, any theory of where animals came from has to take *Trichoplax* into account.

The search for relics has consumed a lot of time and paper, but if this requires a justification I can offer one. A relic might or might not be the lineal descendant of an ancient type that has retained its ancestral simplicity: *Trichoplax* might be, *Xenoturbella* probably is not, myxozoans certainly are not. Regardless of their history, however, their present existence demonstrates that their body plan is viable for a particular way of life in competition with the rest of the world. Any scheme that posits a similar body plan as an ancestor, or as one of a chain of ancestors, cannot be dismissed as unrealistic. Conversely, any scheme that relies on an imaginary body plan that is not displayed by any modern relic is vulnerable to the criticism that no such creature could possibly survive in the real world. This is (I imagine) why relics which resemble some imaginary type are so assiduously sought, and also why even remotely possible candidates are often rather optimistically interpreted.

> It is a capital mistake to theorize before one has data. Insensibly one begins to twist facts to suit theories, instead of theories to suit facts.
>
> —Arthur Conan Doyle, 'A Scandal in Bohemia', in *The Adventures of Sherlock Holmes.*

Well-authenticated types which are simple without clearly having been simplified are scarce and tend to drive the discussion. Speculation is inevitably channelled by types whose present existence suggests that very similar types could have existed in the remote past. Perhaps an animal very like *Trichoplax* lies at the root of metazoan ancestry, but being small and fragile was never fossilized. This is not unreasonable, but the time that has elapsed since this ancestor lived is the same for *Trichoplax* as for (say) a tiger shark, and *Trichoplax* may have been just as extensively modified. Relics may offer a clue, but only fossils are conclusive.

# 6
# The Magic Molecule

By the 1950s, a century of comparative morphology had led to a broad consensus among biologists, mostly working on specimens in museums, about the route of animal evolution. At the same time there was another group of biologists, or biochemists really, who worked in university departments, where their laboratories usually had no specimens at all. The two groups worked in different buildings and their coffee breaks, lunch hours and social circles were quite separate. They did not mingle much and their scientific interests scarcely overlapped. Fifty years later, the laboratory had overthrown the museum, and the whole question of animal evolution had to be reviewed.

One of the laboratory scientists was Fred Sanger, who worked in a Cambridge lab in the 1950s on the structure of insulin, mostly because it was the only protein you could buy in bulk and reasonably pure from the local pharmacy. He invented a method to uncover the sequence of amino acids that made up the protein molecule, which incidentally showed that insulin, and by extension every other protein, really did consist of a single unique sequence rather than a vague mixture of amino acids. This was the decade in which DNA was found to consist of a definite sequence of nucleotides, a sequence which in some way corresponded to the sequence of amino acids in proteins. The genetic code linking the two was worked out in the 1960s, and in the following decade Sanger found a way of uncovering the nucleotide sequence of a stretch of DNA. Everyone knows of the romantic and controversial story of how Watson and Crick discovered the structure of DNA, but fewer know the less exciting story of how Fred Sanger found out how to discover the precise sequence of proteins and nucleic acids. It was Sanger, however, who provided the techniques that eventually superseded the morphological approach with a new and quite different way of evaluating the relationships of animals and the route of animal evolution.

## Gene-based trees

In 1959, when Libbie Hyman was putting the finishing touches to *Smaller Coelomate Groups*, a rather elderly postdoc (he was 37, on leave from CNRS in France) called Emile Zuckerkandl arrived in Linus Pauling's lab at CalTech. By 1967, when she was penning the sad Foreword to her next and final volume, comparative morphology was beginning to look rather old-fashioned. Zuckerkandl (in partnership with Pauling) had devised a completely different approach to evolution and systematics based on proteins such as haemoglobin. In his first experiments, he broke up the haemoglobin with digestive enzymes and used a combination of chromatography and electrophoresis to visualize the two-dimensional pattern of fragments with differing size and net electrical charge. Human haemoglobin gave a pattern very similar to chimpanzee,

*The Sea-Change*. Graham Bell, Oxford University Press. © Graham Bell (2026).
DOI: 10.1093/9780197797464.003.0008

rather less similar to rhesus monkey, still less similar to cattle and pigs, and quite different from fish and sharks. The structure of the haemoglobin molecule, he concluded, reflects the ancestry of a species, being most similar in closely related species and gradually diverging in species with more distant common ancestors. In principle, this provides a method for classifying organisms quite distinct from morphology, although it wasn't clear how to measure similarity from the pattern of spots on a sheet of paper. This problem was solved when it became possible (and cheap) to determine the precise sequence of amino-acids in a protein, using the method invented by Fred Sanger. By 1964, Zuckerkandl and Pauling, in an article with the prophetic title of 'Molecules as Documents of Evolutionary History' could write that 'the most rational, universal, and Informative molecular phylogeny will be built on semantophoretic molecules alone'. By 'semantophoretic' they meant molecules that represent the structure of genes or their transcripts. It was already clear that the ultimate document would be the genes themselves, the sequence of nucleotides in DNA, but this could not be read until 1977, when Sanger's second invention opened the door to a new method of finding out how organisms are related, based on the genes themselves.

The advantage is obvious: only the sequence of nucleotides in DNA is transmitted directly from parent to offspring. It is the purest source of evidence about ancestry. It has other advantages too. To start with, it is universal: all organisms, unicellular and multicellular, use DNA for information storage. If each site is a character, with the nucleotide occupying that site as the character state, then genes furnish thousands of characters, rather than the few dozen provided by morphology. The characters are precisely defined but subject to continual change, so they can be used effectively to allocate species to different groups. No judgment is necessary, because the interminable debates about germ layers and coeloms could be sidestepped by an unambiguous sequence of nucleotides. It was even possible in principle (I am not suggesting this was ever the case in practice) for someone with no knowledge of animals whatsoever to construct a phylogenetic tree from labelled tubes of DNA alone.

The translation of a sequence of nucleotides into a phylogenetic tree involves three main stages. The first is simply to identify an homologous gene in the sample of species that has been chosen. Then the sequences of the gene in different species must then be aligned if they differ in length because of inserted or deleted nucleotides, so that homologous nucleotides can be compared. Finally, some technique (almost always a computer algorithm) must be devised to sort the sample into sister groups to make a phylogenetic tree. The molecule chosen for the earliest attempts to apply the molecular approach to animals as a whole was a subunit of ribosomal RNA, because it is essential for protein synthesis and occurs in all cellular organisms, prokaryote or eukaryote. (The RNA translated from the DNA template has a precisely corresponding nucleotide sequence, so sequencing the RNA product is equivalent to sequencing the DNA gene itself.) This yielded dramatic but sometimes controversial results. Katherine Field and her colleagues at Indiana University, for example, sequenced more than 1,000 nucleotides of ribosomal RNA from a diverse sample of animals to conclude that cnidarians and bilaterians descended from different protist ancestors, that flatworms are ancestral to coelomates, that annelids and arthropods are not closely related, and that deuterostomes are dubious. Morphologists, cell biologists and even biochemists combined to disagree. The conclusions go against a

specified. To overcome these complications it is absolutely necessary to use computer-driven methods arranged in a 'pipeline' to get from the raw data of gene sequences to the final consensus tree. However, there is no general agreement on which pipeline is optimal (or even whether there is a single best pipeline), so every analysis uses a somewhat different procedure. This has had the unfortunate consequence that the average zoologist, outside the small circle of molecular phylogeneticists, is unable to judge the merits of any particular tree. Journals nowadays insist that not only the raw data but also the details of all procedures that have been used to analyse it should be made available on public data bases, but in practice the details are so complicated that it would be extremely difficult to check the results. Nevertheless, the advantages of using tens of thousands of nucleotides from hundreds of species are so overwhelming that this approach is now used by most specialists.

Oddly enough, the ascendency of gene-based phylogenetic trees has resulted in at least a partial reversion to the sort of phylogenetic speculation that we associate with the generation of comparative biologists which followed Haeckel. The ease and cheapness of genomic sequencing has led to a plethora of hypotheses about animal evolution, often contradictory and usually with scant regard for the original function, the immediate selective advantage, of new body plans. To paraphrase the oft-repeated claim of Theodosius Dobzhansky, nothing in evolution makes sense except in the light of morphology. It is only when development, phylogenetics and morphology are welded together that we shall begin to get a clear picture of early animal evolution.

It was hoped, no doubt, that the whole-genome approach would quickly lead to a convincing clarification of animal phylogeny, but instead it led almost immediately to the entirely unexpected conclusion that the earliest branch splitting from the tree was not the sponges, as almost everyone had thought, but a new contender for the title of ultimate ancestor, the ctenophores. Sponges separated early, before cnidarians, but the lineage leading to ctenophores separated earlier still. This heretical proposition was mostly due to Mark Martindale's lab group, then based in Hawaii, and was published in 2008 under the innocuous title 'Broad Phylogenomic Sampling Improves Resolution of the Animal Tree of Life'. This study relied on short gene transcripts, but 5 years later a whole-genome survey reinforced the original conclusion. If this had been black-box genomics run by people who didn't really know about animals it might have been laughed off, but anyone who said that Mark Martindale doesn't really know about animals would be laughed off instead.

This has some startling implications. If the first split in the animal tree separated ctenophores from everything else, followed by a second split that separated sponges from everything else, then we are forced to choose between two unpalatable options: either muscle and nerve evolved twice (in ctenophores and cnidarians) or both have been lost without trace in sponges. Being unpalatable, both options were rejected by many. Morphologists such as Claus Nielsen of the Copenhagen Museum were, of course, strongly opposed and pointed out that the ctenophore-first theory ran counter to practically everything that is known about animal body plans. It was not only the morphologists who objected, however. An influential group of molecular systematists, including Hervé Philippe of the Université de Montréal, claimed that the seemingly early divergence of ctenophores was an artefact, and published analyses based on equally extensive data showing that sponges branched first, with ctenophores grouped with cnidarians as the traditional Coelenterata. The argument still rumbles on, but when doctors disagree I shall not be foolish enough to offer an opinion.

A decade after the introduction of automated short-read sequencing, Gonzalo Giribet's group at Harvard felt able to review the progress that had been made possible by sequencing hundreds of genes from representatives of every animal phylum (Figure 6.1). They confirm the two main innovations of earlier molecular systematists, the division of protostomes into spiralians and ecdysozoans, and the relegation of flatworms (except acoels) to the humble position of rather highly derived protostomes. Otherwise, the groups set up by the old-school morphologists are either retained (such as protostomes and deuterostomes) or even re-formed after having

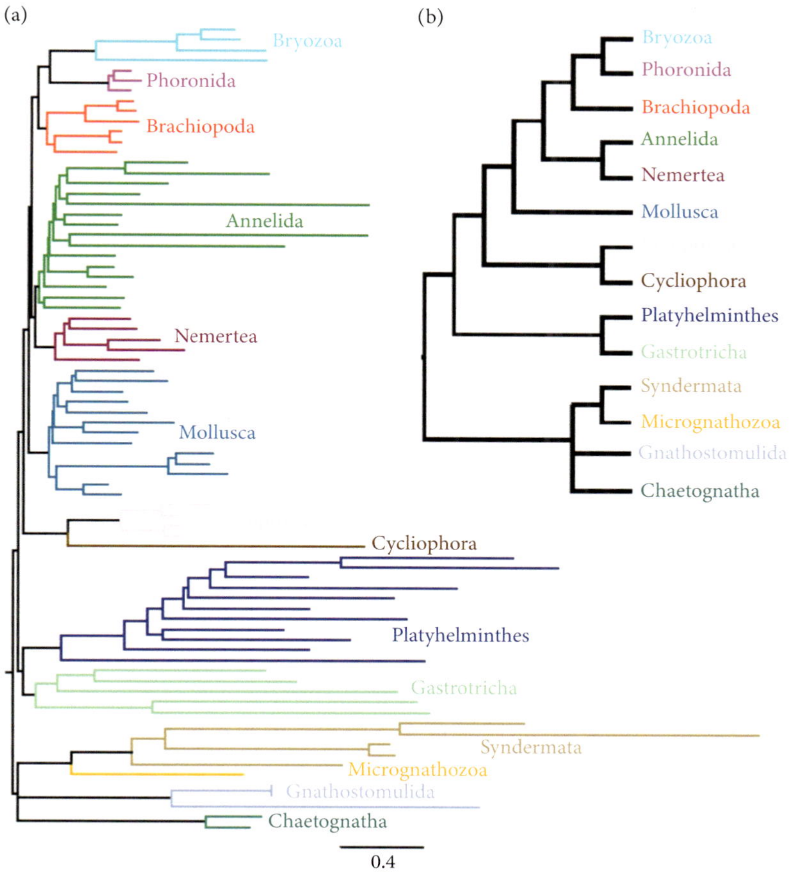


**Figure 6.1** The new animal phylogeny

Reproduced from Laumer, C.E., Fernández, R., Lemer, S., Combosch, D., Kocot, K.M., Riesgo, A., Andrade, S.C., Sterrer, W., Sørensen, M.V. and Giribet, G., 2019. Revisiting metazoan phylogeny with genomic sampling of all phyla. *Proceedings of the Royal Society B, 286*(1906), p.20190831, under a Creative Commons license.

been dismantled (such as lophophorates), or else relationships remain dubious or tentative. The sponge-first vs. ctenophore-first debate is not firmly resolved, and even the branching order of the major spiralian lineages is uncertain. Systematists working with particular groups of animals often assert that some favoured phylogenetic tree has been unequivocally demonstrated, but few, if any, of these trees survive the scrutiny of Giribet's team. 'Even with tens of thousands of well-aligned sites, stringently validated orthology [homology], and the most flexible, descriptive site-heterogeneous models available today, different gene sets can give full support to conflicting phylogenies'.

It may be that yet more genes from yet more species will eventually persuade an unambiguous and generally accepted animal phylogeny to precipitate out. On the other hand, trying to work out animal evolution without understanding animal bodies may be a doomed enterprise. At some point, one must ask why the ocean of data has so far yielded only a few small fish. The answer is obvious (and widely recognized): the branches near the base of the animal tree are very short.

## The problem of short branches

Most gene-based phylogenetic trees show not only branching order but also the length of each branch. The length of a branch is proportional to the number of changes (usually in nucleotide sequence) that have occurred along it, from its origin at a given node within the tree to the subsequent node where it splits into two sister lineages. Either or both of these lineages may go on to split repeatedly, each time at a particular node, before ending in one of the species which have been chosen for analysis. The lineages encompassed in this way comprise a common ancestor and all of its descendants (for the sample of species used), and therefore constitute a natural group.

The total number of changes from any node within the tree to the species at the tip of each branch can be partitioned into two components: the number occurring within the group (from the ancestral node to the node at the first split) and the number occurring between the two descendant groups (the number occurring between the node at the first split and the node at the next split, averaged for the two lineages). Repeating the process until the species at the tip of each branch is reached yields a complete accounting of the pattern of genetic change within and between groups, from the inferred state of the ultimate ancestor to the living species, through every intermediate stage. Since groups arise successively, this is equally the pattern of change over time.

If the branches of the tree are all equal in length, an equal number of changes will separate each node from the next. The rate of change is then a constant, and new natural groups, named or not, arise at more or less fixed intervals. As an example of a real phylogenetic tree (or rather, an estimate of a real phylogenetic tree) I have reproduced one of the consensus trees which Giribet's group arrived at as Figure 6.1. It is obvious at a glance that its branches are very unequal indeed. All the branches at the base of the tree, leading eventually to modern phyla, are short, and they are often very short indeed. Most of the branches within phyla are much longer, and some are very long indeed. The phylogenetic tree turns out to be a bush. What does this imply?

The first possibility is that the origin of new phyla involves very little genetic change, whereas the subsequent divergence of species within phyla involves much more. This is not entirely impossible. It might be that there is a strictly limited range of viable body plans, each protected by strong selection and thus resistant to change, even over long periods of time. It is precisely the opposite of what might have been expected, however, and few if any practicing phylogeneticists (I have found none) have favoured this interpretation.

The alternative is that the initial radiation of animals took place very quickly, allowing little time for genetic divergence. This is the generally accepted interpretation of the short branches near the root of the tree. It is, of course, consistent with the fossil evidence of a rapid diversification of body plans in the early Cambrian. It also presents a serious difficulty: perhaps no amount of evidence, no matter how many genes in no matter how many species are scored, will be sufficient to establish unequivocally the initial pattern of branching of the animal tree. How phyla are related to one another is unknowable, at least in detail. This is a counsel of despair that will be rejected by all phylogeneticists, of course. Whatever the limits of the genomic approach may be, however, it has undoubtedly led to the conclusion that the early radiation of animals, establishing the lineages of modern phyla, took place very quickly indeed.

## The molecular clock

Zuckerkandl realized that proteins and genes not only revealed how organisms are related. They could also show how long ago two groups had begun to diverge from their common ancestor, because groups that had diverged long ago would have accumulated more changes than groups that had diverged only recently. The number of changes that have taken place along a branch will be directly proportional to the length of time that has elapsed since its inception. The number of changes along the branch leading to any modern species is therefore an estimate of its age relative to other species.

Genes are more reliable than proteins because the triplet genetic code that translates a nucleotide sequence into an amino-acid sequence is degenerate: most amino acids are specified by two or more nucleotide triplets, so when one nucleotide is substituted for another the altered triplet often codes for exactly the same amino acid as before. Any mutation like this does not modify the protein and is therefore neutral with respect to fitness—it neither increases nor decreases the reproduction of the individual bearing it, so it will neither increase nor decrease in frequency through natural selection. It will change in frequency nonetheless, simply by chance. The individuals which bear the mutation might fall out of trees or win the lottery by pure luck, good or bad, and the frequency of the mutation will rise or fall as a consequence. It is very unlikely, of course, that any particular neutral mutation will have such a sustained run of good luck that every member of the species bears it, but since there are hundreds of thousands of potentially neutral mutations it is almost certain that one or more of them will be lucky enough to displace the original nucleotide. Once this has happened, the least possible modification of the genome has been made, a single unit change. It is a very informative modification, however, because it is almost certainly unique; after all, there are hundreds of thousands of alternatives, each of which had the same probability of becoming established. A combination of several such changes is bound to be unique. Zuckerkandl realized that neutral mutations convey pure phylogenetic information: 'Due to isosemantic [neutral] substitutions, there probably is more evolutionary history inscribed in the base sequence of nucleic acids than in the amino-acid sequence of corresponding polypeptide chains'. The reason is that altering a protein alters the phenotype, which may be favoured by natural selection and lead to the rapid and more or less deterministic shift from one sequence to another. But what is sauce for the goose may be sauce for the gander, so a phenotype which is superior in one lineage may be superior in many, so that the same shift might occur independently in several lineages. It will still convey information about ancestry, sure, but it also conveys information about utility, and the two cannot be disentangled merely from a knowledge of the sequence alone. It is only neutral mutations that carry a pure phylogenetic signal.

Sequences yield only relative ages; to convert these to absolute ages requires knowing the absolute length of time corresponding to the origin of at least one group. The only way of calibrating the molecular clock is to go back to the rocks and find a series of fossils that documents the earliest appearance of a group and can be reliably dated. Several calibration points are better than one, of course, and ideally the calibration points should straddle the sequences. Unfortunately, the origin of animals occurred before any feasible calibration point (except, perhaps, *Bangiomorpha*), so estimating the date of the origin is bound to be a little speculative.

The first large-scale attempt to put Zuckerkandl's principle to work on the problem of animal origins was not published until 1996—30 years later—by a group from the State University of New York at Stony Brook (a famous address for evolutionary biologists) led by Gregory Wray. Wray's group relied mainly on vertebrate fossils to calibrate the divergence of seven universal (or at least widely distributed) genes, such as the gene that encodes cytochrome c. They concluded that bilaterians first appeared over a billion years ago, at the beginning of the Proterozoic rather than near its close. This conclusion was met with some incredulity, at least among palaeontologists, but was confirmed by other groups in the 1990s. The most extreme result was posted by a group led by Stephen Blair Hedges from the NASA Astrobiology Institute in 2004, using many proteins from many species (it's not clear exactly how many of either). They calculated that sponges split from other lineages about 1,350 Mya, with cnidarians and bilaterians separating about 1,300 Mya. The precise timing depends on how it is calculated, but different methods gave very similar results. It also depends on how the sequences are calibrated, which, as in Wray's analysis, used vertebrate fossils showing that synapsids (such as mammals) split from diapsids (such as birds) about 310 Mya. This is scarcely a secure anchor for such an extreme conclusion, being less than a quarter of the supposed time when animals first evolved. The same date, assumed to be known without error, like the day you were born, was used by Blair Hedges and Sudhir Kumar of Penn State to derive the whole course of vertebrate evolution and infer its causes.

This claim brought a vitriolic response from Dan Graur (a rather polemical evolutionary biologist) and William Martin (a very down-to-earth cell biologist), who were reminded of the conclusion of Archbishop Ussher that the world was created on Sunday 23 October 4004 BC. Perhaps they went a little too far by entitling their critique 'Reading the Entrails of Chickens' (I bet that was Graur's idea) but they certainly got everyone's attention. Using calibration points as though they were fixed values leads to estimates which seem to be very precise and accurate whereas in fact they may be neither.

There is also a fundamental and irresoluble problem with the molecular clock. The genetic distance between two species results from a rate multiplied by a time: the rate of evolutionary change and the time since the last common ancestor of the species. There is no way of estimating rate and time independently, which is why fossils are needed for calibration. Modern clockmakers try to get round this problem by using Bayes' theorem, which requires a brief explanation.

Thomas Bayes was a divine like Ussher, although his calculations have worn a little better. His theorem relates conditional and unconditional probabilities, which sounds rather dry but is one of the few areas of statistics capable of raising the blood pressure. To give a relatively innocuous example, suppose that we wish to ascertain the probability that it will rain, given that it is cloudy. We can write this probability as Prob(rain | cloudy), where the vertical line means 'given that'. Bayes' theorem then states that

$$\text{Prob(rain} \mid \text{cloudy)} = \text{Prob(cloudy} \mid \text{rain)} * \text{Prob(rain)/Prob(cloudy)}.$$

Since it is almost certain that it will be cloudy if it's raining, we can set Prob(cloudy | rain) to 1 and rearrange the equation to get

$$\text{Prob(rain)} = \text{Prob(cloudy)} * \text{Prob(rain} \mid \text{cloudy)},$$

which is obviously true. Aside from meteorology, Bayes' theorem can be used in phylogenetics to estimate the divergence times of species in a phylogenetic tree, with an hypothesis about this time predicated on the sequence divergence between the species substituted for Prob(rain | cloudy). The difficulty is that the analogue of Prob(rain), the probability that the time is correct, is unknown. The subterfuge is to make a reasonable guess at this prior probability and then calculate the likelihood that the observed sequence divergence is predicted by the guess. If it is, fine; if not, make a second guess and see whether or not this improves the likelihood. If it does, change the hypothesis (by altering the structure of the tree), make another guess and proceed in this way for as long as it takes. The estimate of the posterior probability (that the hypothesis is correct, given the observed sequence divergence) should converge smoothly towards some limiting value, which is taken to be correct. The fact that rate and time are inextricably confounded in sequence data, however, means that two priors are needed, one for rate and one for time. This has the unfortunate consequence that the posterior estimate of divergence time does not necessarily converge on any fixed value. Even worse, the priors on rate and time affect the posterior estimates of time even if vast amounts of data are available, as they often are nowadays. This can lead to seemingly precise but wildly inaccurate estimates of divergence time. Dating divergence time from the molecular clock is a promising and valuable exercise, but published dates should be viewed with appropriate caution.

As the genomics era matured, it became possible to use 100–200 genes to estimate when animals first appeared. Calibrating the clock with fossils remains a difficulty, although the date can be pushed back a little to the appearance of the first jawed vertebrates, the acanthodians, small freshwater fish that lived about 420 Mya. The most recent surveys have hauled in Wray's estimate a little, dating the origin of animals to about 850–700 Mya. A refreshingly sceptical study by a British group headed by Ziheng Yang at University College London and Phil Donoghue in Bristol, for example, put the origin at about 730 Mya, with cnidarians and bilaterians diverging about 640 Mya. The very latest article, by a team from Bristol headed by Emily Carlisle (and including Phil Donoghue), hauls the origin of metazoans even closer, although only by using calibration points that more or less guarantee this result.

I should add the further caution that none of these studies heeded Zuckerkandl's recommendation to use neutral substitutions. Many surveys even use proteins. There are good reasons for this choice (neutral sites may have been modified repeatedly in distantly-related lineages) but it inevitably incorporates parallel evolution as well as common descent.

Despite the ambiguities and uncertainties of the molecular clock, there is a pronounced discordance between the molecular data and the fossil record. We have to explain why there is a gap of at least 100 My between the first animals (according to the clocks) and the first fossils (according to the rocks). Perhaps something was happening during this period, roughly between the Sturtian and Marinoan glaciations, which has left no trace in the fossil record, or none that we have found so far. We also have to explain the gap, also of about 100 My, between the divergence of bilaterians, which according to the clock took place at about the same time as the Marinoan glaciation, and the first burrows made by bilaterians in the earliest Cambrian, preserved as trace fossils at Fortune Head and many other localities. A modest discordance between fossils and gene-based phylogenies might be expected, but gaps that make up a substantial fraction of geological time call for a special explanation.

A postscript: Emile Zuckerkandl and his wife, Jane, returned to France in 1965, where he was to found a research institute at Montpellier. He became exasperated with the bureaucracy (sounds familiar) and soon returned to the United States, although at first in rather shaky positions. Once hired by Pauling's Institute for Science and Medicine, however, he became adept at bureaucracy himself, and within a few years had become Vice-President. He wrote some brilliant papers on phylogenetics, but became increasingly sidetracked into research on vitamin C and attacks on creationism and social constructivism. During the few years with Pauling at CalTech he had founded molecular phylogenetics and invented the molecular clock, but he did not thrive in the genomics era and is seldom acknowledged, as I think he ought to be, as one of the most influential scientists of the period. Emile Zuckerkandl died in 2013 in Palo Alto, California.

## Evo-devo

About a century before the gene-based trees invented by Zuckerkandl gave birth to the new animal phylogeny, William Bateson was puzzled by one aspect of the new theory of evolution. If Darwin was right, and natural selection acting on slight differences drove only gradual change over long periods of time, how can one explain why familiar animals and plants mostly belong to sharply demarcated species? Bateson was an early geneticist (he coined the term) who was soon to stumble on an obscure account of breeding experiments by the Augustinian monk Gregor Mendel, but before this, in 1894, he published a collation of observations about variation in natural populations. This work was modestly titled *Materials for the Study of Variation* and is frequently cited even today, but its subtitle is less well-known: *Treated with Special Regard to Discontinuity in the Origin of Species.* Although convinced of the effectiveness of natural selection, Bateson proposed to solve the species problem by showing that a great deal of variation was discontinuous and involved dramatic shifts in appearance. For example, he cited an aberrant individual sawfly in which the terminal joint of one of the antennae develops into a foot. A specimen of a longhorn beetle had been described in which the wing-covers had been replaced by a pair of legs. In one shore crab, a mouthpart developed as a grasping claw. All of these are examples of homeosis (he called it 'homœosis', but the diphthong has long been suppressed), which he defined as 'the assumption of one member of a meristic series of the characters proper to other members of the group' ('meristic' means serially repeated, like segments in an annelid or arthropod). It was an arresting observation, but shortly afterwards his research became dominated by the rediscovery of Mendelian genetics and its implications for evolution.

Homeotic transformations were not forgotten in the early years of the 20th century, when the mechanism of Mendelian inheritance was being worked out, but they were sidelined. One of the main labs working on transmission genetics at that time was the Fly Room at Columbia University, which Thomas Hunt Morgan set up in 1908 to screen caged populations of Drosophila, the fruit fly, in a search for variants that showed Mendelian segregation. His student Calvin Bridges started to notice odd flies as early as 1915, and found a spectacular variant with two pairs of fully-developed wings, rather than the usual single pair, in 1923. Bridges realized the implication of variants like these for evolution in general, and Richard Goldschmidt even used them to elaborate a theory of evolution based on the abrupt appearance of radically different body plans. Bateson had been vindicated. Mainstream evolutionary biologists were not impressed, however, and were inclined to dismiss homeotic variants as 'hopeful monsters' (in Goldschmidt's own phrase). It did not help that Morgan himself was careful to distinguish between the mechanism of inheritance and the evolution of populations, which was perfectly correct but tended to unlink transmission genetics from population genetics. The two fields developed in splendid isolation for most of the rest of the 20th century, until a surprising observation was made in Walter Gehring's lab in Basel.

## The *Hox* cluster

By the 1980s a number of homeotic genes had been identified in Drosophila from the striking appearance of flies where they had been lost or changed—legs instead of antennae (the Antennipedia gene, *Antp*), for example, or two pairs of wings (Ultrabithorax, *Ubx*) The most surprising observation was made by Bill McGinnis, a postdoc who had come to the lab from Berkeley. He discovered that *Antp* and *Ubx* shared a long sequence of 180 nucleotides, encoding 60 amino-acids, which he called the 'homeo box'. This hinted that all the homeotic genes of the Drosophila genome might share the same sequence. Perhaps it might even be found in quite unrelated organisms! This is the sort of late-night speculation that usually leads nowhere, but on this occasion led McGinnis to a local bait shop, where he bought mealworms (beetle larvae) and earthworms, took them back to the lab, and extracted DNA from them. This rather eccentric procedure—molecular biologists don't usually get their material from bait shops—showed that these very different animals did indeed share the homeobox sequence with Drosophila. With the publication of this finding in 1984, molecular genetics was reunited with evolution. A flurry of publications in the following decade made it clear that the range of different body plans that morphologists had uncovered was encoded by a common set of genes in all bilaterians. It follows that the origin of animals is tightly bound up with the origin of the *Hox* genes, as they became known. This realization founded the modern field of evolutionary developmental biology, always known fondly as evo-devo. It is a pity that McGinnis and Gehring never did get the Nobel Prize they deserved for such a fundamental advance.

Drosophila has eight *Hox* genes, which form a tightly linked cluster on the same chromosome. The cast of characters is, in order of expression: Labial, *lab*, which controls head development; Proboscidea, *pb*, mouthparts; Deformed, *Dfd*, maxillary and mandibular segments of the head; Sex combs reduced, *Scr*, head and thorax development; Antennipedia, *Antp*, leg and wing development on the second thoracic segment; Ultrabithorax, *Ubx*, leg and haltere (wing rudiment) development on the third thoracic segment; Abdominal-A, *abd-A*, appendage formation on the abdominal segments; and Abdominal-B, *abd-B*, acting in the final segments and tail. This sequence has two remarkable features. In the first place, the order of *Hox* genes on the chromosome mirrors the segments they affect in the body, from front to rear. Secondly, the order of the genes on the chromosome also mirrors the time course of their expression during embryonic development, from earliest to latest. These quite unexpected features have elicited a range of explanations, but almost everyone is agreed that the *Hox* cluster has evolved through successive rounds of duplication and divergence from a single ancestral gene.

## The *Wnt* genes

The *Hox* cluster is by no means the only set of genes that regulate development. One other important class is made up of the many different kinds of *Wnt* (pronounced 'wint') genes which encode glycoproteins which are able to pass a signal from cell to cell by binding to a receptor protein in the plasma membrane. Wnt proteins establish the site of gastrulation, and thereby direct the development of the anterior region (including the future mouth) of polyps and the posterior region of deuterostomes. Indeed, they are primarily responsible for laying down the primary (anterior-posterior or oral-aboral) body axis in both cnidarians and bilaterians, and have a remarkably similar structure in animals as different as hydras and humans. Hox and Wnt proteins are linked in the complex (and at present incompletely understood) web of induction and repression that is the basis of normal development.

## The zootype

Richard Owen is now best remembered as the antagonist of Darwin, but his most lasting memorial is the Natural History Museum, built, largely at his urging, to house the natural history collections of the British Museum. He was probably displeased when a statue of Darwin was installed in a prominent place on the main stairway while he was still Director, but he got his revenge posthumously when it was removed to the museum café and replaced by a statue of himself. Eighty years later, however, the Darwin statue was put back again and this time it was Owen who was relegated. His defeat in the battle of the statues is probably final, but his influence continues to be felt, especially in the crucial distinction he made between homologous and analogous structures. Another term he coined was 'archetype', meaning the idealized arrangement of structures characteristic of members of the same phylum. The term was based on pre-evolutionary concepts developed by morphologists earlier in the 19th century, long before Darwin's time, but it comes close to the notion that monophyletic groups are defined by a unique set of shared derived characters. The archetype was not an ancestor, however, nor need it correspond to any real animal; it was more like the Platonic essence of a particular group, a mental construct that stripped away the idiosyncrasies of individual species until only the irreducible features of the group remained.

The idea of an archetype might be dismissed as an historical curiosity, except that it has recently been revived by a group of Oxford zoologists—Jonathan Slack, Peter Holland and Chris Graham—as the zootype (Figure 6.2). The zootype is the archetype in modern dress; whereas the archetype was a morphological concept, the zootype is based on genes, and in particular on *Hox* genes. It is defined as the spatial pattern of gene expression characteristic of animals. Indeed, it can even be used as a definition of 'animal': an animal is an organism whose development is governed by the zootype. The earliest stages of development may differ widely, even in closely-related species, for example because the egg contains more or less yolk, and the later stages diverge to give rise to adults adapted to different ways of life. In between there is a period called the phylotypic stage during which embryos closely resemble one another. In vertebrates, for example, the phylotypic stage is the tailbud embryo. It is during the phylotypic stage that the body plan of the phylum is first realized, with all the structures in place, and the zootype is most clearly expressed. Unlike the archetype, the

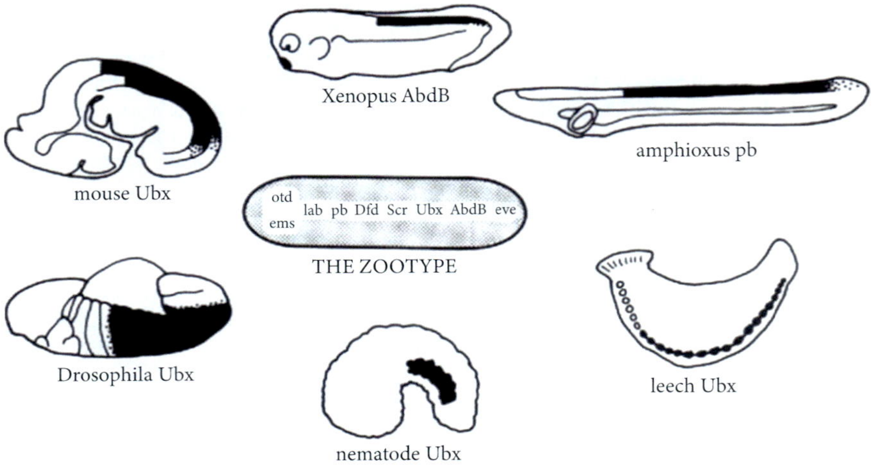


**Figure 6.2** The zootype

Reproduced with permission from Slack, J.M.W., Holland, P.W.H. and Graham, C.F., 1993. The zootype and the phylotypic stage. *Nature*, *361*, 490–492.

zootype explicitly reflects the ancestral state. The ancestor of all animals was the first organism to possess a zootype, and the phylotypic stage for each phylum resembles the ancestral condition.

Not everyone was convinced by the zootype. It is strongly influenced by mice and flies, and it is not clear how far it applies to animals like sponges and cnidarians.

Adult sponges are built around a basic body plan of flagellated chambers fed by pores in the body wall, whereas their larvae are much more diverse. The calciblastula larva is formed by cells moving into the interior of the blastula, with the flagellated cells remaining outside. The amphiblastula has ciliated cells at the forward end and non-ciliated cells at the rear. The disphaerula is formed by invagination, with the cells on the outside remaining ciliated. The parenchymella is formed by the differentiation of cells in a solid-bodied blastula. These are only some of the ways in which the larva can be constructed, and there are several other routes to the adult. Sponges, of course, have no *Hox* genes.

Most cnidarians have a planula larva with extensive differentiation of cell types; some even have a mouth and are capable of sensing, eating and digesting prey. The planula, however, develops from a blastula by many routes, some of them via stages that resemble a plakula or a parenchymella. Cnidarians have *Hox* genes, although only two or three rather than the canonical set of eight found in Drosophila.

The zootype is perhaps best viewed as an attribute, not of animals as a whole, but rather of bilaterians with a full set of *Hox* genes. Nevertheless, the conservatism of the *Hox* cluster provides a mechanism for recapitulation. It is scarcely too much to say that the concept of the zootype restates in genetic terms what Haeckel had expressed in morphological terms. Perhaps that is one reason it has been so contentious.

## The evolution of development

All proteins fold into characteristic structures, such as a sheet, a loop or a helix, because of different kinds of bond between amino acids and because hydrophobic amino acids tend to move to the inside and hydrophilic amino acids to the outside of the molecule. Each kind of protein, specified by its unique amino-acid sequence, therefore has a unique shape. Hox and Hox-like proteins form three helices that fit snugly into the grooves of the DNA helix and thereby control the expression of a particular gene. How tightly it is bound depends on the interaction between the amino-acid sequence of the protein, including amino acids outside the homeobox proper, and the nucleotide sequence of the gene. In this way, homeobox proteins can govern the expression of whole families of genes. The same homeobox protein can act simultaneously to activate some genes while repressing others, and in this way directs the normal course of development, or, when disrupted by mutation, gives rise to homeotic monsters.

All multicellular eukaryotes—perhaps all eukaryotes—have homeobox genes to control gene expression during development. Plants, for example, have a hundred or more, although they are quite different from those of animals. Bacteria and Archaea don't have homeobox genes, but do have genes that encode somewhat similar proteins, from which homeobox genes presumably evolved. Genes containing a homeobox are all capable of regulating others, but three gene families are particularly important for development. The first is the *Hox* cluster itself. The second is *ParaHox*, a group of three or four genes which also forms a cluster and may have originally evolved from some common ancestral gene of *Hox* and *ParaHox*. The third consists of the so-called *NK* genes, which sometimes form a cluster but usually not. (If you look up 'NK gene' you will find many references to Natural Killer cells, which are lymphocytes of the immune system, nothing to do with development; *NK* homeobox genes are named—by themselves—for the two molecular biologists, Nirenberg and Kim, who first described them.) There is an intriguing speculation that these three gene families correspond to the three germ layers of bilaterians: *Hox* controls the development of the ectoderm, *ParaHox* the endoderm, and *NK* the mesoderm. The correspondence is far from perfect, however, and not all embryologists are convinced.

The evolution of the *Hox* cluster by gene duplication irresistibly suggests that the *Hox* and *ParaHox* genes likewise arose by the duplication of an imaginary ancestral gene or cluster of genes given the name of *ProtoHox*. Clearly, *ProtoHox* itself must have evolved from some predecessor, which can be identified from the distribution of *Hox*-like genes among living animals. Sponges are non-polar animals that lack both *Hox* and *ParaHox*, but do have *NK* genes. The ultimate ancestor is thus an *NK* gene, or gene cluster, which duplicated to give rise to *ProtoHox*, which in turn duplicated (or just split) to give rise to *Hox* and *ParaHox* genes. Cnidarians have a small cluster of *Hox* genes and at least two *ParaHox* genes, which direct the formation of the oral-aboral axis and represent this stage. The *Hox* cluster was elaborated further in bilaterians to pattern the anterior-posterior and dorsal-ventral axes by specifying the appendages borne by successive segments, with *ParaHox* governing the development of the through-gut while the *NK* genes became involved in the evolution of the coelom and internal organs. This is a simple progressive theory that reflects the early diversification of animals, incorporates the association of the three classes of development genes with the three germ layers, and is of course consistent with the morphological account of animal evolution.

Unfortunately for simplicity, there is an alternative: animal evolution does not proceed by a steady increase in complexity, but is instead dominated by the loss of genes from a more complete ancestral complement. Sponges may not have either *Hox* or *ParaHox*, but perhaps these were present is some distant ancestor and have since been lost. A group of marine biologists at St. Andrews led by David Ferrier have argued that most of the genes associated with *Hox* and *ParaHox* in animals as distantly related as sea anemones and chordates are also linked in the genome of the demosponge *Amphimedon*. *Hox* and *ParaHox* may be missing now, but their ghosts are revealed by the continued linkage of their neighbours. Ghost genes are not entirely convincing, of course, but in 2014 Ferrier and a group of biologists from Bergen reported finding a real *ParaHox* gene in two calcareous sponges, showing that *ParaHox* at least must have been lost from the demosponge lineage. Furthermore, placozoans seem to have a *ParaHox* gene but no *Hox* gene, which seems inconsistent with the Proto*Hox* theory and implies that *Hox* has been lost from the placozoan genome. More recently, a group from Florida (including Mark Martindale) were puzzled by the lack of similarity between the *Hox*-like genes of cnidarians and bilaterians. Whether or not most of these genes truly descend from the corresponding genes of a common ancestor seems dubious, and might imply that the common ancestor instead had an extensive complement of *Hox*-like genes, many of which were subsequently lost in cnidarian and bilaterian lineages. As a final twist, ctenophores seem to have neither *Hox* nor *ParaHox* genes, which might merely reflect the very early divergence of ctenophores from other lineages—or might equally mean that these genes have since been lost.

The distribution and mode of action of *Hox*-like genes is currently (2024) the focus of intense research, with new discoveries (and new theories) announced every month or so. At present it seems safe to say that the earliest-branching lineages still extant, the sponges, ctenophores, and placozoans, have at most one or two *Hox*-like genes. Cnidarians have canonical *Hox* genes, but not the full set. Bilaterians have the complete suite of *Hox* genes, modified in some phyla by deletion of some genes or expansion of the entire cluster or a mixture of both. The early history of animals must have involved the assembly of *Hox*-like genes to control development, but whether this proceeded mainly by the sequential addition of genes or by gene loss is still an open question.

## Embryos and ancestors

The genes responsible for early development turned out to be much the same in animals as different as mice and fruitflies. This conservatism was taken to prove that the ancestor of bilaterians had a rather complex segmented body featuring a through-gut with mouth and anus, a heart and circulatory system, central nervous system and coelom. (When and how it had acquired these attributes was not specified.) Adam Sedgwick had suggested something along these lines as early as 1884, but the idea was undoubtedly strengthened by the similarity of *Hox*, *ParaHox* and *NK* genes in bilaterians.

At about the same time, however, another branch of genetics was suggesting a completely different ancestry for bilaterians. As the cost of sequencing fell, more and more genes could be fed into the phylogeny hopper, and the branching pattern of some of the more doubtful groups became increasingly stable. In particular, the acoels were firmed up as the earliest branch of the bilaterians, in other words as the sister group of protostomes plus deuterostomes. Acoels are unsegmented and have especially simple bodies. This suggests that the ancestor of bilaterians need not after all be a complex animal with segmentation and a coelom, but could instead have been a much smaller and simpler creature.

A similar ambiguity has many times marked the long journey from Haeckel to Hox. Or rather two journeys, one undertaken by comparative zoologists and the other by phylogeneticists; the two do not always follow the same route nor arrive at the same destination. But there is a third journey that I have scarcely spoken of yet, the journey undertaken by palaeontologists in search of the fossils that will alone provide the hard evidence to settle the controversies among zoologists and phylogeneticists. The purpose of this journey was to discover the animals that left their remains in the supposedly unfossiliferous rock that was laid down before the first trilobite had appeared. These are the fossils that—it was hoped—would reveal the course of early animal evolution and solve Darwin's difficulty. In fact, once the right localities had been found a surprising diversity of fossils was uncovered. In some of these localities, indeed, the rock is crammed with fossils. But they were nothing like what might have been expected.

# 7
# The Early Cambrian Fossils

The first animals were the ancestors of *Spongilla*, *Hydra*, *Mesostoma* and all the rest of the modern fauna. The only direct evidence of what they looked like and how they lived is supplied by the fossil record. This shows us that complex, highly integrated bodies had already evolved early in the Cambrian, so I'll start by describing the types of animal that lived on the sea floor about 20 My after the non-existent golden spike at Fortune Head. Then I'll follow the fossils back in time for 50 My or so, until they fade out altogether. To begin with, the animals are unfamiliar but (for the most part) quite easy to understand. As we move backwards in time, however, the fossils become stranger and stranger.

## Lagerstätten

Most fossils are hard parts, usually shell or bone. It is only in very exceptional circumstances that the whole animal, including its soft tissues, is preserved. The most familiar example is amber, where insects and spiders or even small lizards have been overwhelmed by the resin leaking from coniferous trees. There were no conifers in the Cambrian seas (or anywhere else), of course, and in any case the internal tissues are seldom preserved (they decay before the resin has fully penetrated the animal). Soft parts can certainly be preserved by freezing, and entire mammoths have been excavated (and even on occasion eaten) from the permafrost of Siberia. Even more remarkably, my colleague Lyle White has resuscitated viable bacteria thousands of years old from the permafrost of the Canadian Arctic. It is only the recent series of glaciations that have provided this opportunity, however, and the PreCambrian permafrost has long since melted. We are left with a few rare geological events. Chert is formed when a saturated solution of silica sets to a glassy hardness and can preserve even cellular detail—the Rhynie chert of Aberdeenshire, for example, was precipitated from hot-spring water saturated with silica and has preserved entire communities very early in the evolution of land plants, complete with their associated arthropods and even fungi. More generally, if the soft parts are mineralized by calcium phosphate before they decay they can be preserved as a sort of ghost image of the living animal. This requires very rapid mineralization, of course, but ironically enough the bacteria normally responsible for decomposition may assist the process by forming a sheath over the outside or even the internal organs of the animal. Otherwise, soft parts will be preserved only if the animal is overcome by a mudslide and buried beneath anoxic mud, leaving a fossil which is a mould or cast or impression of its body for future palaeontologists to discover.

*The Sea-Change*. Graham Bell, Oxford University Press. © Graham Bell (2026).
DOI: 10.1093/9780197797464.003.0009

Deposits where the soft parts of animals are preserved were called *Lagerstätten* by the German palaeontologist Dolf Seilacher (of whom much more later). They occur irregularly through geological time. The Rhynie Chert is a lagerstätte; so is the Solnhofen limestone of Bavaria, where *Archaeopteryx* was found, and Beecher's Trilobite Bed, in New York State, where the soft tissues of trilobites have been mineralized by pyrite. There is a rather loose parallel between fossilization and the preparation of specimens for museum collections. At one extreme, mollusc shells and sea urchin tests are common fossils because they do not decompose and are so robust that they only have to be cleaned and boxed to be ready for study. At the other extreme, flimsy hydroids or salps have no hard parts; they just melt back into the water unless they are treated very carefully, and they never become fossils—except in lagerstätten. More generally, the soft parts of any animal soon decompose, leaving only its perdurable skeleton behind. Unfortunately, the soft parts are usually the interesting parts which tell us how it works and what it is related to. They can only be preserved by preventing decomposition, which is done routinely in museums by immersing the animal in alcohol or formaldehyde but seldom happens in nature. The decomposing tissue might be replaced by phosphate, but this works only for very small animals, at most a few millimetres in size. For anything larger, sudden burial beneath an underwater avalanche, so that decomposition is retarded beneath anoxic mud, may leave an impression of some of the soft tissues at least as the mud hardens into rock. If conditions are just right, even fine details such as gills or compound eyes or bundles of chaetae can be preserved. It is those very rare deposits where conditions were just right that form lagerstätten. Moreover, it is only by chance that they are discovered, and geologists are never thick on the ground unless oil is suspected. What is known about the very early history of animals depends to an alarming extent on the juxtaposition of a minority profession with a rare sedimentary event.

Fortunately, there are quite a few lagerstätten in the early Cambrian (Figure 7.1). They are scattered around the world, from tropical beaches in Australia to the frozen coast of northern Greenland, although the best-preserved and most diverse communities have been found at inland sites in China and Canada. The best-known of these is the Burgess Shale, whose fossils have been the subject of many reviews and four complete books. In one of them, *The Crucible of Creation*, Simon Conway Morris imagines the animals that the scientists in a submarine equipped with sampling gear might have been able to see and catch, once their time machine had brought them back 505 My to the muddy sea floor off the northern coast of Laurentia.

The names of the animals they saw will be strange until I describe them a little later. 'Do we have to know the names of all the animals?' is the question asked by many students after a few lectures in my course on animal diversity. 'Ideally', is my answer, 'but if you can't remember them all, at least remember those that I emphasize'. What I don't say is that if I set an exam question referring to an obscure animal I'd only mentioned once I'd be pelted with pickled starfish. The same principle applies here: an animal that I mention several times is probably worth memorizing, even though there is no danger of an exam. All the animals in the next paragraph, for example, will reappear later.

Some of them, such as priapulids, hyoliths and burrowing polychaetes, actually lived in the mud itself. Others lived on the surface of the mud, like brachiopods,

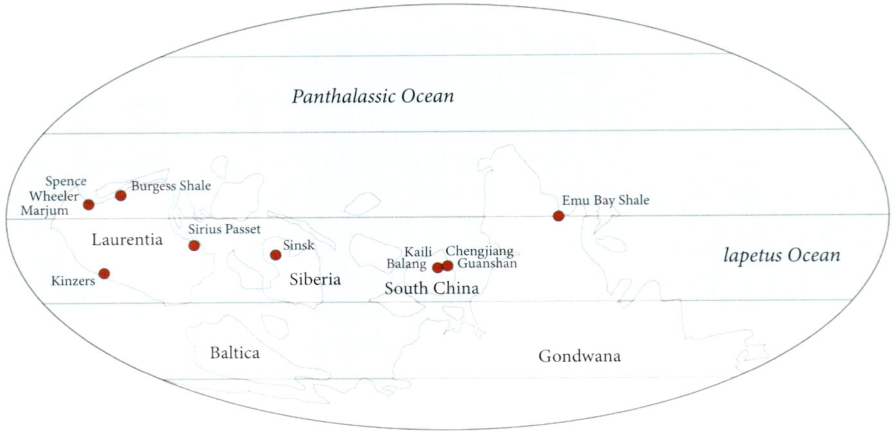


**Figure 7.1** The early Cambrian Lagerstätten

Reproduced with permission from Holmes, J.D., Garcia-Bellido, D.C. and Lee, M.S.Y., 2018. Comparisons between Cambrian Lagerstätten assemblages using multivariate, parsimony and Bayesian methods. *Gondwana Research*, *55*, 30–41.

or grew upwards to harvest the waters above, like sponges and *Chancelloria*. There would have been plenty of arthropods crawling over the seabed, including trilobites, types like *Kunmingella* with a carapace shaped like a Cornish pasty, and great-appendage types like *Leanchoilia*. The crew also saw *Wiwaxia* and motile polychaetes on the sea floor, and found *Anomalocaris* prowling in the open water, along with a variety of floaters and swimmers including ctenophores, chaetognaths and *Nectocaris*. They even saw a chordate, the famous amphioxus look-alike *Pikaia*. They didn't see any obvious molluscs or echinoderms. In short, a familiar cast of characters appeared on the stage, with others lurking in the wings. Without lagerstätte like the Burgess Shale we would know the diverse animal communities of the Cambrian only from trilobites and a scatter of enigmatic tubes, shells and sclerites: Conway Morris has estimated that 95% of the animals living in, on and above the sea floor would have left no fossil remains at all in normal conditions. The lagerstätten alone give us a brilliant picture of the worldwide community of animals that lived in the early Cambrian seas and should enable us to piece together the sequence of events that led to the radiation of body plans and ways of life.

Some 20 My before the Burgess Shale was laid down, a rich community of animals lived on the gently sloping muddy shore of a shallow sea off the coast of the ancient continent of Gondwana. Disaster overtook it one day about 520 Mya, when the whole community was buried by a mudslide. It was quite local; nearby communities were not affected, or at any rate were not fossilized. This is where I shall begin, at a time when animals had begun to diversify and the first trilobites had already evolved. Once the scene is set, I shall trace the fossils back in time for about 50 My, until there is no longer a fossil record of large creatures. Then we have to make sense of what the fossils have told us, which will not be easy because many of them are enigmatic, unexpected, unpredicted.

## The Yangtze Platform, South China, about 520 Mya

The Maotian hill lies south-east of Kunming in Yunnan Province, which borders Vietnam and Laos in southern China. The area around had been mapped by the Geological Survey of Indochina in French colonial times, when trilobites and other early Palaeozoic fossils were described by Jacques Deprat and Henri Mansuy. (They fell out. Mansuy accused Deprat of planting European specimens in his collections, and the scandal that followed sucked in Honoré Lantenois, their boss; Madeleine Colani, their colleague; and the Société Géologique de France. Deprat was disgraced and dismissed, but there are still doubts about his guilt.) Sixty-odd years and a good deal of history later, a graduate student named Hou Xian-guang was looking for fossils in the same area. He had travelled by cart along muddy tracks from Chengjiang, a small town south-east of Kunming (the two are now connected by a fast highway) and had hired a farm worker to break out blocks of shale from a section on the west side of the hill. Splitting these blocks with pick and hammer reveals the fossils, if there are any; it is hard work and often completely unremunerative. On this occasion, however, a July afternoon in 1984, a fossil that seemed almost to be alive was found on the glistening surface of the stone. It was a trilobite, *Naraoia*, known from many other localities. But it is usually only the mineralized exoskeleton that is preserved, whereas this had its soft parts preserved too. It was the first of many; once the exposure had been fully exposed (the geologists who shared digs with Hou Xian-guang thoughtfully blasted it) thousands of specimens could be collected and it was clear that a complete fauna with soft parts preserved had been found. Because they lived during the early Cambrian, about 520–518 Mya, these fossils give a vivid picture of how the radiation of animals had unrolled. A detailed account of the discovery is given in a book, *The Cambrian Fossils of Chengjiang*, illustrated with mouth-watering photographs of the fossils themselves. The site now has UNESCO World Heritage status and houses a large and impressive museum which attracts tens of thousands of visitors every year. Clearly, early Cambrian palaeontology can pull the crowds (although the museum does have dinosaurs too).

### Water filters

The simplest way of life that an animal can follow (and the least interesting, according to generations of zoology students) is to be firmly attached to the sea floor, incapable of crawling or swimming. This is all very well for a plant, which can live on sunlight and minerals, but an animal must be able to eat even if it is completely sessile and cannot search for prey. There are only two ways of solving the dining problem, if we discount exotic options such as farming bacteria or soaking up the dissolved nutrients in seawater. Some sessile animals target large individual prey, using the poisoned harpoons of cnidarians or the hooked spicules of some deep-sea sponges, but these are clearly devices that have evolved from simpler precursors. The alternative is to harvest large numbers of small organisms by creating a current of water and passing it through a filter to remove the edible particles. The only motive power available to create the current is provided by the beat of cilia, and the strength of the current will be proportional to the number of cilia that can be deployed. More cilia will move more water, enabling the animal to eat more and grow larger, always provided that the cost of operating the machinery can be paid by the biomass it captures.

If the cilia are inside the body and the particles, mostly bacteria, are captured by the ciliated cells themselves, the result is a sponge. Sponges will play several roles in this account, but for the time being they can be viewed as animated water filters. The interior of a sponge is occupied by small chambers lined with flagellated cells, which draw a current of water through small openings, the ostia, in the body wall. The flagellated cells catch and eat the bacteria in the current, which continues into the spacious interior of the sponge and then out through an opening at its apex, the osculum. The osculum is much wider than any single ostium, but is narrower than the combined area of all the ostia, so the current speeds up as it passes out of the osculum. The sponge body thus functions as a sort of chimney that increases the draught in the grate (for those who remember open fires). Sponges have a skeleton made up of minute spicules of calcium carbonate or silica or keratin which link up to support the whole apparatus. If you find a sessile animal with spicules and an osculum it is a sponge.

There are a dozen or more different kinds of sponge described from Chengjiang and other sites of more or less the same age, such as Sirius Passet, which I'll describe later. They are not quite the earliest of their kind. Scattered spicules have been found in the lowermost Cambrian of southern China and India, and there are more dubious reports of sponge fossils in older rocks. Odd, when you think about it: sponges being the simplest animals should have been the earliest too, but the fossil evidence for early communities consisting mainly of sponges is decidedly thin. The two classic mid-Cambrian lagerstätten, however, have quite abundant remains of indubitable sponges, complete and fully articulated. Some are low domes that must have lain more or less flat on the seabed, but others are slim vase-shaped animals that towered a foot or more above. Sponges can filter a surprising volume of water quite clear of bacteria, and in doing so would stir the layer just above the seabed. Whether this would have made a lot of difference to the dynamics of ocean bacteria seems doubtful. Waves and currents would stir the sea anyway, and, the average depth of the sea being 4 kilometres or so, even a dense community of sponges would struggle to make much of an inroad into the vast population of bacteria living above it.

There is also an animal called *Allonia* which might or might not be a sponge (Figure 7.2). *Allonia* is a chancelloriid belonging to a group that includes *Chancelloria* from the Burgess Shale. (*Chancelloria* was named for Chancellor Peak in Yoho National Park, which you can see as the CPR train runs through Kicking Horse Pass, which in turn was named for a real chancellor, John Alexander Boyd of the Ontario High Court.) It was a barrel-shaped animal a couple of inches tall whose body was protected by rows of sharp-pointed spicules in clusters a little like the barbs of barbed wire. These clusters are very distinctive and had long been known from the lowermost Cambrian, where they can be found by the thousand. Chancelloriids were clearly very abundant members of these early animal communities. They had an apical orifice, comparable to an osculum, which together with the spicules diagnoses them as sponges, so this is how they were originally described. Their spicules, however,

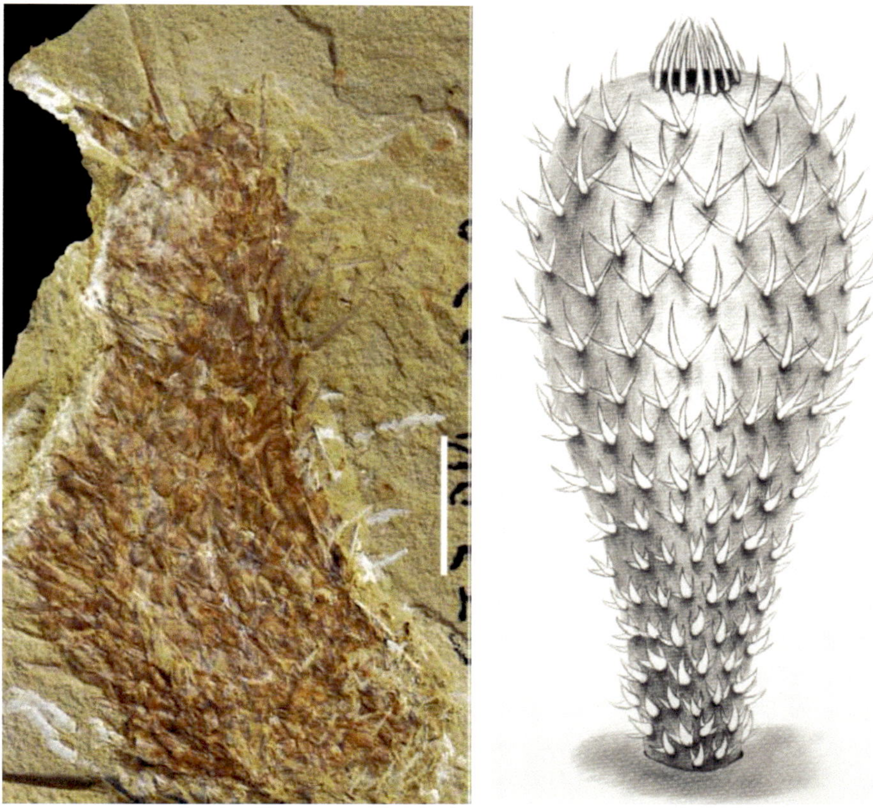

**Figure 7.2** Allonia

Reproduced with permission from Yun, H., Zhang, X. and Li, L., 2018. Chancelloriid *Allonnia erjiensis* sp. nov. from the Chengjiang Lagerstätte of South China. *Journal of Systematic Palaeontology*, *16*(5), pp.435–444.

present a problem. For a start, they are external, in contrast to the internal spicules of sponges, and clearly provide protection rather than support. They are also hollow, whereas sponge spicules are (more or less) solid. In fact, they are strikingly similar to the spearhead-shaped sclerites (scales or spines, more or less) that were the only remains of a contemporary animal called *Halkieria*, which I'll describe in more detail later. This led some very eminent authorities to the bold conclusion that chancelloriids were closely related to halkieriids. Once the entire body of *Halkieria* had been discovered, it was clear that *Chancelloria* and *Halkieria* were completely different kinds of animal that did not resemble one another at all. Others have pointed out that chancelloriids seem to lack the internal structures that were preserved in other Chengjiang animals: no gut, gonads or mesenteries are seen in specimens sectioned by the cleavage of the rock. They also lack ostia; the ostia of most sponges are about 50 μm across and should be readily visible in well-preserved specimens but have never been reported. There is no trace of internal chambers either, so the whole animal is little more than a thin-walled bag with a hole at the top. The sponge theory is not quite dead yet, though; perhaps the body wall of chancelloriids was perforated by tiny ostia only 5 μm or so across, and a single layer of flagellated cells harvested the bacteria as they were pulled in. The chancelloriids would then have been a very simple sponge, of a kind unknown today; but this interpretation relies on the existence of tiny ostia which have never been observed and probably never can be.

## Polyps

Putting the cilia outside, on the surface of the body, seems obvious but leads to a design problem. The ciliated cells themselves could be used to capture particles as before, but the surface area serving a given volume decreases with size, so large animals are impractical. Sweeping the particles into a mouth might work, but only the cilia around the mouth would be useful. The solution is to increase the surface area of the body immediately around the mouth with tentacles, elongate extensions of the body wall that can be fringed with cilia to create a powerful feeding current. The body now consists of a column attached to the seabed at one end with a mouth surrounded by tentacles at the other. This simple kind of animal is a polyp, and its tentacles are the structures that provide the essential ecological link between the sessile polyp and the waters above.

Unarmoured polyps are delicate creatures that are seldom preserved as fossils. The Chengjiang animal *Xianguangia*, however, is certainly a large soft-bodied polyp with a column a couple of inches in length surmounted by a ring of slender, tapering tentacles. Each tentacle has a row of pinnules each bearing fine filaments that resemble cilia, so the whole structure has a feathery appearance. Qiang Ou and his collaborators from the China University of Geosciences and Northwestern University think that *Xianguangia* was a suspension feeder close to the ancestry of modern cnidarians, although it does not fit neatly into any living group.

Other sessile animals are just as puzzling. *Dinomischus* had a long stalk ending in a bulbous calyx bearing a ring of tentacles (Figure 7.3). It was originally supposed to be an ectoproct, an obscure group of living animals which likewise has a calyx with tentacles set on a long stalk, although this gets us little further forwards because the relationships of ectoprocts are disputed and their fossil history is all but unknown. Anyway, as its anatomy became better known this idea was gradually given up and other explanations were proffered—perhaps it was an echinoderm, or belonged to its own ancient and now completely extinct group. The most daring guess is that *Dinomischus* was a ctenophore, even though living ctenophores look completely different. (The 'c' in 'ctenophore' is silent, as is the 'c' in 'cnidarians'. Both names are derived from Greek roots, 'ctenos' meaning 'comb' and 'cnido' meaning 'needle', to the bafflement of undergraduates who no longer enjoy the benefits of a classical education.) Ctenophores may be almost as obscure as ectoprocts, but they are nonetheless large and abundant animals that are often among the dominant predators of the upper waters of the sea. Most are about the size and shape of a cherry or gooseberry, swimming quite fast with a pair of long tentacles trailing behind. Others are larger

**Figure 7.3** Dinomischus

Reproduced from By Han Zeng, http://n2t.net/ark:/65665/3f0719703-0af7-4af6-93a2-af97fab5c74f, CC0, https://commons.wikimedia.org/w/index.php?curid=132284101, under a Creative Commons license CC BY-SA.

and simply engulf anything they encounter, including their smaller relatives, with a huge gaping mouth. One of these is *Mnemiopsis*, a North American ctenophore which was inoculated into the Black Sea by ballast water and has been blamed for the decline of dolphin populations and even reduced oxygen levels. Be that as it may, *Mnemiopsis* and all other pelagic ctenophores have rows of long fused cilia running the length of the body whose beat propels them through the water like the miniature oars of a trireme, or rather, since there are eight rows, an octoreme. Anything less like a stalked suspension-feeder could scarcely be imagined. When you watch them in motion these comb rows scintillate like a night-club strobe light. All ctenophores have comb rows and no other animal has anything similar, so comb rows are the trademark of ctenophores, like the stinging cells of cnidarians or the tube feet of echinoderms: if it has comb rows, it's a ctenophore.

Now, a closer look at the tentacles of *Dinomischus* shows that they are quite complicated structures, each consisting of a central rod from which arises a row of pinnules about 5 mm long furnished with little blades about 200 μm long. This arrangement is rather similar to the tentacle of *Xianguangia*, although with blades rather than filaments. According to Yang Zhao of Kunming and Jakob Vinther of Bristol, these little blades are fused cilia, the pinnules are nascent comb rows, powering a feeding current, and *Dinomischus* was a polyp close to the ancestry of modern ctenophores. Perhaps *Xianguangia* was too. Zhao and Vinther are very convincing and describe other Chengjiang fossils that seem to be polyp-like ctenophores, although a few details of the transition from a sessile stalked animal to a pelagic swimmer still need to be filled in. The switch from feeding to locomotion is not unlikely because any apparatus that can be used for feeding while anchored will serve for locomotion once unmoored, and many animals can use fields of cilia for either feeding or locomotion as occasion serves. (A squeeze of damp moss will usually reveal a few bdelloid rotifers when viewed under a low-power microscope, and you can watch them pulling in a feeding current with their pair of ciliary wheels, or detaching from the strand of algae where they were clinging with their toes and zooming off under ciliary power.) Shifting the fused cilia from the inside of tentacles to the outside of the body is a little more difficult to explain, but it could have happened by wrapping the tentacles around the body and losing the stalk. After all, a sea urchin can be thought of as a starfish with its arms turned upwards and covered in armour plate, the stalk of its distant ancestors having been lost long ago. The pelagic ctenophores of today may have originated as benthic polyps on the floor of the early Cambrian seas, although imagination plays a larger part in this scene than I would like.

## The lophophore

The problem with ciliated tentacles is that they are open structures, so much of the work done by the cilia is wasted on accelerating water that moves away from the mouth. One easily evolvable solution is to fold an array of tentacles into whorls around the mouth, driving a current of water down one side and up the other, capturing particles in sticky mucus that flows into the mouth. This device is called a lophophore and a variety of animals use lophophores today, ceaselessly combing the seas. The tentacles of these lophophorates are hollow, because the central body cavity is extended into their lumen. Among the undoubted lophophorates are brachiopods, whose calcareous bivalved shells are among the earliest Cambrian fossils, as I've mentioned before. Most of the space enclosed by the two valves of the shell of a brachiopod is taken up by an elaborate lophophore, in fact, with the rest of the body squashed into the rear. One of the commonest fossils at Chengjiang is the brachiopod *Heliomedusa*, which is sometimes so well-preserved that its lophophore can be made out as a row of fine lines extending from the valves of the shell. Brachiopods are still around today, although they are much less abundant than they used to be and you are unlikely to find them on the shore. I have found slabs of Ordovician limestone on McGill campus that are crammed with their fossil shells, so numerous that the rock is mostly brachiopod, but I have watched living brachiopods only in aquaria.

The other conspicuous shells at Chengjiang belong to hyoliths, which consist of a hollow conical shell capped by a lid, the operculum, and propped up on two recurved horns, the 'helens', like a sunbather reading a book (Figure 7.4). In some exceptional specimens a U-shaped gut is discernible as a dark trace, although the feeding structures surrounding the mouth have not yet been reported. They have been described in

**Figure 7.4** Haplophrentis

(Top) A specimen from the Burgess Shale, reproduced from the collections of the Royal Ontario Museum.
(Bottom) Reproduced from Qohelet12, Wikimedia Commons, under a Creative Commons license.

fossils from the Burgess Shale, about 20 My later, where a set of rather coarse tentacles, interpreted as a lophophore, protrudes from under the operculum. Now, an elongate animal with a U-shaped gut and lophophore, propped up on its helens, could well be a suspension feeder related to brachiopods, at least according to Joseph Moysiuk of the Royal Ontario Museum. Moreover, some specimens have a cord of tissue at the apex of the shell (the narrow end, opposite the mouth) which resembles the pedicle that brachiopods use to pull themselves back into their burrow. Others disagree. Fan Liu of Northwestern University in Shaanxi Province has described a Chengjiang hyolith whose tentacles do not obviously constitute a lophophore. They are arranged in a straight row, rather than being curved around the mouth, there is no evidence of ciliation, and there are no helens to lift the apparatus above the sediment. The 'pedicle' might be just the crushed remains of the apex of the shell and the U-shaped gut is necessary because the apex of the shell is closed. This suggests a quite different way of life, with unciliated tentacles used to detect and capture small organisms in the sediment.

There are several living animals which forage like this, for example scaphopod molluscs (tusk shells) which likewise have an elongate shell and tentacles that hunt for diatoms and forams in the mud, although scaphopods are buried vertically, mouth down, whereas hyoliths lay flat on the seabed. It is currently an open question whether hyoliths are closely related to brachiopods or belong to a more ancient lineage which diverged early on from other worms, molluscs and lophophorates.

## Legs

Sessile or sedentary animals are in a minority at Chengjiang, where most of the animals that have been found were actively crawling or swimming arthropods, so the astonishing radiation that has led the group to dominate many marine communities was already well under way. Arthropods feed directly on large particles crammed directly into the mouth and have abandoned ciliary feeding, and indeed cilia themselves, except in highly modified form. The very earliest arthropods lived deeper in the Cambrian, perhaps almost at the opening of the period; for example, a bilobed mound (usually likened to a coffee bean) with transverse scratches was originally thought to be a seaweed, and hence named *Rusophycus*, but is now generally interpreted as the impression left by the limbs of an arthropod flopping belly-down on the sediment. Body fossils appeared a little later; Hou Xian-guang was looking for bradoriids, an archaic group of arthropods, when he came across the Chengjiang fossils.

The cuticle of arthropod limbs is tanned by cross-linking the chitin and protein fibres with phenols, which converts the rather floppy limb of a lobopod (I'll talk about them in a minute) into a stiff tube. This sounds like a rather minor modification, but it has fateful consequences. The lobopod limb is little more than a bag to contain the muscles, so it is limited by the length of the muscles used to swing it backwards and forwards. A stiff limb, on the other hand, can be used to extend and augment the force of muscle contraction—the tube itself becomes an important component of the limb apparatus as a whole. Consequently, the shape of the tube will be modified by natural selection to make the limb as effective as possible. Here we run into a snag, because a stiff limb has to resist two kinds of load, the compression load imposed by the weight of the body and the bending load when the limb is flexed. It may fail either in compression (by buckling) or in bending (by breaking). The leg of a table, for example, must resist the compression load of the table top and whatever is piled on it, whereas the handle of a broom must resist the bending load when it is used to sweep. Unfortunately, the best design for either kind of load is not necessarily the best for the other. The load at which a tube fails depends mainly on its length, its diameter and the thickness of its wall, given that a limited amount of material is available to construct it. How its shape is adjusted depends on how the limb is normally used. David Taylor and Jan-Henning Dirks of Trinity College Dublin have calculated the minimal compression or bending load causing failure in thin-walled tubes and compared the results with the shape of arthropod limbs. They found that the shape of the crab merus (the joint of its great pincer closest to the shoulder) resists compression and bending about equally, whereas the tibia of a locust hindlimb is most resistant to bending, which seems biologically reasonable. The shape of an arthropod limb is inevitably a compromise between conflicting demands, and this would have been as true in the early Cambrian as it is today.

The arthropod limb is not just stiffened; it is also jointed, with joints of pliant untanned cuticle separating the rigid sections. A jointed limb is much more versatile than a simple tube. The hind leg of a grasshopper or locust is often used as an example because it is meaty enough to see clearly how it works. It is made up of five articulating sections, called in order from the body wall outwards the coxa (which bolts the limb to the body), trochanter, femur, tibia and tarsus (which is in contact with the ground). Most of the space inside the femur is taken up by a pair of antagonistic muscles: the extensor, which extends the tibia away from the femur when it contracts, and the flexor, which has the opposite effect of bringing the tibia closer to the femur. Both muscles run inside the limb tube and are inserted near the head of the femur and connected to the head of the tibia by a short tendon. When the flexor contracts the insect jumps. (Our hindlimbs work on similar principles but inside-out, with the muscles running outside the bone.)

A jointed limb is also a system of levers which transduces a force applied at one point to an effect at another point. A universal pivot or a set of two orthogonal joints allows the tip of the limb to be pointed or exert a force in any direction. When a grasshopper is crawling the large retractor muscle of the coxa, attached to the inside of its thorax, contracts to move its forelimb backwards by using the joint between thorax and coxa as the fulcrum of a lever. This propels the animal forwards by reaction with the ground. Once the movement is complete the flexor muscle lifts the forelimb from the ground and the extensor then displaces it forwards. Extension of the middle limb and (especially) the hindlimb contribute to the push. There is more to it than this, of course. A grasshopper which moved the limbs on each side of its body sequentially would just fall over, so flexure and extension must be coordinated so that the body is supported by a tripod of limbs at all times. But these are details to be dealt with by appropriate modifications of the nervous system. The essential point is that the jointed limb creates the potential for a whole new range of motile animals to evolve.

Grasshoppers, and terrestrial insects and spiders generally, crawl using relatively simple legs. Marine arthropods such as crustaceans have more complicated two-lobed limbs that combine a crawling leg with a gill. Limbs of this sort seem to have evolved from the gilled lobopods found at Sirius Passet, in which each of the body segments bore both a leg below and a flap of tissue, perhaps a gill, above. Fusion of the two separate structures would produce a biramous two-lobed limb which combined locomotion with respiration, and credible intermediate stages in which a lobopod limb has become united with a (presumably) respiratory flap have now been described from Chengjiang.

A stiffened, jointed, two-lobed limb is in itself a major innovation, but it was only the start of arthropod diversification. The size and shape of a multi-jointed limb can be modified by exaggerating or suppressing some of its component sections so that it can be used not only for crawling or swimming but also for sensing, grasping, stabbing, chewing, holding onto a partner or almost anything else that an animal can get up to. Moreover, the kind of limb that develops on a segment can vary along the length of the body, so that one segment may bear the same kind of limb as the neighbouring segments, or a completely different kind, or no limb at all. Once the jointed limb had evolved, then, arthropods could diversify both from the body wall outwards and from the head backwards. This is not to say that the innumerable combinations of possible limbs are all realized, so that arthropods are a sort of morphological smear. On the contrary, each major group of arthropods is characterized by a particular pattern of limb specialization. Take a lobster, for example. Starting from the head and working backwards, we find a pair of limbs on each segment after the first: antennules for sensing; antennae, also for sensing; mandibles for tearing the food; first and second maxillae for chewing; first, second and third maxillipeds for manipulating processed food; great chelae for grasping; four pairs of pereiopods, the walking legs; five pairs of gilled pleopods, the swimming legs; uropods, which are flattened limbs for bolting away from danger; and, bringing up the rear, the telson, often used as a flap to back away rapidly from a threat. All the Decapoda (lobsters, crabs, crayfish, shrimp, prawns and many others) show the same or a very similar pattern of specialized limbs. Other groups have a different arrangement: in myriapods (millipedes and centipedes), for example, the mouthparts are followed by a long series of walking legs. There is one more refinement: segments bearing different kinds of limb are arranged together in different regions of the body, classically the head, thorax and abdomen. The pattern of limb specialization and arrangement is diagnostic for each major group of arthropods.

## The Chengjiang arthropods

This seems to offer an easy way to describe the Chengjiang arthropods: simply record the form of the limb and the pattern of segment specialization (which highly skilled palaeontologists can do) and relate these to a modern group. In practice this obvious approach doesn't work very well, because most of the Chengjiang arthropods, and perhaps all of them, can't be clearly assigned to any modern group. Take *Kunmingella*, for example, one of the bradoriid arthropods that Hou Xian-guang was looking for when he discovered the site (Figure 7.5). Most of its body is enclosed by a bivalved carapace an inch or two long which is roughly circular in outline when the animal is spread out, as it often is on the rock surface, so in life it would have folded into a shape like a Cornish pasty, with only the antenna at one end and the conical tailpiece at the other poking out. Behind the eyes, which are right at the front as usual, there are 10–12 limb-bearing segments, five in the head and the rest in the trunk. The first bears a long uniramous (unforked) antenna, which presumably explored the environment immediately in front. There is then a series of biramous appendages, each with a leaflike gill branch above the walking leg, followed by two pairs of uniramous limbs to complete the series. This is a rather generalized body plan with limbs (apart from the antenna) that are only weakly distinguished from one another. Related species may have either rather more or even less strongly specialized limbs. Bradoriids superficially resemble ostracods, the little seed shrimp that pullulate in the sea and freshwater, and they were originally thought to be early members of the group. But ostracods are not clearly segmented at all, and besides they have a different sequence of limbs: two pairs of antennae in front, followed by three pairs of mouthparts and then by two pairs of

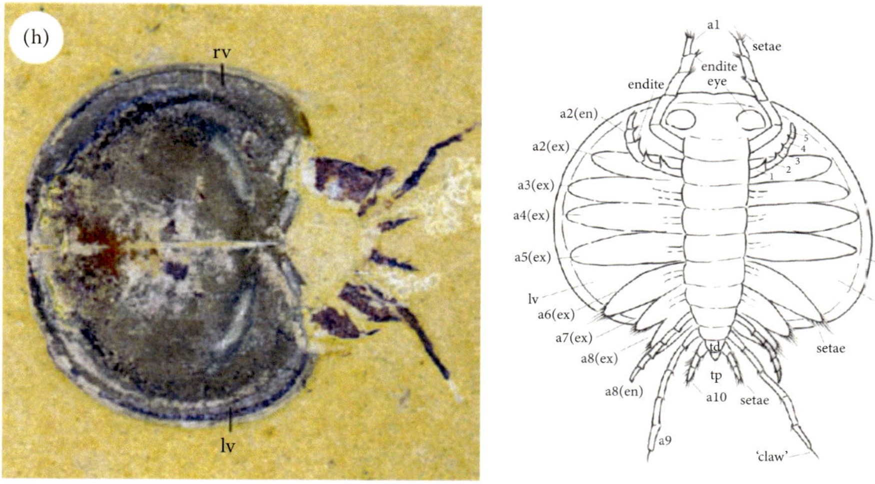


**Figure 7.5** Kunmingella

Reproduced from Hou, X., Williams, M., Siveter, D.J., Siveter, D.J., Aldridge, R.J. and Sansom, R.S., 2010. Soft-part anatomy of the Early Cambrian bivalved arthropods Kunyangella and Kunmingella: significance for the phylogenetic relationships of Bradoriida. *Proceedings of the Royal Society B: Biological Sciences*, *277*(1689), pp.1835–1841.

walking legs. So bradoriids are not ostracods, nor do they belong unambiguously to any other living group. Nowadays they are usually placed somewhere close to the line of descent leading to modern crustaceans, but every new paper has a different point of view.

Wrapping the body in a sort of chitinous shell seems like a good idea and several of the most abundant Chengjiang arthropods are built on this principle. It certainly protects the animal against attack by most predators, provided they are not large enough to simply smash or grind up the carapace. It hampers movement, though, since the limbs barely stick out beyond the edge of the skirt and can do little more than paddle feebly. *Isoxys*, for example, was quite a large animal, an inch or two long, that had a dozen or so segments each bearing a pair of biramous limbs, completely enclosed by a capacious semi-circular carapace. *Waptia* had a similar series of biramous limbs inside its carapace, but also a limbless tail of half a dozen segments that stuck out behind, tipped by a final two-lobed telson. *Canadaspis* also shielded a long series of biramous limbs with its armoured carapace, dragging a tail of several segments ending in a spiny telson. *Branchiocaris* was probably quite similar, although only its carapace has been preserved. The body plan of all these animals was very similar: a pair of stalked eyes and uniramous antennae in front, followed by a series of head and trunk segments bearing rather similar biramous limbs for moving and breathing and protected by a tough bivalved carapace, often with a limbless tail of several segments at the rear, ending in a telson for defence or escape.

*Leanchoilia* was quite different. It had no carapace, although its head region was protected against attack from above by fusing the roof of the first few segments into a solid shield. Its most distinctive feature was a pair of large hinged appendages borne

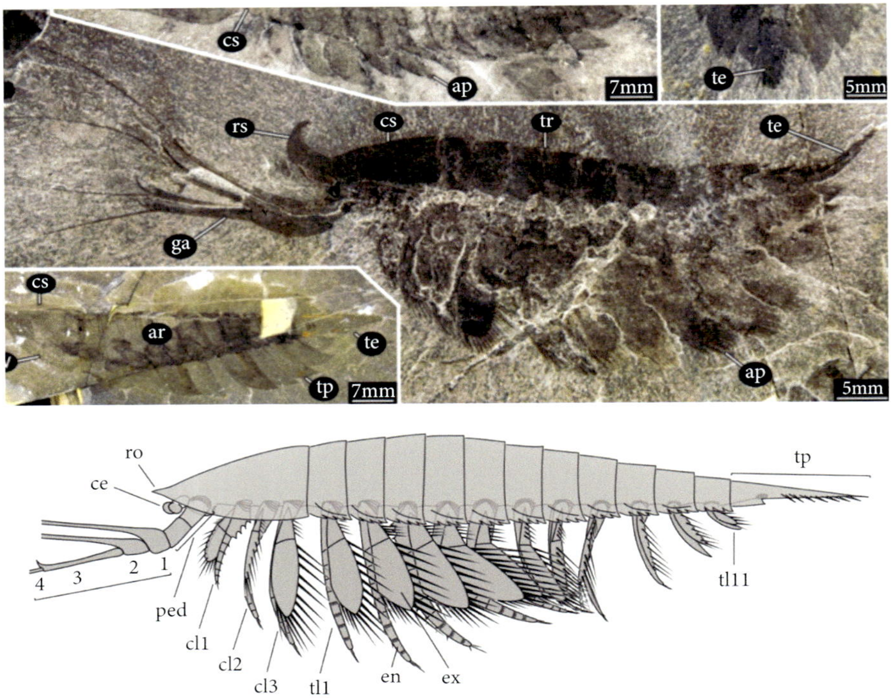


**Figure 7.6** Leanchoilia

(Top) Reproduced with permission from Haug, J.T., Briggs, D.E. and Haug, C., 2012. Morphology and function in the Cambrian Burgess Shale megacheiran arthropod *Leanchoilia superlata* and the application of a descriptive matrix. *BMC Evolutionary Biology*, *12*, pp.1–20. Fig 2D. (Bottom) Reproduced with permission from Liu, Y., Hou, X.G. and Bergström, J., 2007. Chengjiang arthropod *Leanchoilia illecebrosa* (Hou, 1987) reconsidered. *GFF*, *129*(3), pp.263–272. Fig 1.

just behind the eyes and projecting in front of the head (Figure 7.6). Each consists of a stout base and a complex multi-spined head looking rather like a Swiss Army knife. Some spines were long and delicate and might have been sensory, replacing the antennae. Others are more robust and seem to have been apposable, like thumb and forefinger, making pincers to catch prey. Its other appendages are more mundane, a dozen or so pairs of biramous limbs that might have been better adapted for swimming than for crawling, so that *Leanchoilia* might have been an active predator hunting its prey in the open water. Several kinds of these great-appendage arthropods have been found at Chengjiang, although most are quite rare. Some of them, such as *Forfexicaris*, manage to combine the frontal appendage with a bivalved carapace, getting the best of both worlds but presumably restricted to foraging on the surface of the sediment.

*Fuxianhuia* was a less specialized animal (its odd name was given because Chengjiang lies near the Fuxian Lake). It has eyes and antennae in front of a stout head shield that protected the brain, which is actually preserved in some specimens. The next segment bears a pair of spiny uniramous limbs that seem to have been capable of grasping prey, although they are not as complex as the great appendage of *Leanchoilia*. Further back, the trunk consists of a widened section bearing dozens of pairs of biramous appendages, roughly corresponding to a thorax, and a narrower limbless abdomen ending in a flap-like telson. It was presumably a predator or scavenger, crawling across the sea floor in search of soft-bodied prey or dead bodies.

And then, of course, there are the trilobites. It was the abrupt appearance of trilobites in the fossil record that had made Darwin scratch his head. For the geologists of the time, rocks were either post-trilobite and often full of fossils, or pre-trilobite and as barren as cheese. Chengjiang is close to the boundary between these zoological epochs. The trilobites found here are among the earliest found anywhere, but despite a few supposedly primitive features they are unmistakeably trilobites, with a basic body plan that lasted for the rest of the Palaeozoic, long after the other early Cambrian arthropods had disappeared.

## Heads

Trilobites had a pair of jointed antennae that probed the mud ahead as the animal shuffled forwards over the seabed using a long series of biramous limbs, one pair per segment, gill above and walking leg below. In common with the other Chengjiang arthropods they had no specialized mouthparts for processing food. Any edible object they came across was instead torn open by the stout spiny base of their legs and ground against a hard plate, the hypostome, just beneath their mouth. The channel separating the limbs on either side of the body acted, if you like, as an extended jaw that masticated the food and pushed it forwards to the mouth, which pointed backwards to receive it. They chewed their food with their elbows, so to speak, pushing it forwards in the channel formed by the limb bases and mashing it on a hard plate on the floor of the head before swallowing it. Horseshoe crabs still chew their food like this, but in crustaceans and insects the limbs of the first few segments after the antennae have become modified as mouthparts for processing a wide range of foods. Ostracods, for example, have seven pairs of appendages: two pairs of antennae, three pairs of mouthparts (mandible, maxillule and maxilla) and two pairs of walking legs, plus the pointed tip of the abdomen, the furca, which aids the mouthparts. Eating a small annelid, sensed by the antennae, starts with an attack on some vulnerable part of the body such as mouth, anus or gill. The worm is gripped by the mandible palps and the furca and torn apart by a back-and-forth motion. The claws at the tip of the mandible pass chunks of food to the maxillules, which bite pieces off and pass them to the maxillae, which are equipped with teeth that shred the food so that it is thoroughly comminuted when it finally gets to the mouth. Processing the food in this way depends on the coordinated action of several specialized implements, each playing a particular part in a suite of complementary devices.

Arthropod mouthparts are analogous to cutlery, the implements we use for food handling in the kitchen and at the table. In the European tradition, knife, fork and spoon comprise the basic table kit, although they may be supplemented by more specialized implements such as fish knife and fork, eggspoon, soupspoon and butterknife, to say nothing of more exotic devices such as tomato lifter or winkle fork, depending on the meal. In the kitchen we use a wider range of food-processing tools such as carving knife and fork, spatula, corkscrew, eggtimer, and nowadays the essential can-opener. The cultural evolution of these devices can often be followed in detail. The table fork, for example, is a relatively recent innovation. A single-tine fork (such as a toothpick) can be used to impale food items and lift them to the mouth, but they are liable to slide off and you may prick yourself. It is no more satisfactory in the kitchen, where a joint of meat will rotate when you try to carve it. For kitchen use, a two-tine fork permits meat to be carved and has been used since ancient times. It was introduced as a piece of table cutlery in Italy in the 14th century, but was not widely adopted because the food still slips off easily and the tines are too widely spaced to lift small items of food (and it was felt to be a little effeminate at the time). The three-tine fork introduced in Germany in the 18th century was more efficient, but was rapidly displaced by a four-tine version which held food securely and could double as a scoop. Five-tine versions have been made but seem to offer little if any improvement; the only innovation that has stuck is to curve the tines, making a more effective scoop.

Different implements must be manipulated in different ways, may be used differently (such as British and American uses of knife and fork), and tend to shift the balance between food handling at the table and food preparation in the kitchen (whether the table knife is sharp enough to cut meat, for example). If we include non-European cultures these differences are exaggerated—chopsticks, for example, require radically different techniques both at the table and in the kitchen. In a similar way, the size and shape of the mouthparts of any given crustacean are adapted to its diet and the balance between food preparation (typically by the mandibles) and food manipulation (typically by the maxillae).

The specialization of the mouthparts involves reconstructing the head as a distinct multi-segmented region of the body equipped with brain, eyes, antennae and the mouthparts themselves. Understanding the structure and evolution of the arthropod head requires a lifetime of study, so I'm grateful (as you will be) that it lies outside this narrative, except to note that the crucial events took place later in the Cambrian, with the Chengjiang arthropods being only an early stage in the whole process.

## Eyes

It is a slightly odd thought that although we refer routinely to 'the dawn of animal life' there was actually nothing around to see it. Some algae can sense the light and even move towards it, but that's their limit. Only animals have eyes able to see a sunrise. The simplest light-sensing device is a dap of pigment whose conformation is altered when struck by photons, plus a signalling system that transduces this change to produce an effect somewhere else. Algae can do this. Add a neuron which is potentiated by the signal and you have the beginnings of an eye, able to collect the light coming from different directions, focus the light rays to produce an image, and pass the information to a brain to create a picture of the world immediately outside. Not all animals have image-forming eyes, and those that do are compelled by physics to follow a few basic designs for focusing light, just as we are restricted to a few basic designs for constructing optical instruments. The simplest is a pinhole camera, which I daresay everyone made at school: a pinhole in an opaque sheet held up at a window will produce an upside-down image of the surroundings on a screen behind. The device has been known since ancient times but is rare among animals, even though a simple version could be made by curling over the edges of a pigmented cup until they nearly met in the middle. The usual example is the eye of the pearly nautilus, although whether it enables the animal to capture active prey, rather than scavenging dead bodies, seems doubtful. A biconvex lens of the sort used in spectacles, cameras and microscopes will focus light by refraction and has evolved repeatedly in lineages of molluscs, vertebrates and other animals. A concave mirror will also focus light; mirrors are used in telescopes and the eyes of molluscs such as scallops. Pinhole, lens and mirror can each be deployed as a single large unit (such as our own eyes) or as a battery of small units which combined give a pointilliste image of the surroundings which is then merged into a single coherent image by the brain (step back from a picture by Seurat, squint a little, and you will get a similar result).

Tardigrades are the minute 'water bears' that clamber among the stem and leaves of moss and aquatic plants. They probably branched off early from the arthropod main line and have a particularly simple kind of eye, with pigment, signal and nerve impulse embodied separately in a small group of cells on the surface of the brain. Lobopods such as the velvet worm, *Peripatus*, have long segmented bodies and look a little like large caterpillars. They are also still around, although you are likely to come across them only in the damp litter on the floor of tropical forests. Lobopods have large and complex eyes, comprising cornea, lens, retina and a short optical nerve tract leading to the brain. Real arthropods are quite different: they have compound eyes, with hundreds or thousands of long narrow cones, the individual units called ommatidia. The light receptor of each ommatidium is the rhabdom. Light is focused, first by a corneal lens and then by a crystalline cone, onto the tip of the rhabdom, which generates a signal transmitted to the brain by the neuron at its base. The wall of each ommatidium is dark; only light coming from a certain direction is perceived, so that the eye as a whole is able to put together a complete picture of the surroundings. Compound eyes evolved very quickly; even the earliest trilobites have sophisticated compound eyes, and the Chengjiang arthropods would have had acute vision.

## Fins

A burrowing worm lives in a one-dimensional world in which the only direction is forwards. An arthropod that crawls with limbs can explore the two-dimensional sea floor. The third dimension of the open water above the sea floor is more difficult to reach. A minute animal can swim with cilia, although its efforts are likely to be feeble compared with current or tide. The force needed to accelerate a massive body can only be supplied by muscles, and must be transduced in some way to provide locomotion. The force transducers are fins, sheets of tissue extending from the body wall which move the body forwards by reaction when pressed against the water. Fins can be either active or passive. An active laterally placed fin is supplied with muscles that move it backwards and forwards and is constructed in such a manner (like a one-way hinge, for example) that the backward stroke is a power stroke whereas the forward stroke is a weaker recovery stroke. (Think of how your hands move when swimming breast-stroke; you are using them as active fins.) If muscles are used instead to change the shape of the body, usually by throwing it into a curve, a passive longitudinal fin increases the area of the body surface applied to the water and thereby increases the forward component of the thrust that is generated.

*Anomalocaris*, for example, was similar to the yard-long horror of the Burgess Shale, with massive jointed appendages just behind its eyes, ready to grab an unfortunate priapulid and cram it into the camera-shutter mouth behind. The name that has been adopted for the group that it belongs to is Radiodonta, a Latin-Greek fusion that refers to the radial arrangement of plates around the mouth, like the spokes of a wheel. The body segments, however, are provided not with jointed limbs but with a dozen or so pairs of overlapping flaps. In some specimens a block of fibrous material has been found near the base of these flaps and is likely to have been the muscle that moved them. The flaps, then, were active fins used in swimming to propel the animal through the water. How they are related to the limbs of arthropods like *Kunmingella* or *Fuxianhuia* is far from clear. Radiodonts had filamentous outgrowths on each segment, further up near the dorsal midline, which may have acted as gills. The flaps could then have been lobopod walking legs modified for swimming, and the radiodonts branched from the arthropod stem before the evolution of the biramous limb. This is the current consensus, I think, although other interpretations remain possible. Whatever the route that was taken, the flaps took the radiodonts into the third dimension, powering them through the open water above the seabed.

## Gill slits

Brachiopods and arthropods are examples of two solutions to the problem of getting enough food: suspension feeding to capture a large number of small items using cilia, or predation to capture a small number of large items using limbs. It is not surprising that lineages should evolve to specialize in one or the other. What is rather surprising (but characteristically biological) is that a second kind of animal was evolving at the same time along similar lines.

The lophophore is an effective device for suspension feeding, but a more radical solution is to internalize the whole apparatus, putting the current-generating cilia and the filter inside the body. This is not quite straightforward. Assuming that the water stream enters through the mouth (otherwise a new entrance would be needed) it must follow a passageway and leave somehow. The gut is not an option unless food particles are captured by cells of the gut lining, in which case you end up with an inferior kind of sponge. More realistically the stream must leave the alimentary tract soon after entering it, through pores which enable food particles to be trapped. This can be done by carving narrow slits in the wall of the pharynx, the internal gill slits or more formally the pharyngeal slits, separated by bars stiffened with cartilage. The bars support a delicate network of sticky mucus, secreted by a gland, the endostyle, on the floor of the pharynx. As the water stream passes through the slits the particles it bears are caught in this mucus net. The food-laden sheet of mucus can then be rolled up and passed to the stomach and lower part of the gut where it is digested. The depleted stream of water passes out of the pharynx through the slits and enters the 'atrium', a spacious cavity in the interior of the animal. It must then be provided with an exit, of course, which is done by piercing the body wall with exit pores, the external gill slits, allowing water to flow out. Once in place, this rather complicated apparatus is quite economical, because putting the cilia into a tube creates a sort of turbine that makes the work they do more efficient. The water stream can also be used for respiration as well as nutrition, which is why the slits are (sometimes confusingly) called gill slits.

But although it may be economical it is certainly complicated, so how did it evolve? It could not have evolved in the first place as a feeding device, because the water, having passed through the slits in the pharynx, would have nowhere to go: providing exit holes beforehand is not allowed, since evolution has no foresight. However, a capacious body cavity containing fluid circulated by cilia would support the respiration needed by large animals, as I've already mentioned, so we can begin by assuming that a large hollow animal has already evolved. It would have to be fairly placid, perhaps a rather sedentary kind of worm. Opening the cavity to the outside through a pore would allow the fluid to be replenished and this would improve gas exchange and allow a higher level of activity. It's risky because it also opens a route from the outside for invaders to gain access to the interior of the body, but if the benefit of increased activity outweighs the drawback of infection it would be worthwhile. The system can then be tinkered with by separating specialized incurrent from excurrent pores (by orienting the beat of cilia appropriately) or by elaborating specialized tufts of highly vascularized tissue as internal gills, perhaps close to the incurrent pores as gill pouches. Regardless of these potential modifications, once a ciliated cavity with an opening to the outside is in place, irrigating it via pores in the pharynx wall with seawater taken in through the mouth becomes feasible. This would provide an efficient respiratory system based on a continuous stream of water from which enough oxygen could be extracted by blood vessels in the cavity wall, the pharyngeal pores or gill tufts to support active muscular movement. It is only at this point that the pores in the wall of the pharynx could be modified as slits and used as a food collection device, subsequently upgraded with a mucus net. In other words, the package of pharyngeal slits + atrium + outlet might have evolved as a device for extracting oxygen from a continuous stream of water before being modified as a device that would also extract food particles from the stream.

Perforating the pharynx for suspension feeding is incompatible with putting the cilia outside on a structure like a lophophore, so any lineage has to adopt either one or the other. Quite early in animal evolution, therefore, two quite different lineages of filter-feeders evolved: the protostomes, with tentacles bearing external cilia, and the deuterostomes, with internal cilia and a perforated pharynx. The protostomes include not only brachiopods and arthropods but also most of the familiar animals—snails, insects, worms, flatworms and many others. The deuterostomes are an odd bunch that includes some rather unfamiliar animals that use a ciliary pump to power an apparatus of pharyngeal slits + atrium + outlet. In particular, the sessile tunicates you may have seen as bag-like animals about the size of a fist on dock pilings or harbour walls not only filter their food using gill slits, but actually consist of little more than a hard-working cilia-powered pharynx inside their tunic, like a colander packed in a leathery sack and equipped with inlet and outlet hoses.

The more familiar deuterostomes are echinoderms and vertebrates, neither of which fits the stereotype. Echinoderms have lost all trace of pharyngeal slits, land vertebrates don't need them because they don't work in air, and if you're a zoologist (or an angler) you'll point out that fish neither breathe nor feed in the way I've described. Fish gills develop as pouches from the pharynx which elongate and grow outwards until they meet and fuse with the epithelium of the body wall, so there is an open channel from mouth to the exterior without any need for an atrium. The pharyngeal slits are stiffened by jointed rods of cartilage forming a gill arch which bears gill rakers (short prongs like the teeth of a rake) on one side and gill filaments (with a rich blood supply) on the other. As the stream of water flows through the slit (propelled by muscle power rather than cilia) the gill filaments take up oxygen from it and the gill rakers catch any small organisms it contains. The details are complicated, but the system itself is simpler because the pharynx is connected directly to the exterior.

There has been a lot of argument about what the earliest deuterostomes looked like. It's an argument about ancestry: did deuterostomes evolve from simpler to more complex by the addition of features, or from more complex to simpler by the removal of features? The basic architecture of fish gills is simpler than the colander of tunicates, but no fish have yet been found in the lower Cambrian and it seems very likely that the fish system has been simplified from some more complicated ancestor. Whether or not some of the more bizarre fossils found at Chengjiang and other sites of about the same age show us what this ancestor looked like, and even whether or not they were deuterostomes, has provided much entertainment.

For example, *Vetulicola* seems at first glance to be one of the many arthropods which has a carapace covering its head and a tail sticking out behind, something like *Kunmingella*, and this is how they were originally described. A closer look, however, raises some doubts (Figure 7.7). The trunk is indeed strikingly segmented and looks just like the hind end of a shrimp, but it does not have the jointed limbs that would have been expected—has no limbs at all, in fact, and no antennae or eyes either. The head end is a rather robust box with a wide opening, presumably the mouth, at the front. It is not a carapace with an open seam at the bottom where the legs poke through, but instead a more or less uniform hull except for a groove on each side which is punctuated by a series of apertures, each roughly oval in shape and about 2–5 mm wide. Degan Shu and his colleagues at Northwest University at Xi'An think that these are external gill slits, which implies that *Vetulicola* was an early deuterostome. The pharynx itself has not been preserved; the head end is filled with featureless sediment (leading to the absurd conclusion that *Vetulicola* was a deposit feeder), so the decisive evidence of pharyngeal slits is missing.

We are left with a puzzle: was *Vetulicola* an arthropod which had evolved gill pouches (and lost jointed limbs) or a deuterostome which had evolved a segmented hind end? It seems to be a fusion of quite different kinds of animal, like the Chimaera of Greek mythology which had a lion's head and a serpent's tail connected by the body of a goat. One of the most secure conclusions of evolutionary biology, however, is that

**Figure 7.7** Vetulicola

(Top) Reproduced with permission from Aldridge, R.J., Xian-Guang, H.O.U., Siveter, D.J., Siveter, D.J. and Gabbott, S.E., 2007. The systematics and phylogenetic relationships of vetulicolians. *Palaeontology*, *50*(1), pp.131–168. Plate 1, Fig 1. Scale bar is 5 mm. (Bottom) Reproduced with permission from Li, Y., Williams, M., Harvey, T.H., Wei, F., Zhao, Y., Guo, J., Gabbott, S., Fletcher, T., Hou, X. and Cong, P., 2020. Symbiotic fouling of Vetulicola, an early Cambrian nektonic animal. *Communications Biology*, *3*(1), pp.1–9. Fig 6.

ancestors (or 'missing links') are not chimaeras. The common ancestor of arthropods and molluscs, for example, did not have both jointed limbs and a calcareous shell; on the contrary, it had neither. Richard Aldridge and his colleagues at Leicester and Kunming have reviewed the evidence for both the arthropod and the deuterostome interpretations, as well as a wider range of speculation. Henry Gee thought that the segmented tail presaged the evolution of the vertebrate trunk, so that *Vetulicola* was a deuterostome close to the ancestry of chordates, tetrapods and, eventually, us. But Aldridge and co. end by admitting defeat: 'it may prove impossible to resolve their phylogenetic position using evidence from vetulicolian fossils alone'.

Besides being fairly large overall, *Vetulicola* had a circular mouth about 1 cm in diameter, considerably larger for its size than a fish such as a perch. It also had a fairly broad tail fin, probably vertical in position. This could have been used as a passive fin because the segments of the trunk would have allowed it to be flexed from side to side, driving the animal forwards. The wide mouth and propulsive tail suggest that *Vetulicola* might have been an actively swimming midwater filter feeder whose movement rammed water into its mouth to complement (or replace) the action of cilia in generating the current from which it extracted its food. It is not a rare animal; several hundred specimens have been found at Chengjiang, and the lower Cambrian seas must have swarmed with these strange creatures, which despite an abundance of fossils still resist our attempts to fit them into some familiar group.

## The notochord

Passive fins will aid movement only if the body can shift fairly rapidly from one shape to another. Merely contracting longitudinal muscles on one side then the other won't do this because the body would just be scrunched up. Suppose instead that a flexible incompressible rod runs the length of the body, so that alternately shortening one side then the other bends the rod, and hence the body, first into a J-shape then into a Ն-shape. This would result in a sculling motion with a forward component, so the animal would be able to swim. This flexible rod is the notochord, made of cartilage, which supplies the 'chord' in Chordata. We are chordates, so the search for the earliest unequivocal appearance of a notochord has been particularly stormy, with claims and rebuttals flying backwards and forwards. In our own backbone the cartilage is eventually replaced by bone as a long series of vertebrae, each moved by a separate block of muscle. These V-shaped blocks, the myotomes, evolved to flex the notochord and are one of the most characteristic features of chordates. Swinging a tail from side to side would be hampered, though, if the gut tube extended all the way to the end of the body, so the course of the gut has been modified so that it ends a little short. This creates a distinct tail region, posterior to the anus, which can be further modified as a propeller. Animals with a notochord can swim even if they have no fins, but a caudal fin provides a blade to the oar, so to speak, and makes it possible to swim faster and more powerfully.

We can imagine an animal that has evolved along these lines. It would have a spindle-shaped body equipped with pharyngeal slits + atrium + outlet for respiration, complemented by a mucus-secreting gland to aid feeding, plus a notochord and tail fin powered by myotomes for locomotion. We do not have to use our imagination, however, because creatures like this can be found by the million in the sand and gravel of shallow seas. They are the amphioxus or lancelet *Branchiostoma*, which has been used by generations of zoology students as the exemplar of a particularly simple kind of chordate. It has all the attributes that I've listed, plus a few more that are shared by other chordates. It spends most of its life half-buried in the sand, gently inhaling a stream of seawater, but is capable of short bursts of a wriggling kind of swimming if it is disturbed. It is not an open-water swimmer because it lacks the dorsal and pectoral fins that fish use to minimize roll, pitch and yaw, but it has long been taken to illustrate an early stage in the evolution of the fish body. Naturally, the

Chengjiang shale was scrutinized closely for anything resembling amphioxus, and in 1999 Degan Shu and his colleagues (more or less the same team that first described *Vetulicola*) announced that they had found two candidates, which they named as *Myllokunmingia* and *Haikouichthys*.

The two are very similar. Each was initially described from a single specimen, but since then many more have been found, and *Haikouichthys* (as I'll call the two considered together) must have been a fairly common animal. It was only about an inch long, the size of a male guppy or a yearling stickleback. It had no bones, teeth or other mineralized structures, so the fossils are soft-tissue imprints only (Figure 7.8). There was a row of gill pouches on either side of its body associated with faint structures interpreted as gill arches and gill filaments. Perhaps it could be so small because there was a direct route from the mouth to the outside, like a fish. How it fed is a matter of speculation; it is often portrayed as a scavenger, but I think it could just as easily have been a filter-feeder. It did have an undeniable notochord, probably formed of separate articulated blocks of cartilage, powered by muscle chevrons. Locomotion was aided by a high dorsal fin reinforced by rays, although there was no distinct tail fin. *Haikouichthys* is usually assigned to the craniates, which are vertebrates which have a bony or cartilaginous box to protect their brain, and it has sometimes been called 'the first fish'. Unfortunately, neither brain nor skull have been preserved, although large prominent eyes have been, and seem to imply that it must have been fairly brainy. (There may be a degree of anthropomorphism here: many brainy animals lack eyes altogether, whereas box jellyfish have large complex eyes but no brain.)

**Figure 7.8** Myllokunmingia

Reproduced from Degan Shu, Northwest University, Xi'an, China. https://commons.wikimedia.org/wiki/File:Myllokunmingia_big_cropped.jpg

However, the combination of gill openings, notochord and myomeres makes it a card-carrying chordate at least.

There is obviously a wide gap between *Vetulicola* and *Haikouichthys* which might have been bridged by *Yunnanozoon*, a guppy-size animal that is very common at Chengjiang. Behind its mouth there is a series of half-a-dozen or so pairs of semi-circular structures bearing short projections that might have been gills supported by cartilaginous gill arches. A dark streak running down the body was originally interpreted as a notochord, but sceptics have since insisted that it may have been a gut, or something else entirely. Most of the rest of its body bore a prominent dorsal fin which seems to have been distinctly segmented in a manner distantly reminiscent of the segmented tail of *Vetulicola*. The current climate of opinion is that the yunnanozoans were deuterostomes, although perhaps not chordates, and the matter is sufficiently dark and dubious that I'll leave it at that.

## Visceral and somatic

All the animals that I have described were individuals, sharply distinct from one another. The mechanisms that are necessary to protect regular development against somatic mutations must already have been in place, so that highly integrated individuals could evolve. Although animals such as arthropods have highly integrated bodies, one cannot repress the feeling that they are far from completely integrated. Alfred Sherwood Romer expressed this disquiet in an essay written in the 1970s: 'there seems to be a somewhat imperfect welding, functionally and structurally, of two somewhat distinct beings'. He was referring to vertebrates—he was a very distinguished zoologist whose book *The Vertebrate Body* has been part of my mental wallpaper since undergraduate days—but his remark applies to arthropods too. The two distinct beings are visceral and somatic. The visceral being is the alimentary tract, the internal organs, and their blood supply and innervation. It is responsible for the endless complicated processes of metabolism, assimilation and rejection, and goes about its business without fuss until something goes wrong, a blood clot or a tumour, and the individual dies. The visceral being is dark, inward, vegetative and involuntary. The somatic being consists of the sense organs, the brain and the main nerve tracts, and the limbs that move the body. It is responsible for locomotion and manipulation, and if something goes wrong the individual is usually disabled but not dead. The somatic being is light, outward, sentient and apprehensive. Visceral and somatic beings together constitute the individual.

The distinction between visceral and somatic beings was not a new idea, although Romer was later chiefly responsible for promulgating it. It seems to have originated with Marie François Xavier Bichat, a rising star of French medicine in Napoleonic times until his tragically early death at the age of 30. He wrote (in my very free translation from the French):

> The physiological processes of animals fall into two clearly distinct categories. On the one hand there is a continual succession of assimilation and excretion by which the appropriate molecules are ceaselessly incorporated into the body while others

are rejected. This category is solely internal, but an animal, unlike a plant, also perceives the world around it, acts on its sensations, and often gives voice to its desires, its complaints, its pleasures and its pains.

Bichat was clearly thinking of air-breathing tetrapods, but, as with Romer, the concept can be generalized to animals as a whole. A polyp is wholly visceral; the spiny-headed worms show the first signs of a somatic element; the chimeric *Vetulicola* reminded Henry Gee of the original crude separation of the two; and finally the arthropods and chordates are dual animals with distinct and highly developed visceral and somatic beings, albeit somewhat imperfectly welded together, as Romer puts it.

The concept has a decidedly antique flavour and you are unlikely to find it in modern textbooks, but it summarizes a good deal of zoology. In particular, it gives a clear account of the evolution of highly cephalized bilaterians such as arthropods and vertebrates as the stepwise addition of components of the somatic being (sense organs, specialized limbs moved by striated muscle, nerve cords and a central nervous system), besides the occasional reversion, in echinoderms for example, to a purely visceral state. The first animals were no doubt exclusively visceral, as sponges are today. By the base of the Cambrian, burrowing worms must have had a head at least, and thus the beginnings of a somatic being. Among the arthropods of Chengjiang the head region had become more highly differentiated, equipped with sense organs and a brain. It was during the Cambrian that the distinction between visceral and somatic beings was first drawn and became an integral part of early animal evolution.

The distinction between visceral and somatic pain has been retained by modern neurobiology. The concept has even received unexpected support from molecular biology, since the transcription factors which regulate pain are quite different in visceral and somatic neurons. The separation between visceral and somatic is by no means complete—there are doubtful cases, and others where both systems contribute—but it seems useful to make the distinction to understand the main lines of evolution that led to the Chengjiang animals.

An advance made by one component of the somatic being is likely to stimulate corresponding advances in the others. Take an improvement in visual acuity, for example. This will necessarily make it profitable to increase the processing ability of the brain, and at the same time enables more precise movements to be made. A shift towards more specialized limbs will favour enlarging parts of the brain to control them, while the appropriate sensory cues may be modified. Increased brain size will accommodate enhanced sensory perception and allow greater coordination of movement. Modifying any one component thereby creates a demand that can be met by modifying another. The visceral being has no feedback loop to promote evolutionary change in this way. Perhaps the rapid evolution of the somatic being explains why highly cephalized animals such as trilobites, with a distinct head equipped with sense organs and mouthparts, could have evolved only about 20 million years after the worm burrows that marked the beginning of the Cambrian. The somatic being first began to evolve among the animals that lived in this brief interval.

## The Cambrian shore

In Cambrian times most of the land, and thus most of the seashore, was in the southern hemisphere. There was a large continent to the south and west, Gondwana, that extended from the South Pole to the Equator. In later periods it broke up and drifted slowly north to form Africa, Australia, South America and southern Asia. A smaller land mass, Laurentia, lay to the east, and now forms the core of North America. Siberia and Baltica, named for their current positions, were large islands (roughly the size of modern Greenland) between the two continents, Siberia to the north of Baltica. The northern hemisphere was mostly open ocean, and presumably deep water.

All four land masses were surrounded by the relatively shallow seas covering the continental slope, which is where the Cambrian lagerstätten were created when underwater mud slides buried a whole community of animals. Some (like Chengjiang) were on gently sloping mud; others (such as Sirius Passet and the Burgess Shale) lay at the base of steep local escarpments which must have been high-risk environments. There is a ring of sites around these old continents where soft tissue was preserved, spanning a period of some 15 million years between Chengjiang and the Burgess Shale.

### Gondwana: The no-shows at Chengjiang

The community at Chengjiang includes most elements of a distinctive Cambrian fauna, but it does not include all. Presence is more obvious than absence: it is only natural to describe the animals that have been found at Chengjiang rather than those that might have been expected but have not been found, or at any rate not yet. Absences can be quite striking, though, and in particular it is puzzling that so few molluscs or echinoderms have been found. Both are familiar groups with hard parts—the plates or ossicles of echinoderms and the shell of molluscs—that are readily preserved elsewhere but have not been found at Chengjiang. It can't be that conditions were (for some mysterious reason) unsuitable for preserving calcareous structures, since brachiopod and hyolith shells are quite common. Absence of evidence, after a thorough search, is rather good evidence of absence: if no fossils of a particular kind have been found, while others abound, it's likely that this kind of animal simply wasn't there. There is an alternative explanation, though: perhaps it was really present, and even abundant, but unrecognizable.

Unrecognizable echinoderms, in particular, have a rich literature. Living echinoderms (crinoids, starfish, brittle stars, sea urchins and sea cucumbers) are deuterostomes with a body plan so different from that of any other group that it's difficult to understand how it could have evolved from a worm-like animal with gill slits. Instead of limbs or fins, echinoderms use thousands of small tube feet for moving around. These are slender columns about 10–20 mm long that poke through the body wall as the terminal parts of an elaborate system of canals for bringing in seawater and conveying it under pressure (powered by cilia) to all parts of the body. The canal system is pentaradiate, as reflected by the five arms of a starfish. The body is strengthened by calcite ossicles with a foamy or mesh-like structure called stereom that saves on material while remaining strong and tough. Any pentaradiate animal with a water vascular system, tube feet and stereom is an echinoderm. Alas, the converse is not true, for tube feet do not often fossilize and the ancestors of modern echinoderms must have lacked pentaradial symmetry or stereom or both. There was in fact an extensive radiation of echinoderms with bilateral or triradial symmetry in the early Palaeozoic, easily recognizable from their distinctive mineralized exoskeletons but leaving no descendants today. One very early echinoderm, for example, is *Ctenoimbricata*, a disc-shaped bilaterally symmetrical and fully plated animal about an inch across found in deposits in Spain from the middle Cambrian (509–505 Mya).

Earlier candidates are a little murky. *Vetulicystis* consists of two parts, a globose theca or calyx about 5 mm across and a short broad tail. The division of the body into two parts recalls *Vetulicola*, but on the other hand the 'tail' might have been a stalk, attaching the calyx to the sea floor. The animal is not mineralized (its body seems to be covered by a cuticle) and there is no sign of pentameral symmetry. There is no particular reason to dismiss it as a very early echinoderm ancestor, nor any particular reason to accept it. Likewise, a stalked animal bearing a more or less globular theca encased in calcareous plates has been found in the Purella zone of the Yanjiahe Formation in Hubei. *Yanjiahella* might be a very early echinoderm, but this has been vigorously disputed. *Vetulicystis* and *Yanjiahella* exemplify the difficulty of identifying an ancestral lineage (or something close to the ancestral lineage) before any of the distinctive characteristics of a group have evolved.

*Cotyledion* had a rather larger cup-shaped calyx about an inch across borne on an undeniable stalk and armoured with ovoid plates or sclerites. The whole animal might have stood as high as 40 mm above the sea floor. The calyx is usually filled with featureless mud but was fringed with short projections, perhaps the bases of tentacles. It has a superficial similarity to a stalked crinoid, but it might just as readily have been a polyp.

*Eldonia* is a large disc-shaped animal that was originally described (by Charles Walcott over a century ago) as a holothurian, a sea-cucumber. The centre of the animal was occupied by a large coiled structure interpreted as the gut, and the edge by about 80 long rays interpreted as radial canals whose function is unknown. A set of strap-like tentacles surrounds the mouth, or what is presumed to have been the mouth, in some specimens. That's about all; it is not even clear which way up the animal lived, whether it was a pelagic hunter like a scyphozoan jellyfish, tentacles down, or lay on the bottom to ambush prey, tentacles up like a sea anemone. It doesn't look like any sort of echinoderm.

Molluscs are even more of a problem, since presumptive shell-bearing molluscs are prominent in earlier communities but fail to put in an appearance at Chengjiang. There are, however, two animals that might have been molluscs: *Wiwaxia* and *Nectocaris*. *Wiwaxia* was a small slug-like animal that combined the defences of armadillo and hedgehog, tiled with horny plates and armed with a double row of sharp upwards-pointing sclerites to deter any attack from above. The sclerites were initially compared with the chaetae of annelids, which would mean that *Wiwaxia* was an early polychaete, but second thoughts put it among the molluscs, or perhaps in a group close to the ancestry of both polychaetes and molluscs. The argument was decided, I think, by Jean-Bernard Caron of the Royal Ontario Museum, who showed that the mouthparts of *Wiwaxia* (and *Odontogriphus*, which was similar but lacked the impressive dorsal armour of *Wiwaxia*) were a pair of rakes with a strong resemblance to the radula used by molluscs to scrape the thin layer of slime from the surface of rocks or seaweed.

*Nectocaris* is more difficult to place. It had a fusiform body equipped with lateral fins and a head bearing prominent eyes, a pair of long tentacles and a peculiar nozzle or siphon. Jean-Bernard Caron and his colleague Martin Smith think that it was an early cephalopod, and it certainly looks like a squid, complete with grasping arms for catching prey and a siphon for jet propulsion. Others have denied that *Nectocaris* is a cephalopod or even a mollusc, pointing out that true squid did not appear until about 100 My later and interpreting *Nectocaris* as an example of convergent evolution. It was a midwater predator with grasping arms, certainly, but its internal anatomy was unlike a squid and the 'siphon' may have been just an eversible pharynx. I shall not try to be wise when doctors disagree so strongly.

## Gondwana: the finds at Jinyangkou, Balang and Emu Bay

The initial finds at the Maotian hill were only the first of a wave of discoveries across the Yangtze Platform which is still continuing, and no doubt several more localities will be added before this book is published. Two sites in particular have yielded hundreds of fossils whose preservation rivals that at Chengjiang: the Shuijingtuo Formation near the town of Jinyangkou in Hubei Province, about 1,000 km to the north-east of Kunming, and the Kaili Formation near Balang village in Guizhou Province, about 700 km to the east. The Qingjiang biota (named for the river that flows through Jinyangkou) is about the same age as Chengjiang, whereas the Kaili biota is about 10 My younger.

Any account of a new fossil locality is likely to emphasize how it is different, and in particular how it is different from nearby sites of about the same age. After all, nobody gets prestige and promotion (let alone funding) for finding more or less the same sort of fossils in another place. Many of the animals of the Qingjiang assemblage, for example, are cnidarians, both polyps and medusae, exquisitely preserved and quite unmistakable, whereas these are rare at Chengjiang. Moreover, the Qingjiang cnidarians seem to be different: Xingliang Zhang and his colleagues at Northwest University, who announced the discovery of the new locality, describe a polyp, as yet without a name, that had stout naked tentacles, quite unlike *Xianguangia* from Chengjiang, which appears to have tentacles fringed with cilia. It is tempting to conclude (Zhang does not go so far) that this animal used cnidocytes, the stinging cells of all living cnidarians, to capture large prey, rather than using cilia for suspension feeding.

The fossils of the Kaili Formation were actually found by a team of trilobite-hunters from Guizhou University led by Yuanlong Zhao a couple of years before Hou Xianguang split the first block of shale concealing *Naraoia* at Chengjiang. They published, too—but only in the house journal, and in Chinese, so it was not until the significance of the Chengjiang fossils became widely recognized that the outcrop at Balang was thoroughly investigated. One conspicuous difference between the two is the prominence of echinoderms among the Kaili fossils (and in other Gondwana localities, such as Morocco). Most of them are eocrinoids, sessile animals covered with calcareous plates that used long arms for suspension feeding, much like the living crinoids to which they may be distantly related. They are one of the many groups of echinoderms, often very different in appearance, that begin to show at about this time, once their calcite exoskeleton had evolved.

The differences between the assemblages found at Chengjiang, Jinyangkou and Balang might be attributable to ecological conditions (the Qingjiang community probably lived in deeper water), age (the Kaili community is about 10 My younger), preservation or simply chance. An outside observer, however, is more likely to be struck by their similarity than by their differences. All three are dominated by arthropods, including trilobites, bivalved bradoriids like *Kunmingella* and great-appendage types like *Leanchoilia*, with lots of sponges (plus *Chancelloria*) and a scattering of priapulids, lobopods, radiodonts, hyoliths and brachiopods.

Much the same mix of types (together with vetulicolians) has been found at Emu Bay, which nowadays lies on the north shore of Kangaroo Island, across the bay from Adelaide in South Australia. In the lower Cambrian it was the muddy seabed off the east coast of Gondwana, close to the Equator and separated from the cluster of Yangtze Platform sites by the breadth of the old continent. A mudslide buried the local community of animals and turned into a fine-grained mudstone that preserved their tissues in fine detail. The preservation of eyes in particular can be literally spectacular. Just eyes, in fact—the detached compound eyes of a large arthropod, perhaps falling from its head after it had died and decayed a little. Trilobites have compound eyes too, of course, with calcite lenses that would have given them quite acute vision. The non-trilobite Emu Bay eyes are more like modern crustaceans and insects, in that they are not mineralized and have thousands of ommatidia packed in a hexagonal grid. Moreover, the lenses are larger towards the centre of the eye, where its surface is almost flat. According to the Australian team who described them, these eyes would have had a 'bright zone' of enhanced sensitivity, akin to the fovea of our own eyes, which allowed them to be used to hunt for prey (or avoid predators) even in dim light. It is quite remarkable that such an advanced design should have evolved in the early Cambrian, about 514 Mya.

## Laurentia

The old continent of Laurentia, now forming most of North America, straddled the Equator to the west of Gondwana. Lagerstätten have been found around its shores, although they are younger than the Gondwana sites, shading into the middle Cambrian: the Kinzers Formation in Pennsylvania (about 512 Mya), the Mount Cap Formation in the Mackenzie Mountains of northwest Canada (about 510 Mya), and the Spence Shale of Utah (about 506 Mya).

Kinzers is a small town in Paradise (which is, of course, located in the south-eastern corner of Pennsylvania) that has given its name to the lower Cambrian Kinzers Formation. The oldest member of the formation is a fine-grained mudstone that outcrops in the Conestoga Valley, between Lancaster and Hanover, and has long been known for its superbly preserved trilobites, which James Hall first described in the year the *Origin* was published. Apart from trilobites, the most impressive fossils are the large grasping appendages of radiodonts. Among the less impressive were small coiled tubes called *Pelagiella*, but why these caused such a fuss will have to wait until later, when I talk about the odd fauna found on the Siberian Platform.

The Mount Cap fossils were originally found in rock cores from the oil patch west of Great Bear Lake, in the Northwest Territories of Canada, although they were later found in outcrops along the Little Bear River, about 100 km south of Norman Wells. Most of them are carbonaceous fragments and isolated sclerites similar to those found in Sweden (and described by the same Cambridge team) which are unlikely to arouse much enthusiasm outside a small group of specialists. Among them, however, are the earliest remains of diet, of what the animals of the early Cambrian actually ate, in the form of faecal pellets and strings, known to palaeontologists as coprolites. Most of these contain only homogeneous material digested by an unknown animal eating unknown prey, but a few contain fragments of brachiopod and must have been produced by something capable of crunching a hard calcareous shell. Others are packed with the cells of green algae and were voided by a suspension feeder eating acritarchs. These minute blobs and threads, modest as they are, give us the first direct evidence of Cambrian ecology.

The mudstones of the Spence Shale outcrop in the Wellsville Mountains, a few miles north of Brigham City, and their exceptionally well-preserved fossils have been known for over a century. For the most part, the fauna is similar to that of other sites of comparable age, dominated by trilobites and other arthropods, in company with lobopods, brachiopods, priapulids, hyoliths, sponges and hemichordates. On the other hand, it has lots of echinoderms, complete with calcareous plates. The most abundant is *Ctenocystis*, an animal without a stalk or appendages of any kind (like *Ctenoimbricata*, but unlike the eocrinoids which are also found in these deposits) and only an inch long at best. It is protected by plates of stereom, so there is no doubt that it is an echinoderm, but instead of the five-fold symmetry of all living echinoderms it was bilaterally symmetrical, looking a little like a small battered moccasin as it lay on (or in) the mud of the sea floor. Its mouth lay at the centre of a circle of rake-like plates, the 'ctenoid apparatus', which presumably acted as a coarse filter for the stream of water entering the mouth. It seems likely that pharyngeal slits provided the fine filter, since it is undoubtedly a deuterostome, but the pharynx itself has not been preserved, so this is just a guess. What is beyond doubt is that the Spence Shale shows that the early radiation of echinoderms was well under way by the middle of the Cambrian despite the paucity of fossils at other sites.

## Siberia

Something happened in Siberia (then in the southern hemisphere of course) about 515–513 Mya that blotted out much of the burgeoning mid-Cambrian communities. Andrey Zhuravlev, who has written extensively about this period, thinks that it was recurrent bouts of anoxia, perhaps caused by the decomposition of phytoplankton blooms. The rocks of the Sinsk Formation, in the eastern flank of the Siberian Platform, are largely made up of laminated black shales, the lamination suggesting that burrowers were absent and the black colour indicating a high organic carbon content, the two adding up to a poorly oxygenated or even anoxic seafloor. The consequence was a severe reduction in the diversity of the early Cambrian fauna. It was the first mass extinction of the Phanerozoic, and although not as catastrophic as some later events about half the early Cambrian genera were never seen again.

Archaeocyaths were especially hard hit; about 90% of them disappeared and they never really recovered from the blow. They were sessile filter-feeders whose body plan is usually likened to two conical colanders stacked one inside the other, with the outer having smaller holes and the inner larger ones (Figure 7.9). The two are kept apart by a rather complicated set of struts and flanges, and the entire body is supported by irregular crystals of calcite which locked together to make a firm skeleton. Water flowed through the holes into a central cavity and left through a wide apical aperture. The whole structure looked like a little ice-cream cone, just an inch or two high, anchored to a rock. They are very common as fossils and were first described from Labrador and Siberia in the mid-19th century, but what sort of animal they were has been very controversial among the small and vexed group of specialists who have studied them over the years. Stephen Rowland of the University of Nevada has written an amusing account of their arguments. They were at first supposed to be a new kind of coral with porous walls, although some denied they were animals at all. Perhaps they were

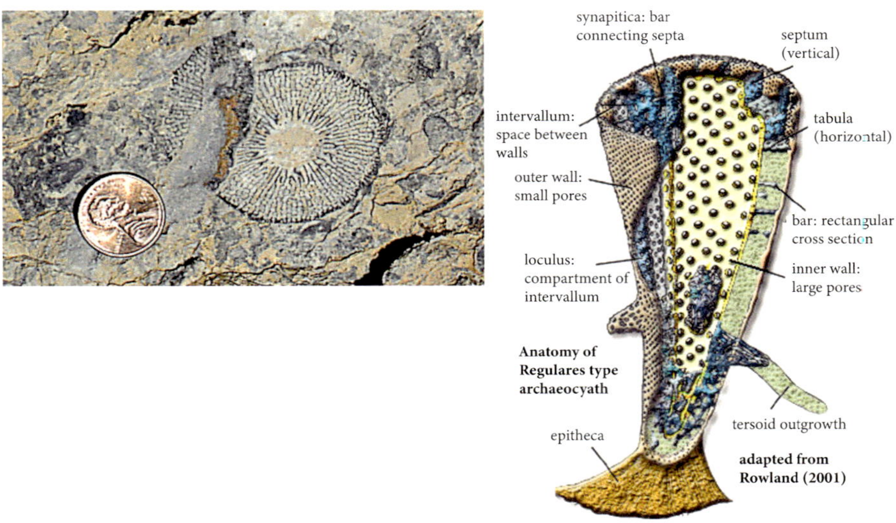


**Figure 7.9** Archaeocyatha

(Left) Reproduced with permission from UC Berkeley, https://ucmp.berkeley.edu/porifera/archaeo.html. (Right) Reproduced with permission from Rowland, S.M., 2001. Archaeocyaths—a history of phylogenetic interpretation, *Journal of Paleontology*, *75*(6), pp.1065–1078, Cambridge University Press.

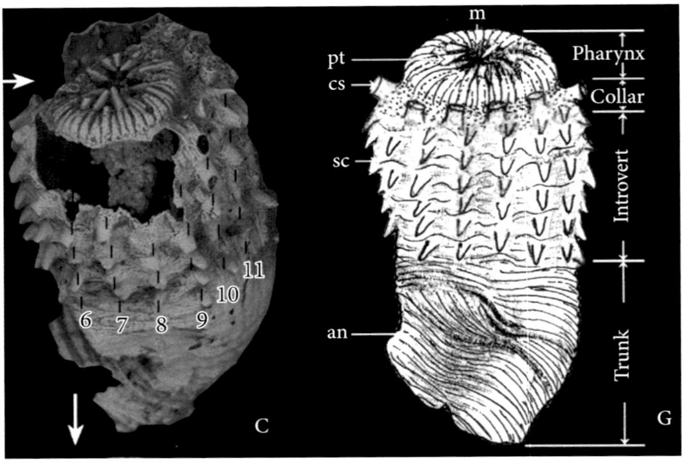


**Figure 7.10** Eopriapulites

Reproduced with permission from Liu, Y., Xiao, S., Shao, T., Broce, J. and Zhang, H., 2014. The oldest known priapulid-like scalidophoran animal and its implications for the early evolution of cycloneuralians and ecdysozoans. *Evolution & Development*, *16*(3), pp.155–165. Fig 2 C and G.

will impale any small organism when it is pulled in by the retractor muscles. A small priapulid will eat correspondingly small prey, diatoms or ciliates perhaps, but larger forms are raptorial predators able to catch and consume any soft-bodied worm they come across, including of course their own kind.

If there were predators around in the Cambrian mud large enough to eat a small worm the priapulids would have needed some sort of protection. Instead of a mineralized exoskeleton priapulids have evolved a fibrous cuticle in which protein and chitin are interwoven to serve as a thin and rather feeble anti-stab vest. A cuticle makes the problem of growth even worse, however, because enlarging the body by adding material to the outside of the cuticle, like adding material to the margin of a shell, would not work because it is the inside of the cuticle that restricts growth. The ancestors of priapulids came up with a drastic solution: simply throw away the old vest and make a new one, a size larger. Shedding the old cuticle is far from easy, but it can be done by first detaching the old cuticle from the body surface and then peeling it off inside-out like a sock, the animal wriggling as it does so to scrape it off against the sand. Some of the pieces of cuticle from Kuanchuanpu show that this solution had been found in the early Cambrian—they might have been just dispersed fragments, except that they are inside-out.

The closest living relatives of priapulids are the kinorhynchs, although you might not think so to look at them. Kinorhynchs are all minute millimetre-size animals which live in marine sands. They are rather spiky-looking animals whose body is clearly divided breadthwise into ring-like sections, and they might be taken for a miniature shrimp except that they have no legs. What they do have is an introvert and an open body plan much like a priapulid. Like the smallest priapulids they are worthy but rather obscure members of the community of minute animals that crawl through the mud and detritus of the sea floor, usually found by zoologists who look hard enough but never dominant. Amazingly, the same Kuanchuanpu deposits where *Eopriapulites* was found have also yielded the phosphatized husk of an entire kinorhynch, *Eokinorhynchus*. It is astonishing, not only that quite complex (although very small) animals had already evolved by then, but also that sister groups that have survived to the present day had already diverged by early Cambrian times.

## Puzzles

There are also some fossils that are more difficult to place because they don't have any unique feature that shows unequivocally which group they belong to. There is a minute phosphatized animal from Kuanchuanpu, for example, which has a slender stalk and an inflated calyx supported by internal strips of tissue, the mesenteries, and ringed by a circle of bumps which are interpreted as tentacle bases. *Eolympia* has been interpreted as a polyp related to sea anemones and corals. It does not have the stony or horny sheath of modern corals (which did not evolve until much later), but some of these small tubular fossils, collectively known as carinachitids, might be the ancestors of other kinds of cnidarian such as jellyfish, which have a polyp living in a mineralized tube as part of their life cycle.

*Saccorhytus* is a minute fossil from the Kuanchuanpu fauna which was originally identified as a deuterostome because it had a row of apertures that were interpreted as gill slits. It would have been the simplest kind of deuterostome, little more than an animated pharynx lying on the seabed. Closer examination has shown that these apertures are probably just the broken bases of spines, however, showing convincingly that *Saccorhytus* is another example of the priapulid-like animals already known from Kuanchuanpu. Anyway, it seems much too small to have been a deuterostome.

## An odd egg

Some of the most abundant fossils in the Kuanchuanpu formation are phosphatized embryos, so abundant that they can be strung together in a developmental sequence from eggs to adults. I shall call them *Olivooides*, the olive-shaped egg, assuming that the many other names which have been used (*Pseudooides*, *Punctatus*, *Pyrgites*, *Quadrapyrgites*) are developmental stages or closely related animals in the same group. You might think that a complete developmental sequence would identify the organism beyond doubt, but it has been surprisingly difficult to pin down *Olivooides*. The cleavage of the egg results in a hollow ball of cells covered with star-shaped spiny outgrowths that develops an opening that is, oddly enough, five-sided, reflecting an internal five-fold symmetry. The embryo then hatches, pushing through the envelope that covers it, as a blunt cone-shaped animal ('Punctatus') bearing a series of strongly marked transverse ridges which retains a pentagonal opening at its wider end. What is it? Perhaps it was a priapulid or a kinorhynch, since the adults are found in the same deposits; but *Olivooides* has no introvert at any stage, which seems to rule out this possibility. The pentameral symmetry is reminiscent of echinoderms, but *Olivooides* lacks both the characteristic hydraulic system and the peculiar calcite endoskeleton—which evolved before pentameral symmetry—of starfish and sea urchins, so that doesn't seem very likely either. The best guess is that *Olivooides* was a jellyfish, although since it was not in the least gelatinous this conclusion needs a little unpacking.

Most scyphozoans have a life cycle that alternates between a large pelagic medusa, the jellyfish that is familiar to most people from being washed up on beaches, and a much smaller and more cryptic polyp, known only to zoologists. The polyp was discovered living in a sponge by George Allman, a well-known naturalist of his day, while on holiday in Antibes in the 1870s, although it is more often dredged up from deeper water attached to a hard surface such as a nodule or an empty shell. Allman named it *Stephanoscyphus* because he did not realize—how could he?—that it was the polyp phase of a medusa that had already been described, belonging to a rather odd group of deep-sea jellyfish, the coronates or crown jellyfish. The most distinctive feature of the polyp is the elongate chitinous tube, that it makes to live in. This tube is divided into sections separated by ridges and looks much like the Punctatus stage of *Olivooides*. Moreover, rather similar tubes belonging to a group called the conulariids are quite common in the Kuanchuanpu deposits and are usually regarded as cnidarians. Furthermore, *Stephanoscyphus* produces young star-shaped medusae by transverse budding, so that its column comes to look like a stack of asterisks—and a similar star-shaped structure has been found in the same beds as *Olivooides*. It is all very circumstantial, no doubt: the resemblance of *Olivooides* to *Stephanoscyphus* might be superficial, the assignment of conulariids as Cnidaria is little more than a guess and there is no proof that the juvenile medusa, if that is what it was, is connected in any way to *Olivooides*. Nevertheless, until someone comes up with a better fit *Olivooides* looks like a jellyfish.

## The Siberian Platform, Russia: 540 Mya

The Siberian Platform, like the Yangtze Platform, is a raft of old rock that lay south of the Equator in Cambrian times but later drifted north to its present position, Rather than being in the middle of the Middle Kingdom, however, with an equable climate and fertile soil, it lies at the frozen limit of Eurasia, extending from Lake Baikal to the shore of the Arctic Ocean. The ancient rocks of the Platform are now overlain by a jumble of later deposits, including some laid down in the earliest Cambrian; indeed, the Russian candidate for the base of the Cambrian came from the Aldan River in the south-east sector of the Platform. The very lowest stage of the Cambrian is the Fortunian, or if you are Russian the Manykayan, or if you are very Russian the Nemakit-Daldynian; all of these names denote more or less the same period, right at the beginning of the Cambrian. Fossils had been found in deposits of this age by painstakingly scrutinizing the rock surface, but they were few, small and obscure, nothing to compare with the splendid array of trilobites to be found only a few metres above. It was not until a simple chemical trick had been discovered that the diversity of animals that lived in the earliest Cambrian could be appreciated. What sort of animals they were, though, turned out to be hard to say.

Animals can make skeletons out of either calcium carbonate or calcium phosphate, depending on some complicated chemistry involving the availability of phosphate and bicarbonate ions, the oxygenation and pH of the seawater, and probably the activity of the local microbes too. Moreover, an originally calcium carbonate skeleton may become coated with a thin veneer of phosphate after the animal that made it has died. These rather mundane facts have an interesting consequence, because calcium carbonate is soluble in weak organic acids whereas calcium phosphate is not. Consequently, soaking pieces of limestone in vinegar will dissolve the matrix while leaving any phosphate-based fossils it contains intact. This is much easier than searching through thin slices of rock for the occasional microfossil. When it was first applied to the early Cambrian limestones of Siberia in the 1960s by Alexey Rozanov and Vladimir Missarzhevsky it produced bucketfuls of minute and extremely puzzling fossils—well, thimblefuls at any rate, enough to show that the early Cambrian seas were already populated by a rich fauna of small animals long before the first trilobite crawled across the ocean floor. But if you spilled them out onto a table top, they would have looked like the contents of a fairy's button-box, with tiny buttons, long curved needles, miniature cups, scraps of lace, and a dozen other things that fairies might find useful.

The oldest, right down near the base of the Cambrian, are the tubes called *Anabarites* and the tusks called *Protohertzina*, which I'll describe in more detail later. The variety of structures, all small, increased over time until early in the middle of the Cambrian, when most of them vanished after the Sinsk mass extinction. The survivors of this characteristic if enigmatic collection of fossils lingered on for tens of millions of years into the next period, the Ordovician, but they never again dominated animal communities to the extent they did in the earliest Cambrian. Perhaps their apparent decline was in part an artefact because phosphatization became less prevalent, but the sheer morphological variety of form before the Sinsk event was never seen again. Missarzhevsky later coined the term 'small shelly fossils' to describe them, and despite protests that many are neither small nor shells the label has stuck. What kind of animals they were has been much more difficult to work out, and it is only after decades of work, and a few lucky finds, that some of them have been pinned down. The small shellies (to be even more informal) have been found all over the world, but most of them have come from the Siberian Platform.

## Swords and shields

The small shelly fauna might appear to be just a litter of stony fragments, but it begins to make sense once it has been sorted into swords and shields. The swords are usually elongate structures ending in a sharp point, like a needle. The shields are plates worn on the outside of the body that are able to resist or deflect a sword thrust. The swords and shields found on the Siberian Platform show that animals had already begun to eat one another, and thus become divided into predators and prey, in the early Cambrian.

I once went to a seminar given by Martin Grant, a physicist who was at that time Dean of Science at McGill. He was explaining that it was impossible, in principle, to tear a sheet of paper because the intermolecular bonds between the cellulose fibres are simply too strong. An elementary calculation (elementary for a physicist) showed that you would need earthmoving machinery to overcome them. In practice, of course, it is dead easy, as he demonstrated by ripping up his notes. So what accounts for the discordance between principle and practice? The answer he gave is 'cracks': any real material incorporates minute flaws which yield under slight pressure and start a crack. Once started, the crack tends to elongate because even a modest pull generates a large force at the tip of the crack which tears the fibres apart. This is important if you are designing a shield, because it gives you a choice of two options: you can either make it hard, to prevent a crack from forming, or you can make it tough, to prevent a crack from spreading.

The ideal material for an exoskeleton would be both hard and tough, like the jointed metal armour worn by mediaeval knights. The modern equivalent would be tank armour. Metals can be worked only at temperatures that living organisms cannot withstand, however, so this solution is impracticable and evolving an exoskeleton must negotiate a compromise between hardness and toughness. Really hard materials are difficult for animals to fabricate, and most plump for either calcium carbonate (snail shells, for example) or, better, calcium phosphate (our bones) or, better still, opaline silicate (glass sponges), all of which use materials which are readily available in seawater. Calcium carbonate is not very hard; depending on how the molecules are arranged it rates only 3 for calcite on the Moh scale (which arranges materials in the order they will scratch one another, from 0 for talc to 10 for diamond) and about 4 for aragonite; snail shells have layers of both. As a thin layer it is easily broken—eggshells are famously fragile—but a thick layer will resist most stabbing or crushing predators. Thick plate armour is expensive to make, however, and is apt to be cumbersome, as any mediaeval knight would tell you, especially if he had just fallen from his horse.

The alternative is to fabricate a tough material, usually a composite of fibres laid down in an amorphous matrix. The anti-stab vests worn by police, for example, are usually made of aramid fibres (cross-linked aromatic polyamide molecules) woven in a thermoplastic matrix. The anti-stab cuticle worn by animals such as arthropods and nematodes have a similar design, consisting of cross-linked chitin molecules complexed with protein. Some animals try to get the best of both worlds by mineralizing their cuticle with calcium carbonate, like a crab carapace, but inevitably give up speed and flexibility.

Armour is necessary when there are predators about. A predator might use projectiles to overcome these defences, but animals usually don't because there is no point in killing prey that could then be eaten by some other animal that happened to be closer. The exceptions are harpoons tethered by a filament, such as the stinging cells of jellyfish and sea anemones, which enable the prey to be reeled in and eaten before any competitor arrives on the scene. Pushing weapons, swords and knives, such as the fangs of *Protohertzina*, are much more common. They will penetrate armour if a sufficiently great force is applied to a sufficiently small area: the force is supplied by muscle contraction accelerating a tusk ending in a pointed tip.

## The mysterious small shellies

The shields are more numerous and more diverse than the swords (Figure 7.11). For example, very early in the Cambrian a minute bowl called *Purella* appeared. It is the shape of a low cone, with thick armour protecting the animal underneath. It is almost universally reckoned to be a mollusc, although it is much smaller than most later molluscs, being barely 1 mm across (the smallest snails alive today are about the same size). For comparison, *Purella* would have been attacked by animals bearing swords that were also about 1 mm long, so the early Cambrian seafloor must have featured relatively large predators, perhaps 1 cm or so long, hunting much smaller soft-bodied prey. The low shell of *Purella* would have protected it effectively, especially since a predator could have struck it only a glancing blow.

A shell provides protection, but it also creates a problem: how is the animal inside to grow bigger while continuing to be protected by an inflexible shell? The two possible solutions are to grow broader or to grow taller. The simpler option is to add material to the edge of the shell, which is what limpets do. The animal inside is roofed all the time, and by contracting the muscles of its foot it can lift slightly so that atmospheric pressure will stick it to the rock, which is why limpets are so difficult to dislodge. One disadvantage is that the potentially vulnerable area beneath the shell expands proportionately. This can be avoided by growing taller, so that a large animal can be enclosed within a shell that has a relatively narrow aperture. But nothing is perfect, and a cone-shaped shell will tend to topple over if it becomes too long. This can be ameliorated in turn by coiling the shell, so that its centre of gravity remains over the body of the animal it is protecting. The result is a gastropod with the familiar coiled-cone shell of snails everywhere. All stages in this progression can be found among

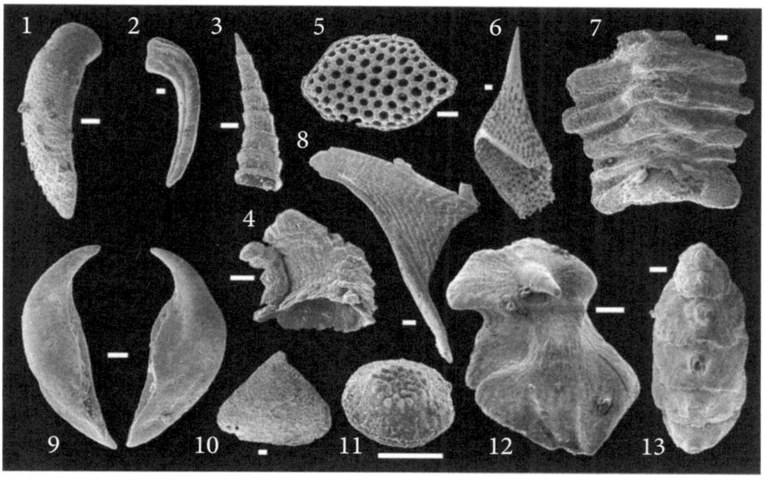


**Figure 7.11** Small shelly fossils. Scale bars are 0.1 mm

Reproduced with permission from Bengtson, S., 2004. Early skeletal fossils. *The Paleontological Society Papers, 10*, pp.67–78, Fig 1.

the small shellies, from the low cone of *Purella* to long straight cones to cones with a pronounced hook to fully coiled forms such as *Aldanella*, found on the banks of the Aldan River in the south-east part of the Platform. It took about 10 million years of gradually increasing diversification to reach this stage.

Among the more modest of the fossils found at Kinzers are the small coiled shells of *Pelagiella*, which appear on the Siberian Platform as part of the small shelly fauna at more or less the time that *Aldanella*, with a very similar appearance, fades out. They look just like small gastropods, with a tightly coiled shell, and this is what they were taken to be until Roger Thomas of the local Franklin & Marshall College took a closer look at the Kinzers material and saw a pair of fan-like bundles of rods, about 2 mm long, protruding from the aperture of the shell. These are most un-molluscan and instead resemble the hair-like chaetae of polychaete annelids. Thomas and his collaborators concluded that *Pelagiella* is not a gastropod shell at all, but instead the coiled tube of a marine worm, similar and perhaps related to the chalky tubes that can today be seen scribbled on seaweeds and rocks in the intertidal. Their argument is very convincing, but it throws the conventional picture of mollusc ancestry into disarray if *Aldanella* and *Pelagiella* are no longer accepted as gastropods. Ed Landing of the New York State Museum has finessed this difficulty by splitting *Pelagiella* into two genera, one of which (*Pelagiella sensu stricto*) has an oval aperture and really is the shell of a gastropod, whereas the other ('*Pseudopelagiella*') has an angular aperture and is a polychaete tube. If he is right, the gastropod lineage can be traced back to the small shellies of the early Cambrian such as *Aldanella*, and the polychaete lineage even further, to the straight tubes left by *Sabellidites* a little below the base of the Cambrian. Not everyone agrees that he is right, however: Jerzy Dzik of the Palaeontological Institute in Warsaw claimed to have found traces of chaetae inside specimens of *Aldanella* and suggested that both *Aldanella* and *Pelagiella* were coiled hyoliths. This seems unlikely (the 'chaetae' are much too small), but it would explain why large snails are conspicuously missing from the Cambrian lagerstätten.

As some shells were becoming more and more coiled the small shellies were diversifying in other ways too, both on the Siberian Platform and elsewhere. Hyoliths are common almost everywhere, although shell, lid and helens are often found separately, which can complicate matters when trying to match them up or even to recognize them—the opercula look like mollusc shells to me, and even specialists can be led astray. *Lapworthella* was named for Charles Lapworth, whose Ordovician Period eventually ended the long and bitter Cambrian controversy. The fossil is a rather blunt straight cone, or a sharply pointed cone, or a slightly curved cone—a very variable cone, in short, which suggested that the fossils might be individual sclerites of a series running down the side of an elongate animal to protect it from attack. ('Sclerite' is a handy term that can be applied to just about any hardened body part without knowing what it is for.) *Lapworthella* is one of the tommotiids, a loose grouping of animals with shells made of calcium phosphate, some of which might be the remote ancestors of brachiopods, The shells of *Halkieria* look rather molluscan but are found together with blades or spines that seem to be part of the skeleton of the same animal. The sclerites of *Microdictyon* are lacy plates that look a little like minute doilies. The star-shaped sclerites belong to *Chancelloria*. The spherical object is *Olivooides*. *Fomitchella* is a sharply pointed cone which seems to be the spine of a conodont or a protoconodont, something like *Protohertzina*. Blade-shaped sclerites, straight or curved or S-shaped, are *Siphogonuchites*. And there are many more.

Some of these remains are complete individuals, such as hyoliths and *Olivooides*, but most of them seem to be fragments: the cones of *Lapworthella*, the blades of *Siphogonuchites*, the delicate traceries of *Microdictyon* and the daisy chains of *Chancelloria* look more like elements of a larger and more complex structure than the complete armature of a small animal. They are akin to chain mail, made of interlinked steel rings, which was the mediaeval version of an anti-stab vest. A coat of mail is much lighter and more flexible than plate armour, but it was laborious to make when each ring had to be individually annealed, and was correspondingly expensive to buy. It will also create a problem for the palaeontologists of the distant future, when they come across the isolated steel rings, to infer the shape of the whole coat, let alone the shape of the animal inside.

This ancient puzzle has a counterpart in more recent remains. The museum where I work has large collections of snail shells, hundreds of thousands of them. We have them because the people who have spent a good part of their lives putting together a collection of cone shells or cowries find that they are growing old and none of the children have any interest in shells, so they donate the lot to their local museum (In return for a tax receipt, of course). It is easy to see the attraction: many of the shells are very beautiful objects, more like gemstones than hardened animal secretions. At the same time, shells are lustrous because they have been cleaned and polished, justifying the sneer that they are snails with the interesting bits missed out. You might never be tempted to spend money on snail shells, but strolling along any sandy beach you will come across any number of shells cleaned and polished by a jeweller's paste of sand and water. They look much like the small shellies, in fact, except that they are much bigger, or are the broken fragments of bigger shells. One thing you will seldom come across are the interesting bits, the soft parts of the animal which show how its body worked. These will have been consumed long ago by scavengers and fungi and bacteria so that no trace of them remains. You can still reconstruct the living animal by finding live specimens, but the small shellies have been dead half a billion years and their shells and sclerites are all we have to go on.

## The *Halkieria* animal

Greenland is a long way from Siberia, but is likewise situated in the Arctic and if anything is even colder. Geology can be difficult when the rocks are covered by ice and fossil collecting is possible only at sites where the snow melts in the summer. Nevertheless, it was on the inhospitable north coast of Greenland that the *Halkieria* animal was finally found in the thin strip of land that borders the permanent ice. On the shore of J.P. Koch Fjord, on the southern edge of the Nansen Land peninsula, there is a small cliff of layered mudstone that has weathered out into a talus slope where fossils were first collected in 1984 by Tony Higgins and N.J. Soper, two British geologists working for the Grønlands Geologiske Undersogelse. The site is called Sirius Passet, an intensely romantic name that has a rather muddled origin in the Cold War. In the 1940s and 1950s both Canada and Denmark were keen to assert their rather sketchy sovereignty in the high Arctic against possible Soviet incursions. Denmark, which governs Greenland, set up Operation Resolut in 1950 as a dogsled patrol that created an armed presence in the north. It was soon realized, however, that there was already a weather station and airstrip at Resolute Bay in northern Canada, likewise set up to deter any Soviet claim to the Arctic islands. To avoid any deplorable misunderstanding, the patrol was then renamed Sirius, after the brightest star in the sky, so Sirius Passet is a pass on the route of the Sirius patrol, the Slædepatruljen Sirius. The patrol still uses dogs, which do not break down or need fuel dumps.

The mudstone cliff, and some nearby localities that were discovered later, are in the earliest part of the Buen formation, just before it shades into deeper water deposits. In the lower Cambrian, about 518 Mya (roughly the same age as Chengjiang, and about 20 My after the small shellies appeared), the area around Sirius Passet was a shelf sea off the continent of Laurentia, a little south of the Equator. The animals were originally preserved as flat films of carbon, although the carbon was later burned off during the formation of the North Greenland hills, and only their impression remains. The impression is enough, however, to show the outline of the body of the animals, entire and articulated, much more informative than the scattered sclerites and shells which was all that was previously known about many of them.

Nevertheless, *Halkieria* is still something of a puzzle (Figure 7.12). The animal resembled an armoured slug up to three inches or so in length. Its body was roofed by a layer of flat plates bordered by rows of pointed blades, with curved sickle-shaped sclerites on the outer edges. There were hundreds of these sclerites, which overlapped to form a screen that it must have been difficult for any predator to break through. There were also two large convex shells at either end of the body, each with an 'umbo' (by analogy with the hinge area of clams, where the two shells are joined together) pointing towards the centre of the body. *Halkieria* had no limbs, and how it moved is something of a mystery, although a mollusc-like muscular foot is an obvious possibility. Since the carbon film has melted away it cannot be dissected and its internal structure is unknown. What is it? The original description from the 1960s places it tentatively among hyoliths, and it has subsequently been assigned, with different degrees of confidence, to brachiopods or molluscs, perhaps an early version of chitons. It has often been compared with *Wiwaxia*, an armoured dome a couple of inches long with prominent upwards-pointing sclerites which is well known from later deposits such as the Burgess Shale. This does not help much, though, since *Wiwaxia* itself might have been an annelid or a mollusc or something ancestral to either or both. Some experts have thrown in the towel by creating a new group, the Sachitida, not assigned to a phylum, but this solution will not satisfy everyone.

More to the point, perhaps, the purpose of the shed of sclerites that covers most of the body is obvious enough, but what are the pair of large shells for? They must tell us something important about the way of life it led, but nobody seems to know what it is. My favourite guess was made by Stefan Bengtson, the sage of Uppsala, who surmised that they were used to plug the ends of a U-shaped burrow. This would explain why the fossils are often curved, especially at the anterior end, but no convincing burrows have yet been found at Sirius Passet or anywhere else.

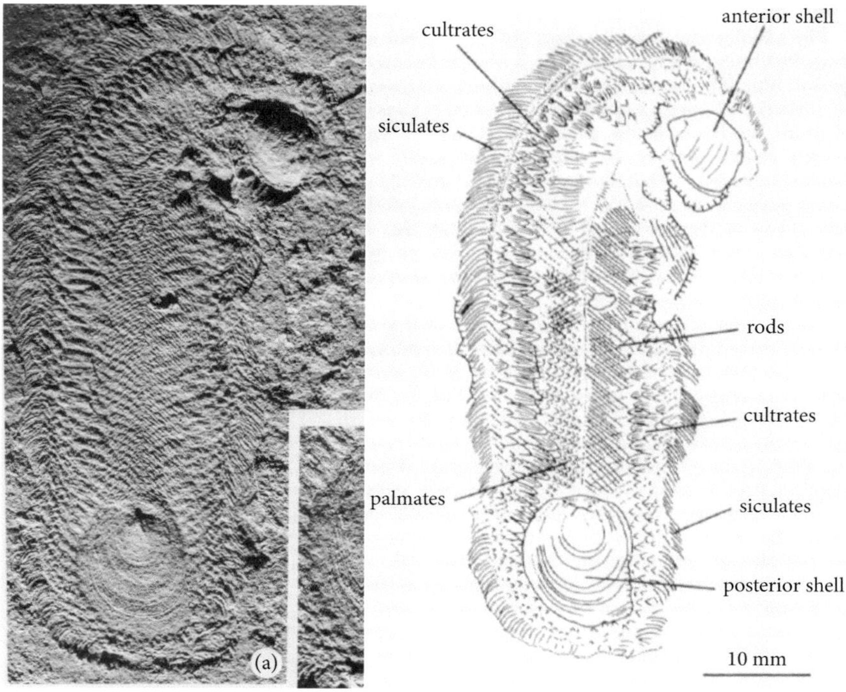


**Figure 7.12** Halkieria

Reproduced from Morris, S.C. and Peel, J.S., 1995. Articulated halkieriids from the Lower Cambrian of North Greenland and their role in early protostome evolution. *Philosophical Transactions of the Royal Society of London. Series B: Biological Sciences*, *347*(1321), pp.305–358.

## The *Microdictyon* animal

The second body fossil found at Sirius Passet was even more startling. *Microdictyon* was already well known from the characteristic plates, each a fine calcareous filigree forming a circle or rhomboid about half a millimetre across, that have been found by the thousand in a score of localities throughout the world as a prominent member of the small shelly fauna. The whole animal came as a surprise, showing that the plates were arranged along the body of an elongate animal like the row of shields slung outside the hull of a Viking longship (Figure 7.13). Even more surprisingly, the animal was segmented, with each segment bearing not only a shield but also a rather bulbous limb shaped a little like those on the sausage dogs that skilled people make out of long balloons at fairgrounds. The form of the limb immediately identified the group: *Microdictyon* was a lobopod.

There are lobopods still around. When you are outdoors in the countryside you are seldom more than a few steps away from a tardigrade, and you may actually be standing on one. You won't notice, however, since tardigrades are only a fraction of a millimetre long. They live in moss, puddles, damp soil and anywhere there is a thin film of water, where they stumble around on four pairs of stumpy legs sucking the

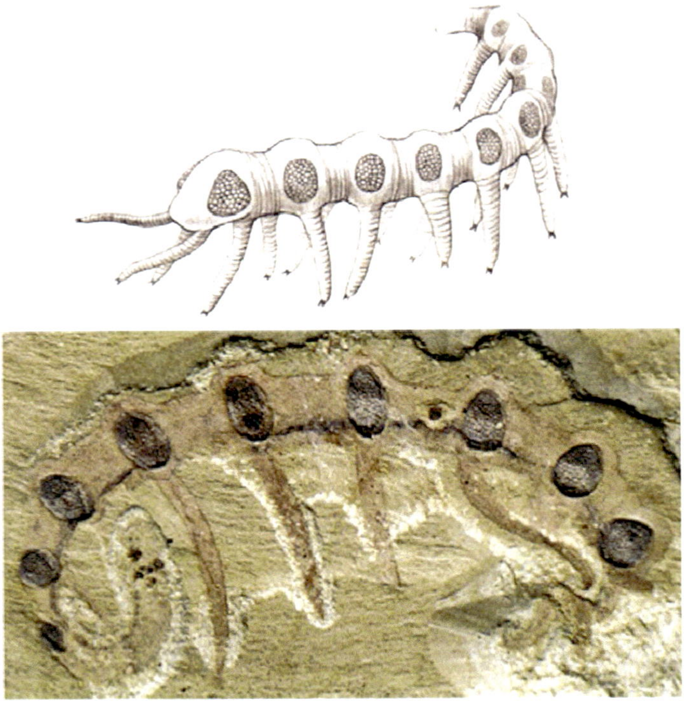

**Figure 7.13** Microdictyon

Reproduced with permission from Hou, X., Siveter, D.J., Siveter, D.J., Aldridge, R.J., Cong, P., Gabbott, S.E., Ma, X., Purnell, M.A. and Williams, M, 2007. *The Cambrian fossils of Chengjiang China, the flowering of early animal life*. Oxford: Blackwell.

juices from plant cells. The other kind of living lobopod is very different. *Peripatus* and its relatives, the onychophorans, are much larger animals, several inches long, which crawl through the leaf litter of moist tropical forests on a whole train of stumpy legs, a little like a large caterpillar. (The similarity should not be pressed too far. Donald Williamson, a marine biologist at Liverpool, proposed that caterpillars actually evolved from onychophorans by a sort of hybridization. This idea has been justly forgotten.) Onychophorans are carnivores which can capture animals the size of a beetle or woodlouse using a stream of sticky saliva jetted from a pair of nozzles on their head. Onychophorans are similar in size and structure (less the side armour) to *Microdictyon* and have similar limbs, but the improbable route they took from the bottom of the sea to the floor of tropical forests is unlikely ever to be traced.

Above all, lobopods are segmented animals. Below the head, their body is divided crosswise into a series of compartments, separated by flaps of tissue, each of which has its own dedicated muscle groups, blood supply, nerves and waste disposal system. Segmentation allows local muscle control and may have evolved originally because it aids active burrowing, an idea that I'll discuss later. Once in place, it facilitates the evolution of ever more complex body plans because each segment is a quasi-autonomous module of the body as a whole. This greatly simplifies the developmental program needed to specify a series of limbs, for example. Instead of having to encode the complicated instructions for placing a limb here, then another limb there, a third limb somewhere else, and so forth, it can simply specify that the limb package be expressed in every segment. The limb itself develops as an outgrowth of the body wall and is therefore hollow, with muscle strands inside that swing it backwards and forwards. If the forward stroke is through water and the limb grips a firm surface during the backward stroke then swinging the leg to and fro will move the animal forwards. A train of legs (three pairs is the minimum) avoids belly-flops and (given the coordination made possible in segmented animals by local nerve and muscle control) results in a steady crawl across the seabed or forest floor. Optional extras such as claws to increase traction can be added later.

The importance of segmentation can scarcely be overestimated; rapid crawling on hard surfaces has evolved only in segmented animals, with a few weird exceptions such as sea pigs and mudskippers, plus of course the glaring exception of land vertebrates such as ourselves, whose limbs evolved much later by modifying the bony fin-bases of crossopterygian fishes. *Microdictyon* is therefore a convenient marker for an important event in animal evolution, the prior evolution of a segmented body that led to limb-based propulsion.

## *Anabarites*

The very earliest Cambrian fossils of the Siberian platform, long before large animals like *Halkieria* and *Microdictyon* appeared, are slim tubes a few millimetres long and less than a millimetre in diameter, open at their broad apex and coming to a blunt point at the other end (Figure 7.14). They can be recognized by a characteristic tri-lobed plan, so that a transverse section looks something like a clover leaf. Why they should have adopted this odd construction is unknown, although the simplest idea, that it prevented them from rolling around, has a good deal to be said for it. They have been named as *Anabarites*, although what sort of animal secreted them is also unknown. They were originally taken to be the serpulid worms whose white scribblings can be seen by the hundred on intertidal rocks and seaweeds, but this theory seems to have no takers nowadays. Most specialists think that they were scyphozoans, a group of cnidarians that includes most of the familiar jellyfish found in all oceans throughout the world. The life cycle of scyphozoans alternates between a medusa, the jellyfish itself, and a small benthic polyp which is much more cryptic. (Raising jellyfish in the lab and working out their complete life cycle is quite difficult, so polyp and medusa of the same species have often been given different names. Generations of zoologists have devoted their careers to connecting the two.) It is true that one group of scyphozoans, the coronate medusae that float upside-down in deep water, secrete a tube in their polyp phase, although this is made of chitin rather than calcium carbonate. Until—if—their soft parts are discovered it seems safest to corral *Anabarites* as a distinct kind of animal, perhaps belonging to a living group and perhaps not, among the very first animals with hard parts.

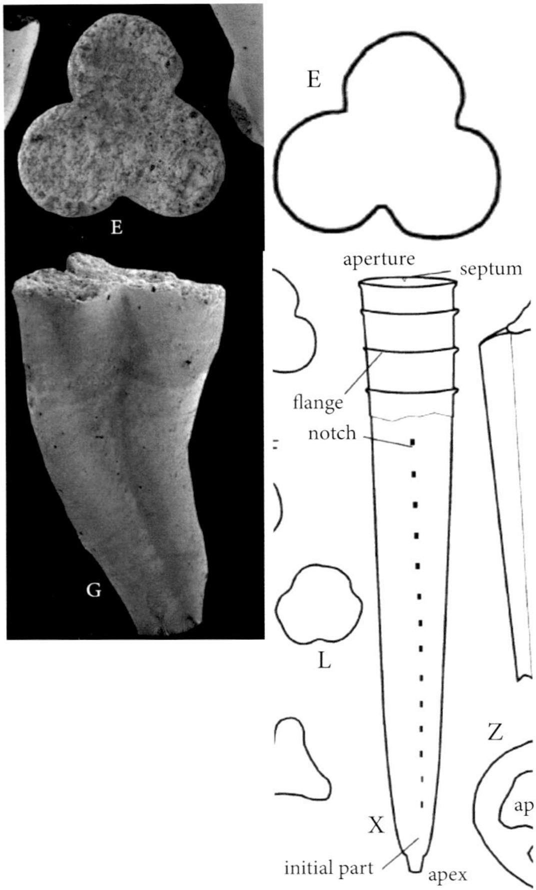


**Figure 7.14** Anabarites

Reproduced with permission from Kouchinsky, A., Bengtson, S., Feng, W., Kutygin, R. and Val'kov, A., 2009. The Lower Cambrian fossil anabaritids: affinities, occurrences and systematics. *Journal of Systematic Palaeontology*, *7*(3), pp.241–298.

## *Protohertzina*

The tubes of *Anabarites* are accompanied by the curved tusks of a protoconodont, *Protohertzina*, in the lowermost Cambrian deposits of Siberia (Figure 7.15). Conodonts themselves are just what they say: conical teeth, varying from a simple curved tusk to a more complicated structure looking like a set of false teeth for Dracula. From the Cambrian onwards, a great variety of these miniature teeth—they are mostly only a millimetre or so in size—litters the rocks throughout the 300 My or so of the Palaeozoic. They were discovered long ago, but only became important once it was realized that their distinctive and continually shifting morphology made it easy to recognize rocks of the same age in widely separated localities. There are two kinds of geologist,

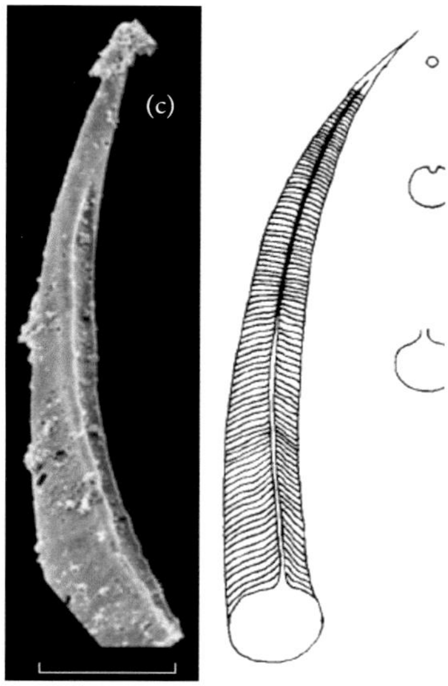


**Figure 7.15** Protohertzina

Reproduced with permission from Demidenko, Y.E., 2006. New Cambrian lobopods and chaetognaths of the Siberian Platform. *Paleontological Journal, 40*, pp.234–243.

I am told: poor geologists, who study rocks and fossils in university departments, and rich geologists, who conduct mineral exploration in companies, especially oil companies. When conodonts began to be used in stratigraphy the arcane details of a particularly obscure group of microfossils suddenly became of gripping interest and no doubt some poor geologists became rich overnight. Dinosaurs are more dramatic, but conodonts are more profitable.

Conodonts are found only as disarticulated teeth made of apatite (like our bones) which were presumably part of the jaw apparatus of some extinct predatory animal. What kind of animal remained a mystery until body fossils were discovered in the Shrimp Bed, a mid-Palaeozoic deposit exposed on the shore of the Firth of Forth a few miles from the centre of Edinburgh. They turned out to be eel-shaped creatures with staring eyes, V-shaped muscle blocks running alongside the remains of a notochord (perhaps), and a caudal fin supported by rays. In short, conodonts look very much like chordates, and even vertebrates, which would push the ancestry of our own group well back into the Cambrian. Whether or not the conodont animal was really a vertebrate is still controversial, although current learned opinion seems to favour this interpretation.

*Hertzina* is a Cambrian conodont, so this looks straightforward: protoconodonts evolved into conodonts, as their name implies, and consequently *Protohertzina* is an early vertebrate. But names can be misleading, and the experts have now concluded that protoconodonts did not evolve into conodonts after all. Instead, the structure of their spines, when studied in minute detail, has pointed to a very different group of animals, the chaetognaths. These are much less familiar than vertebrates, so this explanation itself needs a little explaining.

Chaetognaths still swarm near the surface of seas around the world, where they are a terror to other small animals in the plankton. They are long narrow animals, usually a few centimetres in size, which are not very bendy, so most of the time they lie fully extended in the water until with a sudden contortion of the body they flick to a new position. They are fully equipped with prominent eyes, a central nervous system, a through-gut with mouth and anus, a secondary body cavity partitioned into three pairs of compartments, paired lateral fins and a tail fin. Their mouth is circled by a fearsome array of spines which can be spread like a basket and snapped shut like a cruel trap. Chaetognaths are ambush predators, lying still until they detect the approach of a small animal and then lunging forwards to impale it. How they are related to other groups of animals has been vigorously debated, with the tide of battle flowing backwards and forwards between opposed camps, but they certainly have a complex and distinctive body plan, and finding them snuggled down at the very base of the Cambrian is a little surprising, to say the least. It is still not quite certain that protoconodonts are chaetognaths—their spines are made of apatite, for example, like conodonts, whereas chaetognath spines are made of hardened chitin—but whatever the outcome of that particular debate the curved tusks of *Protohertzina* show that quite elaborate predatory animals had already evolved when the Cambrian period was only just getting under way.

## *Cloudina*

Oddly enough, tubes appear in the fossil record before burrows, a little below the base of the Cambrian. One of the earliest is *Cloudina*, named for Preston Cloud, which has been found at many localities throughout the world (although not, as far as I know, at Fortune Head). It has a puzzling structure like a stack of funnels, each apparently open at both ends except at the very base (Figure 7.16). It is quite small—usually about a millimetre in diameter and a few millimetres long—but has been preserved because it made its funnels out of calcium carbonate.

This is quite a trick. Modern animals with a calcified wall or shell, such as corals or molluscs, pump calcium and bicarbonate ions across a special kind of membrane into an internal space, where the concentration eventually reaches saturation and calcium carbonate precipitates out. It is unlikely that the first animals to make hard tubes had such sophisticated machinery, but they would certainly have had the rudiments to build on. The membrane that surrounds living cells is not permeable to ions, but they can nevertheless be pumped in or out through specialized ion channels. Perhaps the word 'pump' is a little misleading: there is no motor impelling a stream of water through pores in the membrane, but rather proteins that span the membrane whose shape allows some ions to pass but not others. Calcium ions convert the electrical signals carried by sodium or potassium ions into cytoplasmic signals that regulate events such as the beating of cilia or the contraction of muscle fibres. Perhaps the many different kinds of channel in modern animals had not yet evolved and *Cloudina* may have lacked muscle, but it would have had the ability to pump calcium across a membrane. To form a calcareous tube it would also need an enclosed space where

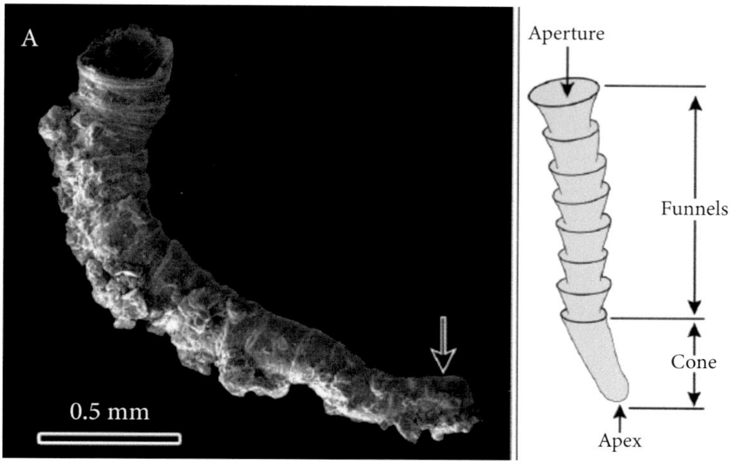


**Figure 7.16** Cloudina

Reproduced with permission from Cai, Y., Cortijo, I., Schiffbauer, J.D. and Hua, H., 2017. Taxonomy of the late Ediacaran index fossil Cloudina and a new similar taxon from South China. *Precambrian Research*, *298*, pp.146–156.

calcium could be concentrated, perhaps simply through being embedded in a mat. Exactly how *Cloudina* formed its tube remains speculative, however, since almost all specimens have been washed out of place and dumped elsewhere by currents.

The funnel-in-funnel structure of *Cloudina* tubes is difficult to explain. It seems unnecessarily complicated, relative to a simple uniform tube, and does not much resemble anything living today. There are some well-preserved specimens which show that new funnels are formed within the parent funnel, presumably by some sort of asexual budding, and this might be the reason for the funnels (and explain how they become calcified).

There were simpler tubes around at more or less the same time, some of them quite large. *Sabellidites*, for example, built a tube that was only a millimetre or two wide but 10 cm or more long out of a composite material consisting of fibres dispersed in a matrix of some sort. There were also animals making burrows, such as *Oldhamia* or the narrow wandering tunnels made by something in the Marenga Formation in Brazil.

The body fossils of a few other calcified organisms have been found just below the base of the Cambrian, but they are just as strange. *Namacalathus* is found together with *Cloudina* and likewise has a wall made of calcium carbonate. Otherwise it looks very different, having a stem rising from the sea floor and topped by a hexagonal bulb with a star-shaped opening in the top and a window in each facet, like an eccentrically designed lighthouse. The stem can be a few centimetres in height and the bulb a centimetre across, so this is quite a large organism. It is difficult to place, and has been likened to a large protist, a cnidarian polyp and even a ctenophore. A recent reconstruction based on CT scans of exceptional specimens in which the original soft tissue was replaced by pyrite shows an internal tube which might be a gut, in which case

*Namacalathus* was certainly an animal and perhaps close to the ancestry of modern groups such as bryozoans and brachiopods.

Quite apart from the unfamiliar body plans of *Cloudina* and *Namacalathus*, however, is the question of why hard outer coverings should have appeared at just this time, a little before the Cambrian period proper opened. Calcification is quite an expensive process, after all, and would not evolve unless it conferred some substantial benefit. Several possibilities have been suggested, but most of them, such as providing a chimney for filter-feeding or simply a waste-disposal system (calcium ions are mildly toxic), are not very convincing. The obvious benefit is protection: the shell of a snail protects its soft body from being swallowed or torn to pieces by a predator. Protection offers a simple and sufficient explanation for the evolution of calcified defensive armour, but one which takes for granted that unknown predators were already abundant. The identity of both predators and prey is uncertain; the attempts to compare tubes such as *Sabellidites* to modern groups such as ventworms are imaginative at best. We do not really know what was happening in these late PreCambrian seas, although from later events it seems that the diversification of animals was already under way.

The Cambrian history of animals is rather adequately documented by the fossils. The earliest volume is murky, and the small shellies still retain some of their mystery, but by the time of the lagerstätten of the Yangtze Platform and north Greenland a diverse community of animals had evolved in the oceans of the world. This community showed that many of the characteristic attributes of animals had already evolved too. But where is the ancestor? Or, rather, where are the ancestors? These new body plans had clearly evolved from much simpler precursors. Even the burrowing worms at the very base of the Cambrian must have been quite sophisticated animals, nothing like the very simple creature we imagine at the root of animal evolution. At least they make it clear that we are looking for at least two ancestors: the ancestor of all animals and the ancestor of something like *Treptichnus*. They also make it clear that we have found neither in the Cambrian. We shall have to look further back in time. What was found in older rocks came as a surprise.

# 8
# The Ediacaran Fossils

The discoveries made by the last generation of palaeontologists have filled in the gap between the beginning of the Cambrian period and the Burgess Shale with a menagerie of unexpected new animals, but these discoveries have not solved the original difficulty. Indeed, they have made it worse. To pursue the investigation we shall have to move back further in time, beyond the base of the Cambrian into the PreCambrian seas that were thought to be devoid of anything more complex than a seaweed.

Earth history before abundant fossils of large animals can be found—the source of Darwin's difficulty—is divided into three long eons, the Hadean (meteorite bombardment, ending about 4 billion years ago), Archaean (hydrological cycle and origin of life) and Proterozoic (beginning with the evolution of oxygenic photosynthesis about 2.5 billion years ago). The Proterozoic was found to be uncomfortably long at 2 billion years or so, and was divided rather roughly into three parts, the latest of which was the NeoProterozoic, beginning a billion years ago. This was also found to be uncomfortably long and was in turn divided into three (geologists are fond of triplets): Tonian (business as usual), Cryogenian (worldwide ice ages between 720 and 635 Mya), and the final stage of the NeoProterozoic, the Vendian or (more properly) the Ediacaran, which began after the last of the Cryogenian glaciations and ended at the base of the Cambrian 541 Mya. It was in the Ediacaran that the long-anticipated PreCambrian fossils were first found. You might expect that they would be obvious precursors of the small shellies, but instead they were large soft-bodied creatures resembling feathers or bowls. Moreover, the first to be recognized came from an area that had been worked over by geologists and naturalists for more than a century.

## The Ring Pit

The mostly flat and fertile core of England is hemmed in by the hills of Scotland and the Lake District to the north and by the hills of Wales to the west. The hills are older than the plain. The Cambrian period, in fact, was first recognized as the lowermost fossiliferous deposits in the Welsh hills, after the acrimonious academic controversy between Adam Sedgwick and Roderick Murchison in Victorian times. Most of England is underlain by much more recent deposits. When I was a schoolboy, for example, my father used to drive me to a nearby ironstone quarry of lower Jurassic age, no more than 200 million years old, where I collected ammonites and belemnites, the extinct relatives of modern squid and octopus. There is one exception, however: an isolated outcrop of old rock north-west of Leicester, a town almost exactly in the centre of the country. This is the Charnwood Forest, where the overlying deposits

*The Sea-Change*. Graham Bell, Oxford University Press. © Graham Bell (2026).
DOI: 10.1093/9780197797464.003.0010

have been eroded away to expose a succession of sandstones and siltstones which lie below, if only just below, the base of the Cambrian. They have been difficult to date precisely, but the crucial horizons are 560–570 million years old, and some formations may be considerably older. They are at any rate indisputably PreCambrian. I often visited Bradgate Park, in the middle of the Forest, as a schoolboy, but I was in search of the elusive palmate newt and took little notice of the rocks. They were in any case just dark grey rocks, apparently devoid of fossils, as was only to be expected, given their PreCambrian age. What I did not know at the time was that a most surprising discovery had been made only a few years before and almost a hundred years after Darwin's despairing admission.

The great advantage of Charnwood is that it lies in a densely populated region with several large towns nearby, providing flocks of professional geologists and amateur fossil-hunters to scour the area for traces of Ediacaran life. Other PreCambrian sites, such as the Flinders Ranges of South Australia and the White Sea coast of northern Russia, are quite remote, but you can take a bus from the centre of Leicester and be in Woodhouse Eaves in less than an hour. There is no shortage of exposed surfaces and loose blocks from drystone field walls, but they are bumpy, dark grey and (to my eyes) as barren as cement. More skilled eyes have found that the fossils can be seen quite clearly when the light slants across the rock at a low angle to accentuate the relief. When latex is used to mould the surfaces a whole new community is revealed, including some abundant species that also occur elsewhere but also others which are rare or unknown anywhere else.

One morning in April 1957, three boys set out for Charnwood to do a little rock-climbing. All three were pupils at the Wyggeston Boys' School in Leicester (which I was myself to enter three years later) and they were headed for an old slate quarry near the village of Woodhouse Eaves. As they scrambled over the rock face, one of them, Richard Blachford, pointed out a curious impression about seven or eight inches long on a sloping surface of grey-green slate. It looked like a feather, or a long narrow leaf, with a series of lateral lobes, each with a fringe of shorter ridges. It was clearly a fossil, although it resembled no familiar organism; but one of the boys, Roger Mason, knew that the local formation lay below the base of the Cambrian and so should contain no fossils at all, familiar or otherwise. Guessing that it might be important, he visited the Geology Department of the local university, which happens to be next to the school (and opposite the cemetery, as our teachers often sardonically remarked). The only member of staff he could find there was Trevor Ford, a young lecturer who had only recently joined the department. Ford was at first rather sceptical, as he admits; after all, the Charnwood formation was widely known to be unfossiliferous, as was to be expected in rocks of that age.

And at this point the discovery might have been forgotten, at least for the time being, like the other sightings of strange markings on rock faces in the quarry. The earliest were made by the quarrymen themselves, who called their quarry the Ring Pit, from the curious circular markings on the surface of the slate they exposed. Shortly after the quarry closed, James Harley, a well-known local naturalist, wrote a letter to a colleague in 1848 that mentioned several different kinds of markings, including both the mysterious circles and more complex structures, probably fronds similar to that found over a century later. This stimulated some interest, but as time went on they became regarded as being inorganic in nature, and forgotten. Other people saw them afterwards, including Tina Negus, who found a frond in 1956 (probably the same one seen the following year by Richard Blachford) and was likewise ignored. What rescued the latest sighting was the old tradition of brass rubbing.

English churches often feature the recumbent effigy in brass of an important (or self-important) figure in local affairs, going back to mediaeval times. A very accurate image can be produced by laying a large sheet of paper over the memorial brass and rubbing it lightly with a wax crayon. Brass rubbing enjoyed a great vogue in Victorian times and although often discouraged (it tends to erode the brass, in time) it continues to be popular and even has its own registered charity, the Monumental Brass Society. Roger Mason applied this technique to the frond in the quarry and showed the result to Trevor Ford, who immediately realized that an important discovery had been made. This was no vague verbal report, but concrete evidence of an indisputably organic fossil where no fossils should exist. Ford drove out to the quarry, with Mason, to inspect the rock face in situ and to take photographs. Fully convinced now, he published a description of the frond (as *Charnia masoni*) and the discs (as *Charniodiscus concentricus*) in September 1958 (Figure 8.1). With hindsight (he comments ruefully), such an important discovery should have been published in a prominent journal with a general circulation, such as *Nature*.

**Figure 8.1** Charnia

(Left) Courtesy of Leicester Museum and Art Gallery, Leicester UK. (Right) Reproduced from Matteo De Stefano/MUSE. This file was uploaded by MUSE—Science Museum of Trento in cooperation with Wikimedia Italia, CC BY-SA 3.0, https://commons.wikimedia.org/w/index.php?curid=48168696, under the Creative Commons Attribution-Share Alike 3.0 Unported license.

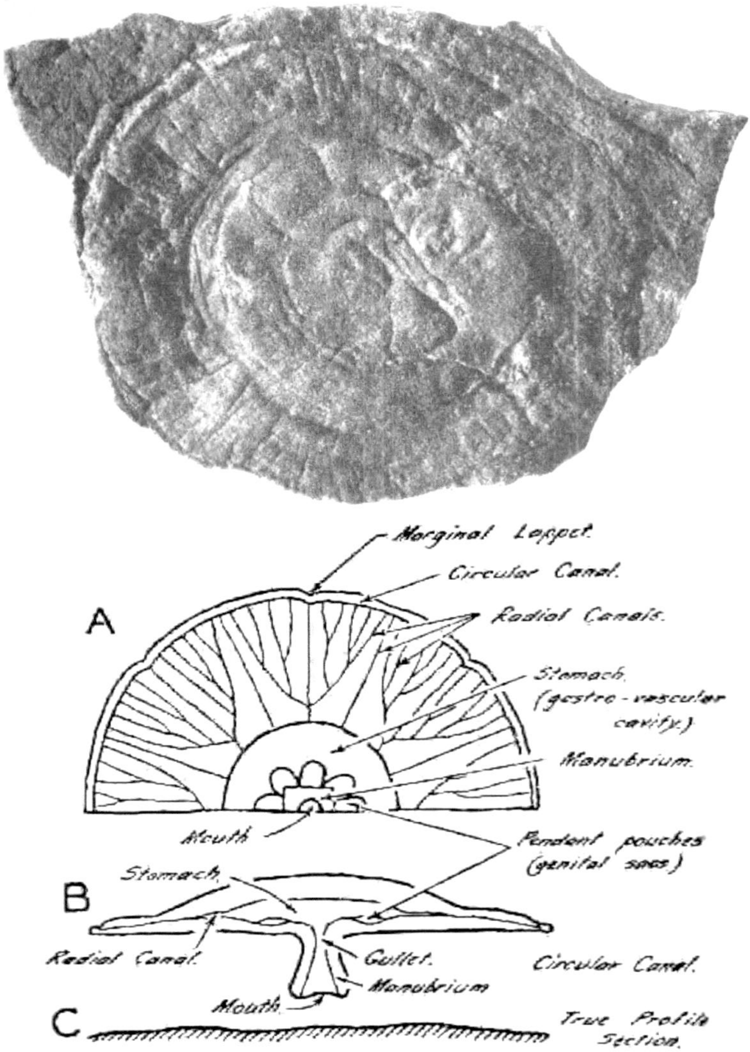


**Figure 8.2** Ediacaria

Reproduced from Sprigg, R.C., 1947. Early Cambrian (?) jellyfishes from the Flinders Ranges, South Australia. *Transactions of the Royal Society of South Australia*, *71*(2), pp.212–224.

The first discoveries of large, clearly multicellular fossils were the fronds, discs and bowls reported from late PreCambrian deposits in England and Australia a lifetime ago, in the 1940s and 1950s, but these were just the beginning of a long series of discoveries from sites around the world. I shall describe the fossils found in Africa (Aus, Namibia), Australia (Flinders Ranges, South Australia), Europe (Charnwood Forest, England), Asia (White Sea coast, Russia) and North America

(Avalon Peninsula, Newfoundland), but there are many other sites. Perhaps the revolutionary claim that fossil remains of large animals could be found well below the base of the Cambrian stimulated geologists everywhere to look more closely at their local PreCambrian rocks, but at all events the decades after 1960 saw a dramatic rise in the number of localities where these fossils could be found. As the discoveries increased, a new and unexpected fauna began to take shape.

## Nama Fm, Namibia, about 543 Mya

Most of southern Namibia is today a hot dry rocky plain, almost as arid as the Sahara Desert, studded with ranges of low mountains. In Ediacaran times the area south and east of the modern town of Aus was a shallow sea in which beds of mudstone, shale and fine-grained sandstone were deposited. The fossils found here date from the latest Ediacaran and include frond-like forms which are known from several other localities, besides forms that are unknown elsewhere, such as the peculiar three-vaned frond *Swartpuntia* which has been found only at Farm Swartpunt, on the road south of Aus. They also include some fossils with a completely different kind of structure, more like a bag or bowl than a frond.

### Bags

The characteristic fossil at Farm Swartpunt is the bag animal *Ernietta*, which looks like the sort of small clay bowl you made aged six: a shallow vessel with a rounded base standing a couple of inches high and measuring about the same across (Figure 8.3). It turns out that the usual fossil is only the base of a larger and more complex organism,

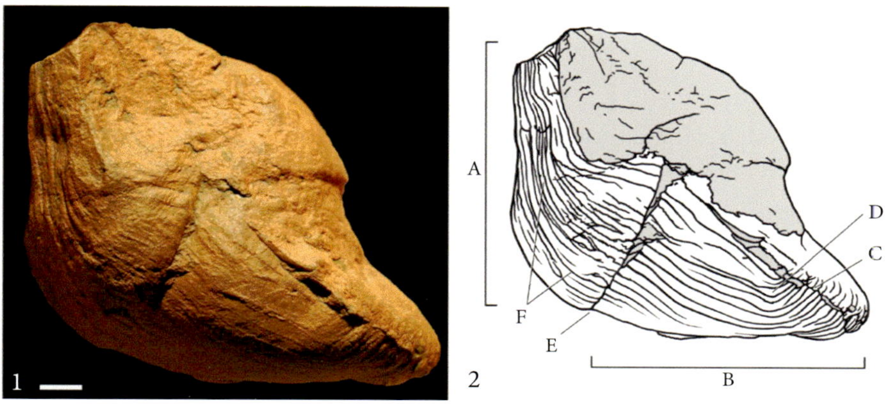


**Figure 8.3** Ernietta. Scale bar is 1 cm

Reproduced with permission from Elliott, D.A., Trusler, P.W., Narbonne, G.M., Vickers-Rich, P., Morton, N., Hall, M., Hoffmann, K.H. and Schneider, G.I., 2016. *Ernietta* from the late Ediacaran Nama Group, Namibia. *Journal of Paleontology*, *90*(6), pp.1017–1026.

although this was not realized until intact specimens were found in an old channel where they had been swept by a mudflow and buried by sediment. *Ernietta* was more or less the shape of a shopping bag half-buried in the seafloor, with walls formed from long tubes, like a palisade (or likely a double palisade), extending upwards into a narrower trunk region before flaring out into a pair of wing-like structures. The tubes met in a zigzag suture along the midline of a roughly oval base and were probably filled with sand lower down and fluid further up. The base of the bag may have been ballasted with more sand, and the wings, held above the sediment surface, used for feeding. There must have been herds of these strange creatures on the shallow sea floor, like the stubs of a drowned heath. How they made a living is a puzzle. One daring guess is that dense aggregations of *Ernietta* generated turbulent flow that drove edible particles into the cavities of their bag-like bodies, so they lived by a sort of socially mediated suspension feeding.

The Namibian fossils are strange enough, but they could be dismissed as an outlier to the main trend of animal evolution, were it not for discoveries in other parts of the world that make it clear that they were part of a distinctive community of large organisms that flourished before the Cambrian period began. The first to be described were slightly earlier in time and on a different continent altogether, the fossils of the Ediacaran Hills in South Australia.

## Flinders Ranges, South Australia, about 550 Mya

Sprigg showed his original jellyfish fossil, now named *Ediacaria*, to Mawson at a meeting of the Australian and New Zealand Association for the Advancement of Science at Adelaide in August 1946 (Figure 8.2). Mawson was not impressed, and thought that the markings in the rock might have been caused by purely inorganic processes such as shrinking and cracking. He seems to have changed his mind after Sprigg's political tea-party made it clear that *Ediacaria* was not an isolated oddity, but he was then getting close to retirement. Sprigg did not push hard for his Ediacara fossils, either; he had a day job, after all, with the South Australia Geological Survey. When he resigned from this job in 1954 it was to set up as a geological consultant and found the oil exploration company SANTOS, which later discovered huge reserves of natural gas in the Cooper Basin. Ediacara was by now collecting dust on the back shelf.

### Martin Glaessner

The man who, more than anyone else, was responsible for the widespread recognition of a rich diversity of large marine organisms in the PreCambrian was no young Australian iconoclast. Indeed, he was neither young nor Australian. Martin Glaessner had been born in the Austro-Hungarian empire, and was old enough to remember, when a small boy, seeing Emperor Franz Josef parading along a street in Vienna. His career began, like so many others, by working in the local museum, in his case the very grand Nuturhistorisches Museum in the Maria-Theresien-Platz. By the age of 25 he had survived the First World War, obtained two doctorates (in law and geology), published a monograph on decapod crustaceans (crabs and their relatives), and spent a year at the British Museum in London. He was then invited to join the new Petroleum Institute of the USSR in Moscow, as part of the drive to recruit thousands more engineers and technicians during the early years of Stalin's drive to industrialize the country. His task was to study the geology of the Caucasus oilfields by mapping the fossils of foraminiferans, which can be found by the thousand in drill cores. He combined this work with marrying Tina Tupikina, before being offered the stark choice of taking out Soviet citizenship or leaving the country. He and his new wife chose to leave, but returned to Austria only a few weeks before the Anschluss of March 1938. He was promptly arrested (he had Jewish ancestry on his father's side) and put to work at the German army barracks. He was saved by the offer of a job from George Lees, the chief geologist of the Anglo-Iranian Oil Company (later British Petroleum). This extricated the couple from Austria to London and eventually to Port Moresby in New Guinea, where Glaessner set up a micropalaeontology lab to map foraminiferans and other microfossils for the company. This was, of course, just in time for the Second World War, although Glaessner's work was considered important enough to excuse him from active service. After the war, he moved first to Melbourne and then, in 1950, to the University of Adelaide, where Mawson had offered him a position as Senior Lecturer. Having lived precariously in some troubled areas of the world, Glaessner may have welcomed the security of Adelaide; at all events, he stayed there for the rest of his career. He was actually present at the Adelaide meeting where Sprigg and Mawson discussed *Ediacaria*, very soon after being appointed (no doubt by some sleight of hand, since his position was Special Lecturer) at the University of Melbourne. Like Mawson, he was unimpressed, or at any rate not impressed enough to shift away from micropalaeontology, the field where he had built his reputation.

The event that moved the Ediacara fossils from the back shelf to the front seems to have been a trip by two amateur fossil collectors, Hans Mincham and Ben Flounders, to the site in 1957. They found dozens of specimens, which they showed to Glaessner when they returned to Adelaide. He immediately became very interested (perhaps micropalaeontology was losing its first allure) and expressed this by the usual academic procedure of assigning students in his lab, Mary Wade and Brian Daily, to the new project. Following up, the South Australia Museum organized a trip to the Ediacara site (including the two amateurs and Glaessner's two students) in March 1958. They collected *truckloads* of fossils. It was now clear that what Sprigg had discovered was not merely a few isolated specimens, but instead a whole community of strange and previously unknown organisms. The news soon spread, and, with the fossil out of the bag at last, nobody could doubt that large, complex individuals lived in PreCambrian times, long before the first trilobites appeared on the scene.

The government reacted promptly—it was still possible for governments to react promptly in the 1950s—by declaring the site a fossil reserve in May, which closed it to collectors (including, ironically enough, Sprigg himself, who had plans to mine the site for fossils). Its real importance was still not clearly recognized, however.

Glaessner himself first described the material found by Mincham and Flounders as coming from the basal Cambrian, and it was not until May 1959 that he published a letter to *Nature* arguing for a late PreCambrian date. Once this had been accepted there were two sites at opposite sides of the world, in England and South Australia, which showed conclusively that the fossil record of large multicellular organisms extended below the base of the Cambrian. But what were they?

## The Ediacarans as ancestors

Martin Glaessner was a strong personality whose insistence on clarity and rigour, stemming from his education and experience, did not always chime with the more casual habits of his colleagues. There is an undertow of resentment that perhaps reflects this. Two of his colleagues at Adelaide, for example, wrote long afterwards that Sprigg's ideas were ignored or rejected until "they were taken up (or taken over) by a distinguished European palaeontologist". (The 'distinguished' is particularly cutting.) Nevertheless, Glaessner was the academic heavyweight who more than anyone else clearly established that whole communities of large marine organisms flourished before the Cambrian radiation of animals. He had no doubt what they were, either—the early ancestors of modern animals, of course, primitive things like jellyfish. Glaessner firmly assigned the Ediacara biota, bar a few problematic forms, to living groups of animals, mostly cnidarians but including annelids and arthropods. *Charnia*, for example, was probably a sea pen, a colonial soft coral that is still abundant in modern seas; I have seen splendid living sea pens in the Vancouver Aquarium. This was the point of view that he put forward in *The Dawn of Animal Life*, a very influential book that he published in 1984 (long after his retirement in 1971, at the statutory age of 65) to summarize what was known about PreCambrian animals.

Many of the palaeontologists who have worked on the sandstone flags of Ediacara and nearby sites in the Flinders Ranges of South Australia have followed the forceful advocacy by Martin Glaessner and interpreted them as early representatives of groups of animals still in existence. The jellyfish *Ediacaria* was joined by fronds similar to those found 10 years later in England, which were likewise interpreted as sea pens. There was a sponge, *Palaeophragmodictya*, with a dome-shaped body from which long rays extended. Reg had a fossil named after him, *Spriggina*, which seemed to be a polychaete annelid, judging from its shield-shaped head and the long row of paired segments, or appendages, running down either side of its body (Figure 8.4). Mark McMenamin of Mount Holyoke College, who has written many articles on the Ediacaran fossils, thought it was more likely to be an arthropod close to the ancestry of trilobites. *Dickinsonia* and *Kimberella*, which I'll describe in the White Sea section, are also found in the Flinders outcrops—*Dickinsonia*, indeed, was first found there, and named by Reg Sprigg after his boss Ben Dickinson in the Bureau of Mines. There seems to be general agreement that at least the big fossils found in the Flinders Ranges can be assigned to the lineages of groups of animals alive today.

**Figure 8.4** Spriggina

(Top) Reproduced from Daderot—Own work, CC0, https://commons.wikimedia.org/w/index.php?curid=24340282, under a Creative Commons license CC0 1.0 Universal Public Domain Dedication, https://creativecommons.org/publicdomain/zero/1.0/deed.en. (Bottom) Reproduced from Matteo De Stefano/MUSE. This file was uploaded by MUSE—Science Museum of Trento in cooperation with Wikimedia Italia., CC BY-SA 3.0, https://commons.wikimedia.org/w/index.php?curid=48168696, under the Creative Commons Attribution-Share Alike 3.0 Unported license.

## Pit-and-ridge creatures

There are also some smaller fossils, mostly only an inch or so long and sometimes much less, which can likewise be placed in living groups. *Parvancorina* has a central ridge which expands at one end into a semicircle, making a shape like an anchor (Figure 8.5).

Glaessner originally interpreted *Parvancorina* as an arthropod, and his diagnosis was strengthened by the discovery of a fossil called *Skarnia* in the Kaili biota with a similar shape but also the remains of jointed limbs. *Tribrachidium* has three curved ridges radiating from a central point, rather like the three legs of the Isle of Man flag. It

**Figure 8.5** Parvancorina

(Left) Reproduced with permission from Coutts, F.J., Bradshaw, C.J., García-Bellido, D.C. and Gehling, J.G., 2018. Evidence of sensory-driven behavior in the Ediacaran organism *Parvancorina*: implications and autecological interpretations. *Gondwana Research*, *55*, pp.21–29. (Right) Reproduced from Matteo De Stefano/MUSE. This file was uploaded by MUSE—Science Museum of Trento in cooperation with Wikimedia Italia., CC BY-SA 3.0, https://commons.wikimedia.org/w/index.php?curid=48168696, under the Creative Commons Attribution-Share Alike 3.0 Unported license.

has been identified as a sponge, a cnidarian and an echinoderm by different authors. *Arkarua* has five ridges radiating from a central mound, or perhaps pit, which has inevitably suggested an association with echinoderms, albeit echinoderms without a water vascular system or even a mouth.

There is an alternative interpretation of the structure of these strange creatures, however. Perhaps they are not related to modern groups at all, but rather represent a quite different body plan based on ridges that channel a feeding stream. Dolf Seilacher imagined them as living like sponges, crouched on the surface of the microbial mat. The fine channels led the stream to the centre, where edible particles were removed by choanocytes lining flagellated chambers, and the broad channels carried the exhaust stream to the outside. An alternative to the hypothetical flagellated chambers is that the ridges simply directed direct water flow into the central pits, where the current is slowed and particles can settle out of suspension. The body plan is simply the number of pits: two in *Parvancorina*, three in *Tribrachidium*, and five in *Arkarua*. How the particles were captured by cells in the pits is uncertain. Perhaps they were taken up directly by phagocytosis once they had settled under gravity. Phagocytosis is a very general feature of eukaryote cells, which unlike bacteria can engulf solid particles by an infolding of the plasma membrane at the surface of the cell and draw them into the interior in a little membrane bag that turns into a digestive vacuole. This is basically how sponges feed, so it is perfectly credible that *Tribrachidium* and other pit-and-ridge creatures would have had the same ability.

## The White Sea coast, Russia, about 555 Mya

The White Sea lies just below the Arctic Circle in north-western Russia, hemmed in to the north by the Kola Peninsula, shaped like the head of a gigantic sperm whale. It has supported a fishery for cod, herring and salmon since mediaeval times, based on the port of Arkhangelsk (Archangel) at the southern end of Dvina Bay, where the Dvina River reaches salt water. The fishery used to shift from the coast to the west of Arkhangelsk in the summer to the east in the winter, and they are still known as the Summer Coast (Letrniy Bereg) and the Winter Coast (Zimnii Bereg). The Summer coast has wide sandy beaches where you can sunbathe if you time your visit carefully; July is the only reliably snow-free month at this latitude. Both coasts have deposits of Ediacaran age, but it was the crumbling cliffs of the Winter Coast where a whole new Ediacaran fauna was discovered.

Science is taken seriously in Russia, where the National Academy of Sciences, founded under the Tsars and expanded in Soviet times, ran a thousand institutes and employed over 100,000 people at its peak. (It reported directly to the Central Executive Committee under Stalin, but has now been demoted by Vladimir Putin as a government agency under the Minister of Science.) Well, science is taken seriously everywhere, but in Russia even palaeontology is taken seriously. Within the Academy, a Laboratory for PreCambrian Organisms was set up by Boris Sokolov after he became an academician in 1968. It was Sokolov and his eventual successor, Mikhail Fedonkin, who first exposed the richness of the Winter Coast biota. This included many of the forms that had already been found at Ediacara. But there is also a host of new and quite different kinds of organism, which have turned out to be just as puzzling.

### *Dickinsonia*

*Dickinsonia* is preserved as a broad flat oval, somewhat pointed at both ends, like an old Zulu shield (Figure 8.6). A small individual is only the size of a fingernail, while a large one would fit into the palm of your hand. Regardless of size, its main feature—well, almost its only feature—is a series of raised ridges, separated by grooves, which extends from the midline to the margin down the length of the body. Like all the White Sea fossils, *Dickinsonia* is preserved as an impression of the upper surface of its body on the overlying sediment.

It is a very distinctive fossil that is abundant both on the Winter Coast and in South Australia, so there is no doubt that it was a prominent member of the seafloor community for several million years. It is also a very simple structure that will bear almost any interpretation that imagination can supply. Reg Sprigg, describing it for the first time, thought it might be a coelenterate, but wasn't sure. Martin Glaessner was quite sure that it was an annelid worm, with the ridges indicating a segmented body. More recently, a group of authors (including Jim Gehling) analysed specimens of different sizes in great detail and concluded that growth

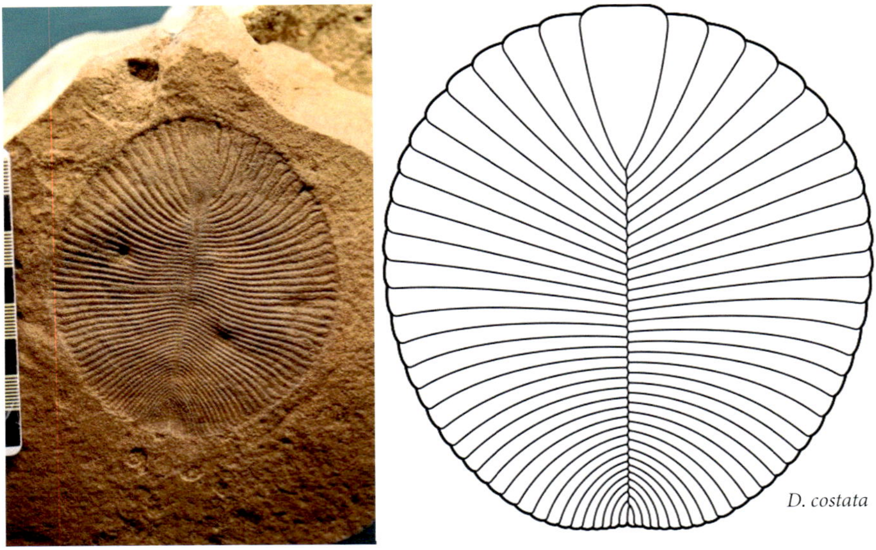


**Figure 8.6** Dickinsonia

(Left) Reproduced from Hoekzema, R.S., Brasier, M.D., Dunn, F.S. and Liu, A.G., 2017. Quantitative study of developmental biology confirms *Dickinsonia* as a metazoan. *Proceedings of the Royal Society B: Biological Sciences*, *284*(1862), p.20171348, under a Creative Commons license. (Right) Reproduced from Aleksey Nagovitsyn (User: Alnagov), https://commons.wikimedia.org/w/index.php?curid=9006157 under a Creative Commons license CC BY-SA 3.0.

occurred by adding units to the posterior end of the body, the characteristic way in which many animals grow. An equally authoritative group found that it grew in a quite different way, but likewise concluded that it was an animal. A third group extracted lipids from the fossils and found them similar to those made by animals. The current consensus, then, is that Sprigg and Glaessner were correct in principle even if they were wrong in detail: *Dickinsonia* was an animal. But what sort of animal?

If the ridges which meet at the midline are truly segments then they are very odd segments, because there should be no midline. The bodies of animals like earthworms or centipedes have a long series of similar segments which extend over the whole width of the body. Worms don't have a seam. The ridges also don't line up, but are (usually) staggered alternately right and left. Fedonkin calls this arrangement 'glide reflection' and uses it to define a group of Proarticulata which includes *Dickinsonia* and was the ancestor of segmented bilaterians such as annelids and arthropods. Another member of his laboratory, Andrey Ivantsov, has explained how food particles were swept by cilia into the grooves between the ridges and from there to the digestive tract, represented by a pair of elongated sacs running along the main axis of the body. In a related proarticulate called *Yorgia*, the gonads (or structures supposed to be the gonads) lie in these grooves. Proarticulates such as *Dickinsonia* and *Yorgia* were thus bilaterally symmetrical animals with complex internal structures including a body cavity, a gut and a set of gonads. Or alternatively the Russian group has let its imagination run away with itself, gut and gonads are accidents of preservation in a few individuals, and Proarticulata is a fantasy.

## *Keretsa*

One of the strangest fossils found on the Winter Coast is *Keretsa*. It has a prominent half-moon shape at one end, perhaps a head shield, which bears two processes that might be antennae. The other end of the fossil has transverse furrows and what appears to be a segmented trunk, although it is not clear whether the trunk units are arranged in an opposite or an alternate fashion. There are no obvious appendages on the trunk. *Keretsa* looks much like *Parvancorina*, in which case it is a typical member of the Ediacaran community, but Ivantsov thinks it resembles *Naraoia*, an early Cambrian trilobite, in which case it is an arthropod.

## *Kimberella*

The star turn of the Winter Coast is *Kimberella*. It was about the same size as *Dickinsonia*, usually between the length of a fingernail and a hand, but was much less decorative and looked something like a slug with a frilly skirt (Figure 8.7). It had

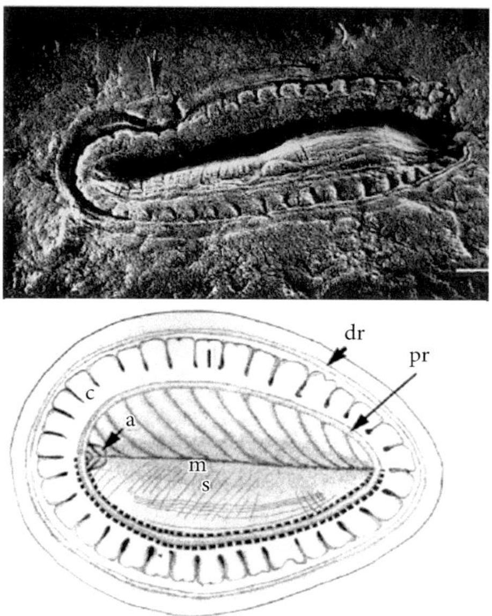


**Figure 8.7** Kimberella

Reproduced with permission from Fedonkin, M.A. and Waggoner, B.M., 1997. The Late Precambrian fossil *Kimberella* is a mollusc-like bilaterian organism. *Nature*, *388*(6645), pp.868–871.

a broad flat ventral surface with a rounded dorsal hump bordered by a row of compartments or a scalloped margin. *Kimberella* was an elongate mound quite unlike the other Ediacarans.

The body fossils are often accompanied by minute scratches in the sandstone that seem to show how it fed on the microbial mat (Figure 8.8). They are very shallow, only a millimetre or two deep, but deep enough to rake up fragments of the mat that could then be swallowed. This has weighty implications.

A rake cannot be powered by cilia or pseudopodia; it needs a pair of antagonistic muscles to move the toothrow of the rake backwards and forwards. Molluscs such as snails and chitons use a device of this sort to scrape a living from the biofilm on rocks. Their rake is called a radula, which usually consists of a battery of hard chitinous teeth mounted on a firm support which is alternately extended and retracted. (A pond snail browsing on the glass side of an aquarium gives an excellent view of the radula in action.) The feeding trails of molluscs are occasionally preserved as a trace fossil, *Radulichnus*, that has been reported from rocks of all ages from the Cambrian to the Pleistocene. But swallowing the fragments that have been scraped up requires a mouth, and a mouth implies a gut, however simple it might be. These feeble scratches on the mat, barely the thickness of a fingernail deep, seem to lead inexorably to the conclusion that *Kimberella* was a bilaterally symmetrical animal equipped with mouth and gut and powered by muscle.

Moreover, *Kimberella* may have left trails. It is not quite out of the question that *Kimberella* was propelled by cilia, but its massive body, shaped like an upturned boat, make this difficult to believe. The alternative is that it was propelled by muscle, perhaps by a series of muscular ripples propagated along its underside, much as snails and slugs move nowadays. Its trail is not a simple continuous trough, perhaps because the mat was too coherent to take an impression. Instead, it is quite short and varies markedly in width. Ivantsov has a rather complicated explanation for

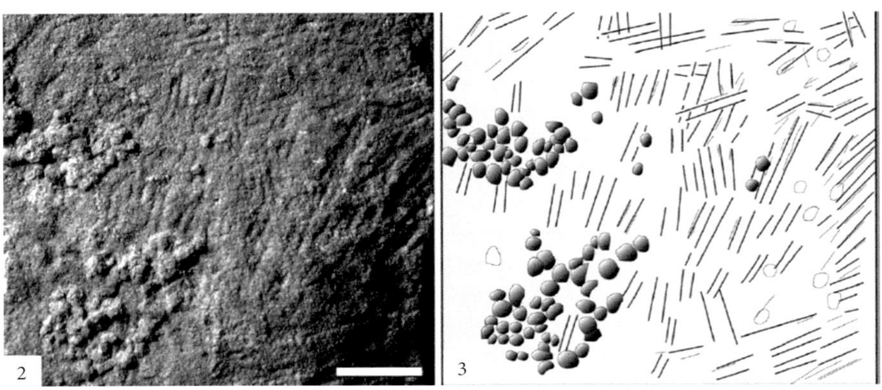


**Figure 8.8** Kimberichnus

Reproduced with permission from Gehling, J.G., Runnegar, B.N. and Droser, M.L., 2014. Scratch traces of large Ediacara bilaterian animals. *Journal of Paleontology*, 88(2), pp.284–298.

this, involving the production of a mucus sheath when the animal sensed that it was about to be buried.

If *Kimberella* looked like a slug, moved like a slug and fed like a slug then perhaps it was a slug, or rather a primitive mollusc with an initial sketch of the diagnostic molluscan combination of a muscular foot and a radula. This is the interpretation, first urged by Mikhail Fedonkin and Benjamin Waggoner, that replaced the original guess of Mary Wade that *Kimberella* was a box jellyfish lying on its side. I think it is now widely accepted. The Ediacaran scratches are not exactly like those left by a radula, however. In particular they tend to be paired, either as single marks or rows or fan-shaped arrays, whereas a typical radula, consisting of many transverse rows of teeth, makes a mark like the head of a rake. This suggested to Jim Gehling that *Kimberella* had only two teeth at the end of an extensible proboscis that could be protruded and moved from side to side (presumably by longitudinal retractor muscles) while grazing on the mat. Pulling on this proboscis in order to dislodge a chunk of mat would also tend to haul an elongate body forwards in a hook-and-grapnel fashion, although progress might be slow (because two teeth do not provide much purchase) and uneven (because the texture of the mat might vary from place to place). This does not sound much like a mollusc—Gehling renamed the scratches *Kimberichnus* to emphasize that they were not made by a radula—but it does imply that a large coelomate bilaterian lived on the sea floor in late Ediacaran times.

## Charnwood Forest, England, about 560 Mya

### Discs

The slate miners named the original site the Ring Pit because of *Charniodiscus*, a more or less circular impression an inch or so across marked with one or two concentric lines and bearing a central stump. What kind of creature it was became clear when very similar discs were found attached to a stalk bearing a frond that looked rather like *Charnia* and equally consisted of a bundle of tubes. The frond seemed to be covered by a sheet of tissue and was bounded by a distinct rim. Organisms with fronds like this are called arboreomorphs, to distinguish them from rangeomorphs like *Charnia*, and may be a quite different group which independently discovered how to elevate a feeding structure above the seabed. If so, they likewise lacked mouth, internal structures and muscles, but presumably shared a similar way of life.

The holdfast is at first glance a rather puzzling structure. An upright frond in a current will create drag, which will uproot the frond and carry it away unless it is firmly anchored in place. A small flat circular disc embedded in loose sediment will not prevent this, any more than a small boat can be securely anchored to a saucer. Seaweeds growing in the rocky intertidal zone have elaborate holdfasts of root-like tangles that clutch the rocks, while sessile animals such as pennatulids which live in loose sediments have long carrot-like holdfasts to resist the current. A tall frond attached to a small disc would soon be swept away in modern seas. Before there were animals to make burrows or plough through the sediment, however, the sea floor would have been a different kind of habitat. A small circular holdfast is perfectly capable of anchoring a frond securely when it is embedded in a microbial mat.

Discs are the most common Ediacaran fossils at many localities, including Charnwood. They are very simple structures, just flat or slightly convex circles or ovals, often with a raised rim, sometimes ornamented with rays or rings and sometimes not, but almost always with some kind of feature, either a ridge or a furrow, in the middle. Their simplicity allows full play to the imagination, and since palaeontologists are very imaginative people they have come up with a rich variety of interpretations. Take *Aspidella*, for example, which is strewn by the thousand over outcrops in downtown St John's, Newfoundland, and was described in the middle of the 19th century, so it is scarcely rare or inaccessible. They are about the size and shape of an old-fashioned wax seal. These are unfamiliar objects nowadays, but before the invention of envelopes a letter was simply a folded sheet of paper with the address written on the back. If the sender wanted to keep it confidential (they usually did) a dollop of hot wax was poured over the flap. Once the wax had set any attempt to remove it would have been obvious, so the recipient of a letter whose wax was intact could be confident that it had not been read—unless, of course, the old wax had been lifted off with a hot knife and replaced by new, matched for size and colour if necessary. To counter this stratagem, the sender could press a seal with a distinctive design into the hot wax, over the flap, making an impression that would foil any malefactor. The seal could be specially made, but any small personal object, such as a signet ring or a cuff-link, would do at a pinch. The adhesive envelope spoiled this fun until it was in turn made redundant by email, but in Victorian times a large oval cuff-link pressed into hot wax would have left an impression looking rather like *Aspidella*. What kind of organism might have left such a mark in the rock?

According to Hans Hofmann, a tall grey cadaverous figure who was associated with the Redpath Museum and left a few specimens of *Aspidella* in our collections, it was not an organism at all, but a purely inorganic structure produced by the 'differential movement of mud'. I have looked at his specimens and I can see what he means; they are pretty obscure (the first problem is often just to find the fossil) and the pattern of bumps and folds could easily be no more than the plop and swirl of mud caused by a bubble of gas escaping from the sediment. Other geologists disagreed, and instead argued that they were looking at the bona fide remains of an organism of some kind. Still, it might be no more than a colony of unicellular microbes such as bacteria, fungi or algae. Some discs, for example, have a somewhat pustular appearance (they were called 'pizza discs' but look to me more like splashes of lumpy porridge) and sometimes appear to be covered by the mat itself, so they might well be microbial colonies of some sort. I like this idea because I have spent many hours staring at the green circles of algal colonies growing on agar in petri dishes, and it would be pleasant to believe that all the while I have been looking at the latest version of these enigmatic fossils. It is true that discs come in a variety of shapes—simple flat coins, flat plates with a raised rim, concentric circles and so forth—but so do microbial colonies. A few years ago I grew a couple of hundred lines of a green alga in the dark for a few hundred generations (they can do this if you give them sodium acetate as a carbon source), pricked a single cell from each into the centre of an agar plate and let it grow in the dark for a few weeks. The ancestral stock, kept in the light, forms the usual low green circular domes which are to all appearances identical. I was astonished to find that the lines grown in the dark grew into a wide variety of colonies—some formed the familiar green domes, but others became high stacks, very thin irregular smears, lumpy blobs and doughnuts. The apparently uniform appearance of the original stock concealed a wide range of cryptic diversity that was revealed only when the environment changed. This would not happen outdoors: intact colonies of microbes are found today only in the shelter of petri dishes in the lab because outdoors they would be eaten by worms or beetles, but they might have grown undisturbed before there were any worms or beetles. It's an attractive idea, but I'm not sure why they suddenly appeared in the Ediacaran and are never found in much older rocks from the long stretch of time when microbes were abundant and animals had yet to appear.

The remaining possibility is that *Aspidella* was a multicellular organism, or at any rate part of one. To visualize what it was, imagine a gigantic layer cake with alternating layers of vanilla and chocolate. Each layer has a scattering of whole hazelnuts, all base down. When the cake is assembled, the hazelnuts near the top of a layer poke into the base of the layer above, whereas those near the base press into the top of the layer below. Alas, the cake is then forgotten—pushed to the back of the cupboard for several million years, until the vanilla and chocolate layers have become hard and solid. The hazelnuts themselves decompose and disappear completely, but the cavities they occupied remain. Within a layer, the cavity would be an exact mould of the outer surface of the original nut. A nut which spans two layers leaves a different print in each. One which is near to the top of a chocolate layer, for example, leaves an impression of its base, together with an impression of its top in the vanilla layer above. Conversely, one near the bottom of the same layer leaves an impression of its top, together with an impression of its base in the vanilla layer below. Given the entire solidified cake, one could work out the structure, or at any rate the outward appearance, of a hazelnut.

In practice, it may be a little more complicated, for example if the crumbs have trickled down from the layer above to the layer below, filling in the cavity, or if the layers themselves can be squeezed or stretched. Add a dash of jargon, though, and you will be able to describe the fossils that are found in alternating bands of shale (originally mud) and sandstone (originally sand), even when the bodies of the organisms that lived there have completely decayed. After heroic detective work by Guy Narbonne, Jim Gehling and others, it has been shown that *Aspidella*—actually about the size of a hazelnut—is preserved on the lower surface of the sandstone layer as a cast in the shale, filled by sand grains trickling from above. Consequently, it is only the base of a structure with a central furrow which is preserved. Its shape is misleading, because the parallel alignment of specimens on the same slab shows that the rock has been deformed; in life *Aspidella* was probably circular or nearly so, with a central pit (Figure 8.9). It would have looked much like *Charniodiscus*, in fact, and because the original *Charniodiscus* is undoubtedly the holdfast of a frond then perhaps *Aspidella* is a holdfast too, attached in life to the base of some unknown frond, rather than being a different kind of organism altogether. This would at any rate reduce the number of mysteries from two to one. The difficulty, of course, is that the upper surface

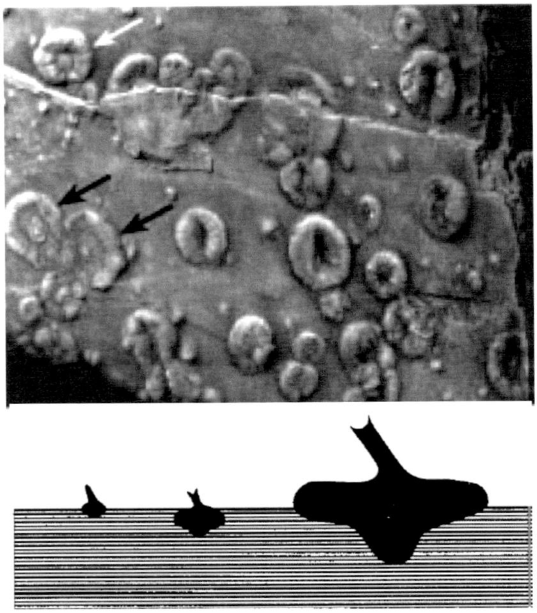

**Figure 8.9** Aspidella

Reproduced with permission from Gehling, J.G., Narbonne, G.M. and Anderson, M.M., 2000. The first named Ediacaran body fossil, *Aspidella terranovica*. *Palaeontology*, *43*(3), pp.427–456.

of *Aspidella*, which should bear the stalk of the frond, is unknown. There are a few rare and rather dubious specimens where the remnant of a stalk-like structure seems to extend from the margin of the disc, but a definitive connection between *Aspidella* and a frond has yet to be demonstrated.

There is another problem, too: why would a holdfast have a pit? A projection of some sort would seem to provide a better anchor, ideally a projection with four flanges at right angles like the vanes of an arrow, but a simple knob might serve. Narbonne and Gehling finesse this difficulty by showing that there is a continuous series of intermediates between the classical type of *Aspidella* and two other kinds of disc-like fossils. One is convex with a central knob, rather like the desk bell that you can press (if you dare) to call for service. The other is flat, bears a series of concentric circles like the growth rings of a tree, and also bears a central knob. Getting from a pit to a knob is not straightforward, but there are some examples which bear a bump inside the central pit. I am not sure how this transition can be justified functionally, but the overall scheme has the great merit of reducing the wilderness of names that have been applied to disc-like Ediacaran fossils to a single name, *Aspidella*, and the multitude of organisms they are supposed to represent to a single interpretation, the holdfast of a frond. A simple interpretation is not necessarily correct, but it is refreshing to sweep away the lichens, fungi, jellyfish, sea anemones, sea pens and other speculations that have consumed so much ink, and instead view all Ediacaran discs as holdfasts unless proven otherwise.

## Avalon Peninsula, Newfoundland, about 565 Mya

South from St. John's lies the Avalon Peninsula, whose east coast is studded with Ediacaran fossil sites, culminating in the spectacular surfaces at Mistaken Point, crammed with fossils. Mistaken Point used to be avoided by sailors, because of the fog. It lies about five miles WSW of Cape Race, the usual landfall for ships crossing the North Atlantic. Heading north from Cape Race along the coast of Newfoundland will bring you to a safe anchorage at St. John's or at one of the harbours along the way. If it is foggy, as it very often is, you may mistake your landfall, and heading north from Mistaken Point brings you onto the pitiless rocks of the Avalon Peninsula. If you are a palaeontologist, on the other hand, these rocks are a magnet, because their broad surfaces, sloping towards the open Atlantic, are crowded with the remains of mysterious creatures that lived on the sea floor 25 million years before the Cambrian period began.

This part of the world has a lot of geology. Fortune Head is just round the corner, across Placentia Bay, and Mistaken Point is only one of several sites on the Avalon Peninsula where these extraordinary fossils can be found. The two are linked by the unmistakable figure of Guy Narbonne, who not only led the team which fixed the base of the Cambrian at Fortune Head but also persuaded UNESCO to recognize Mistaken Point as a World Heritage Site, probably a unique geological double. The fossils themselves were discovered almost 50 years before the UNESCO listing by a graduate student from Memorial University, Shiva Misra, and have been scrutinized very closely by many palaeontologists since Narbonne began to investigate them in the late 1990s. You can see them yourself by booking in for a guided tour (at https://mistakenpoint.ca/) but if you are deterred by the 3 km hike to the site or are short of time there is an excellent display in The Rooms museum at St. John's. Most of the fossils are more or less frond-shaped and are informally known as rangeomorphs (from *Rangea*, the first kind to be described). Their structure has turned out to be so peculiar that to explain it needs a brief detour.

## Fronds

The simplest multicellular structure is the filament, a one-dimensional string of cells joined end-to-end. This can elongate in one of two ways: either the tip cell alone can divide, or the in-between cells can divide. Either way, it just gets longer and longer. But suppose that the in-between cells can divide at an angle to the main direction of growth; the structure is now a branching filament whose shape depends on how growth is regulated. Whether the structure is literally a filament consisting of a single string of cells or a three-dimensional solid rod or hollow tube makes no difference: provided that it grows in a single direction, the shape that emerges will depend only on the rules that regulate growth.

The frond of *Charnia* and other rangeomorphs has a plaited appearance, like a pigtail, because of its manner of growth. Most plants grow only at the leading tip, where the tissue consists of cells dividing actively to elongate the main axis of the shoot. Lateral branches are initiated subsequently as buds and grow outwards from the main axis. The result is a typical tree, with a clearly defined main trunk that gives off branches at intervals. This is called monopodial growth. Other plants (such as some orchids and bamboos) have a different pattern of growth: the terminal bud is suppressed, while each successive lateral branch elongates, before giving rise to another lateral branch to which it transfers the capacity to grow. The result is a zigzag formed by the suppression of lateral branches, without any single main axis. This is called sympodial growth. Not only plants but also colonial animals such as hydroids and bryozoans, in which zooids arise at intervals along a growing stolon, may have either monopodial or sympodial growth. Only the rule matters, not whether the organism is an animal or a plant.

*Charnia* has sympodial growth, and some seaweeds do indeed grow like this (for example, the red seaweed *Dasysiphonia*, which is currently expanding on the rocky shores of New England), but then so do some seed plants (such as orchids) and even some colonial animals (such as hydroids). Clearly, sympodial growth can readily evolve in any organism that grows along a stem or stolon (there are sympodial and monopodial orchids, for example), and cannot tell us what sort of creature *Charnia* was.

The lateral branches of *Charnia* themselves bore lateral branches (of the third order), which themselves bore lateral branches (of the fourth order). The branches themselves seem to have been hollow tubes that communicated with one another (because they become filled with sediment after death) and were stuck together to form the frond.

Now imagine a sympodial organism in which a lateral branch (which we'll call 'primary') continues to grow, having itself given off a lateral branch (which we'll call 'secondary') which likewise grows at the tip. If you draw this rule on a sheet of paper you'll see that it quickly gives rise to quite a complicated pattern. A third order of branching, producing tertiary branches, makes a shape which, spreading out in two dimensions, looks like a particularly intricate doily (and is becoming very difficult to draw). Let us take this shape and try to fold it into three dimensions by sticking the tertiary branches to the secondary, and the secondary to the primary. This will create a fan-shaped blade, whereas classical rangeomorphs are more or less frond-shaped. This creates a packing problem, because there are too many tertiary branches to fit, which we'll solve by making the tertiary branches thinner than the secondary branches, and the secondary thinner than the primary. A judicious thinning rule will allow us to fold the doily into a thick frond. We have now created a reasonable approximation to a classical rangeomorph like *Charnia*: a felted sympodial frond-shaped organism with at least three orders of branching (Figure 8.10).

Or, rather, we have created a reasonable approximation to *Charnia*. Other kinds of rangeomorph consist of a *Charnia*-like structure with added complications. Some have a main axis bearing *Charnia*-like fronds (*Avalofructus*) or bearing lateral branches which themselves bear fronds (*Rangea*). Some have a short stem and a simple frond (*Culmofrons*), or a bushy fan of fronds (*Bradgatia*), or both (*Hyloecullulus*). Some are elongated oval creatures, pointed at both ends, with a row of fronds on both sides, which presumably lay recumbent on the mat or on the surface of the mud (*Fractofusus*). Others are more complicated still, but the general picture

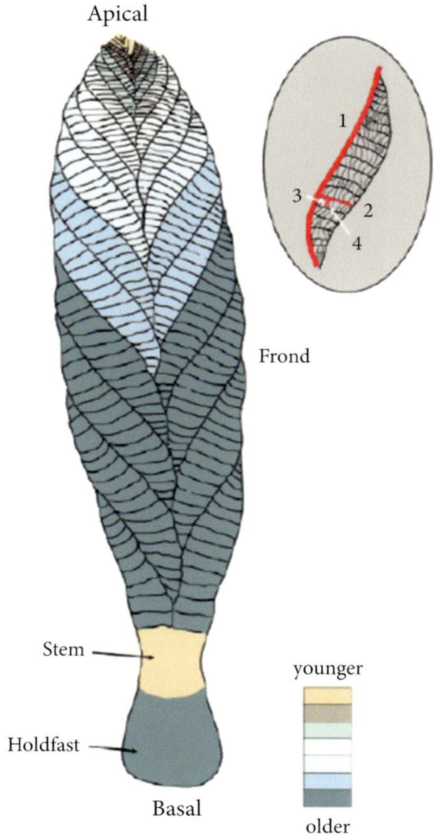


**Figure 8.10** Rangeomorph anatomy

Reproduced with permission from Dunn, F.S., Liu, A.G. and Donoghue, P.C., 2018. Ediacaran developmental biology. *Biological Reviews*, 93(2), pp.914–932.

is clear: by Avalon times, there had been a radiation of rangeomorphs that culminated in a diversity of frond-like body plans whose basic unit is a sympodially branching filament.

There is one further complication: rangeomorph individuals were sometimes connected to one another by thin strands of tissue whose course can still be traced from frond to frond. This suggests that the upright fronds of one kind of rangeomorph growing in a particular small area arose as buds on a stolon running over or just under the surface of the seabed, rather like a clone of sumac. This kind of growth, forming a colony of interconnected individuals, is quite popular, not only among plants but also among seaweeds (such as the dreaded *Caulerpa*) and many kinds of animal (such as hydroids and bryozoans).

To complete your image of the rangeomorph you will need two scales. The first is the scale of the whole organism, which is huge. The fronds splayed out on the Avalon rocks can be a metre or more in length, almost as big as a geologist, although most are much smaller, about the length of a hand. The second is the scale of the smallest resolvable structure. At Mistaken Point the fossils are preserved as impressions beneath a thin layer of volcanic ash, which drifted downwards from a nearby eruption and flattened the organisms on the seabed. The smallest structure that can be resolved is set by the particle size of the ash and is limited to a millimetre or two. At Spaniard's Bay, further north on the peninsula, the same organisms are preserved in mud, which has much smaller particles and allows structures as small as 30 μm (about the thickness of a human hair) to be resolved. This presents an opportunity to study the morphology of a large organism at an almost cellular level of detail. It should be straightforward to say whether or not it is an animal, and, if so, not much more difficult to decide what kind of animal it is. In practice it has been far from straightforward even for the experts to agree on what sort of creature left these cryptic remains in the Avalon rocks.

## Bubbles or seaweeds?

The simplest interpretation is that they are not creatures at all, but some kind of inorganic structure caused by purely physical processes. I've already remarked that this was for long accepted as an explanation of *Aspidella*, which was found nearly a century before the Mistaken Point fronds by Elkanah Billings, the first palaeontologist hired by the Geological Survey of Canada, on Duckworth Street in downtown St John's, of all places. Billings thought that *Aspidella* was the remains of an organism of some kind, but others thought that it looked like the structure formed when a bubble of gas escapes from the sediment and bursts on the surface. I've described all this before. It is a simple and credible explanation, so it is not surprising that the initial description of the Mistaken Point fossils immediately provoked Roland Goldring to protest that they might not be organic at all, like *Aspidella*. Goldring went on to become a very distinguished palaeontologist, one of the leading authorities on trace fossils, but on this occasion he blundered. The Mistaken Point fronds are too uniform yet complex to be anything but the remains of living organisms.

A more conventional interpretation of rangeomorphs is that they were seaweeds, as originally suggested by Trevor Ford for the Charnwood frond, or at any rate some kind of photosynthetic organism (the most outlandish idea is that they were giant marine lichens). This is an attractive idea which gives a simple explanation for the upright frond, which would shade out any prostrate forms, much as trees shade their competitors in forests today. Unfortunately, it seems to be refuted by the geology. The original report refers to ripple-marked surfaces which must have been formed in shallow water, presumably on the authority of the first-named author, M.M. Anderson, who was very familiar with Avalon localities. Anyone who has walked on the Mistaken Point surfaces in stockinged feet will have noticed they are rippled, but this is now attributed to tectonic movement rather than waves. The ripples formed by waves on a shallow sea floor have faded from the literature, replaced by deep dark water in which seaweeds could not grow. Guy Narbonne, who ought to know, is quite explicit: 'Thorough search by generations of sedimentologists has failed to reveal any mud cracks, wave ripples, storm-generated sedimentary surfaces or any other indication of shallow water in any part of the fossiliferous succession'. The sediments were instead deposited in the deep sea, which rules out the possibility that rangeomorphs used light as an energy source.

## Seilacher's heresy

They are not necessarily the remains of multicellular organisms, though. A large single-celled organism might seem to be a contradiction in terms, but there are plenty of living examples. *Acetabularia*, for example, is a unicellular green alga found on the floor of shallow subtropical seas which looks like a little umbrella and grows to several centimetres in height. The main problem with growing larger is that metabolism must increase proportionately to support the extra cytoplasm, which involves making more of all the multifarious proteins that staff the metabolic pathways from the instructions encoded in the DNA of the genome. Making proteins from genes can be speeded up a little, but the only way of supporting any substantial increase in size is to have many copies of the genome. Ciliates do this by delegating the production of their metabolic machinery to a single enormous nucleus that contains hundreds of copies of each gene, a strategy that enables them to grow as large as a metazoan, albeit only a very small one of less than a millimetre or so. Most other large protozoans achieve much the same result by having lots of nuclei, dotted around in the cytoplasm like currants in a bun. The nearest modern equivalent might be xenophyophores (which I'll call xenos for short), which are abundant in the deep sea and are probably related to planktonic foraminiferans (forams) such as *Globigerina*, whose remains form a thick deposit of ooze on the sea floor that covers about one-fifth of the surface of the Earth. Some xenos grow on the seabed as fan-like structures a few centimetres high and look a little like a scale model of a rangeomorph. This led Dolf Seilacher to make an outrageous proposal.

Dolf Seilacher was from all accounts (I never met him) a strong personality who did not suffer fools gladly, if indeed at all. He was quite rude to Professor Goldring, for one. Perhaps he had to have a colourful character because his subject area was so colourless, at least in popular estimation: he was an authority on trace fossils such as burrows and trails. Palaeontology tends to attract colourful characters, or perhaps it just encourages them, but most tend to dig up dinosaurs rather than trying to puzzle out vague scrapes and blurs. Seilacher brought trace fossils alive by picturing them in vivid terms as frozen behaviour. If you ever visit my office in the Redpath Museum, for example, you will climb a rather grand butternut staircase and find, about halfway up, a slab of sandstone that Seilacher once examined. It was deposited in a quiet sea, probably around the low tide mark, towards the end of the Cambrian, long after the time of the Avalon rangeomorphs. If you look carefully you will see that although there are no real fossils—shells or bones—the slab does bear wide shallow grooves, much narrower grooves and lines of pockmarks, which Seilacher used to work out what had happened on this patch of sand some 450 million years ago. The wide shallow grooves have a ribbed appearance and are bordered by a slight mound. These are the trails of two large molluscs, moving on a broad muscular foot, that appeared early on the scene. The trails are crossed by a more complex pattern of narrow, more or less continuous grooves bordered by sharp dints, often in pairs. These are the walking traces of two or perhaps three trilobites: the groove is made by their tail dragging in the sand, and the pockmarks by their legs, which are usually paired structures consisting of an upper gill and a lower walking part. They arrived later, because their footprints are not smeared out by the molluscs, and probably avoided the mollusc trails because they turned away when they met them. You can tell (well, Seilacher could tell) which direction they took because the tail wags slightly, like a trailer behind a car. It is rather like a detective reconstructing a crime scene from the evidence left by tyre tracks and footprints.

Rangeomorphs are not trace fossils, of course, but body fossils, the casts or impressions of the organism itself. Seilacher used his detective skills, honed on trace fossils, to interpret them as quilted organisms consisting of a series of compartments separated by lines of tissue, rather as the panels of a down jacket are separated by lines of sewing. This might explain how a protist could grow into something the size of a rangeomorph frond, but nothing living—nor anything else extinct, as far as we know—has this sort of body plan. It follows that rangeomorphs were some quite different kind of organism, which Seilacher called the Vendozoa, which flourished for some 20 or 30 million years before disappearing completely. The nearest living organisms might be the xenos of the deep sea, although whether xenos are the descendants of rangeomorphs or just resemble them in some ways is debatable. Xenos consist of a branching filament enclosed in an organic tube and can grow to be quite large, forming a ball or fan the size of a hand. They are quite abundant; on the floor of the deep sea they can make up 50% or more of the biomass. Komokiacea ('komokis') also live in the deep sea and have an even simpler structure of branching tubes about a millimetre wide that usually clump that traps sediment to form a mudball. Komokis are easily broken up by most kinds of sampling gear and are also quite abundant; a large fraction of the deep ocean floor may be made up of a skin of komoki tubes.

## Jellyfish and sea pens

Most of the people who have spent their careers puzzling over the Ediacaran fossils are convinced that they are not inorganic structures, fungi or seaweeds, and are not impressed by the notion that they might be gigantic protists. Instead they have concluded that Reg Sprigg and Martin Glaessner were basically correct in viewing them as animals. Most of their original attributions of particular fossils to modern groups such as Scyphozoa (jellyfish) and Pennatulacea (sea pens) have been abandoned, but their central thesis, that we are looking at the earliest known ancestors of animals, has become firmly rooted. Indeed, the most recent book-length survey of all the main localities where these fossils have been found, co-authored by five very experienced palaeontologists (including Guy Narbonne), is uncompromisingly titled *The Rise of Animals*, and its subtitle, *Evolution and Diversification of the*

*Kingdom Animalia*, is just as explicit. Perhaps the case really is closed, but some niggling doubts remain.

Whatever they were, rangeomorphs were quite abundant for several million years, and must have had some way of making a living. Perhaps we should be thinking about them in more ecological terms, asking not which group they belonged to but rather how they lived and in particular what they ate.

## What did the fronds eat?

Some people have suggested that rangeomorphs simply absorbed nutrients from the surrounding seawater. Whether seawater was nutritious enough to support such large organisms is questionable, especially after it had been screened by bacteria. At the surface of the ocean, the concentration of nutrients such as nitrate and phosphate often falls below the limit of detection. A kilometre down, their concentration is much greater, but is still only 50–200 nM. (A nanomole, nM, is one-billionth of the molecular weight of a substance in a litre; the molecular weight of nitrate is 62 g. Hence, 100 nM nitrate = 0.0062 ppm.) In any case, rangeomorphs are the wrong shape. For a given mass of tissue, the efficiency of absorption is governed by the total surface area exposed to the medium, which in turn depends on the dimensionality of the body. The ideal shape is a point with zero dimension—a cloud of isolated unicells, each with their whole surface exposed, like bacteria or yeast. The next best is a one-dimensional filament, which loses surface area because the ends of the cells are joined together. In a two-dimensional network, such as a fungal mycelium, there is likely to be interference between nearby strands, and a two-dimensional sheet is still less efficient because it loses surface area where the sides of the cells are joined. Worst of all is a three-dimensional body in which cells at the surface have only one side exposed, while interior cells are not exposed at all. Large three-dimensional rangeomorphs would not be able to compete for dissolved nutrients with bacteria which had had a couple of billion years to sharpen their game.

Some rangeomorphs at least grew upright, because they were all splayed on the surface of the sediment in the same direction by the ash fall. Why grow upright? One reason might be to grab the sunlight first and shade out any competitors underneath, like trees in a forest, but if they were not photosynthetic this does not work. They might be avoiding benthic predators, but at the time they flourished nothing capable of eating a rangeomorph had yet appeared. The remaining possibility is that they grew tall to exploit a new food source, the bacteria and plankton suspended in the water column out of reach from organisms lying flat on the surface of the sediment. These small particles can be collected actively by using a pump, and an erect posture would have the bonus of preventing the pump from being clogged by sediment. A pump requires power, and since rangeomorphs lacked muscle the only source of power would be cilia, beating in unison to create a stream of water from which the edible particles could be harvested. This is a perfectly feasible way of life that is followed nowadays by animals as different as sponges and bryozoans. A pump alone is not enough, however: there must also be some device for harvesting the stream. Sponges use the collar of their choanocytes to filter out bacteria, while bryozoans march the plankton they capture towards a mouth. Rangeomorphs certainly lacked a mouth, but might have had collar cells, although no trace of them has yet been described. Nor can they have lined internal chambers, since there are none, and placing them on the outside seems inefficient because the available surface area becomes proportionately smaller as the body becomes larger. Altogether, active suspension feeding would not be an attractive way of life for a rangeomorph, even if it were possible.

Passive suspension feeders wait until their prey bumps into them before subduing it. There are many animals that make their living in this way. Cnidarians and ctenophores use armed tentacles to catch quite large prey: cnidarians such as hydroids, jellyfish, sea anemones and corals use lethal batteries of miniature harpoons while ctenophores capture their prey with a sort of glue gun. Some of these animals, such as ctenophores and most sea anemones, are quite large individuals, whereas others, such as most corals and hydroids, form extensive colonies consisting of hundreds of tiny zooids. It is tempting to conclude (and many have) that rangeomorphs were large colonial animals with a similar body plan and way of life to soft corals such as sea pens. The fossils have no trace of zooids, however, and the tubes of the rangeomorph body were in any case too narrow to accommodate them. Without tentacles, the only way to capture plankton is to weaponize the cytoplasm itself.

The cell body of many protozoans is protected by a thick sheath which provides protection but makes it awkward to feed. Forams and xenos solve the problem by perforating the solid wall with a multitude of small holes, each less than a micron across, through which long thin fingers of cytoplasm extend outwards from the surface of the cell into the water beyond. These filopodia are supported and extended or retracted by microtubules, the same sort of microtubules that provide the internal scaffolding of cells. This gives them great flexibility, so the coat of filopodia that surrounds a foram is not a static array of filaments but rather a dynamic network that is continually forming and re-forming, advancing and retreating to search for potential prey—diatoms, other single-celled plankton or even small animals such as copepods. Once the filopodia have recognized an object as being edible they are able to consume it, although I don't know how they do this; they seem to be able to adhere to the surface of their prey, penetrate the cytoplasm and digest it. Their main limitation is that they are very short, only a few tens of microns in length. This restricts them to the boundary layer, the thin skin of water covering the surface of a pebble or the blade of a seaweed (or the surface of a rangeomorph) where the current is slowed by friction. The thickness of the boundary layer depends on the speed of the current, but if the current is weak (less than 1 cm per second, say), as it usually is in the deep sea, it will extend well beyond the reach of filopodia. Particles such as the diatoms that forams are fond of will accumulate in the quiet water of the boundary layer like leaves in an eddy, providing a continually renewed source of food.

The innocent-looking surface of a large foram or xeno is a danger zone for algae and protozoans, which tend to tarry in the boundary layer where they can be attacked by filopodia. Conversely, an upright frond that catches the current is an effective feeding structure for any foram or xeno which is able to grow tall enough. It would be just as effective for rangeomorphs, if they were large foram-like creatures bearing filopodia. The point cannot be proven, because the filopodia themselves are evanescent and the pores they protrude through are far below the limit of resolution for any impression in mud or ash. Still, this kind of passive suspension feeding would be a credible way of life for the lawn of fronds that grew in the mud of Avalon more than half a billion years ago.

## Anhui Province, East China, about 600 Mya

The black shales of the Lantian Formation outcrop in Anhui Province in the eastern part of the Yangtze Block, about 200 km inland from Shanghai. They were laid down about 600 Mya as the sediment of a shallow sea about 50–200 m deep, and preserve striking fossils of quite large multicellular organisms as carbonaceous compressions, quite different from the sandstone casts and moulds of Avalon and the Flinders Ranges.

This was truly the Age of Seaweeds. About a dozen species of seaweed have been described from the shales, growing a few centimetres above the surface of the seabed. They seem to be bona fide seaweeds which have branching or fan-shaped bodies without organs or apertures. It has been suggested, however, that a few of these fossils might be animals. The most likely candidate for animal status is *Lantianella*, preserved as a flat triangular compression that was probably a cone-shaped body in life, supported by filaments running the length of the body and surmounted by crown of about a dozen straps or ribbons (Figure 8.11). It grew several centimetres tall, anchored to the seabed by a small holdfast. The ribbons on top look something like the ring of tentacles surrounding the mouth of a solitary cnidarian polyp, but there is no sign of a mouth or any other diagnostic feature, and despite extravagant claims its status as an animal seems dubious at best. *Lantianella* looks like a seaweed, it lived in a community of seaweeds, and it is reasonable to conclude that it *was* a seaweed.

One of the most striking features of the Lantian biota is the absence of any of the fronds, discs, seals, pit-and-ridge creatures or bags that dominate the Ediacaran biota of Avalon and elsewhere. There is no doubt that they would have been fossilized along with the seaweeds if they had been present. It is always possible, of course, that they were unable to grow here for some reason but flourished elsewhere. Until that has been demonstrated by fossil evidence, however, we have upper and lower bounds for the characteristic Ediacaran biota: it was not present in the Lantian, about 600 Mya, but had appeared at the latest by the oldest fossils at Mistaken Point, about 580 Mya. It is during these 20 My that the Ediacarans must have evolved and spread.

## The hiatus

### The worldwide PreCambrian fauna

As the discoveries mounted, it became clear that the sea floor had been occupied by a distinctive community of large organisms long before the radiation of modern phyla during the early Cambrian. Similar fossils were found at sites all over the world, not

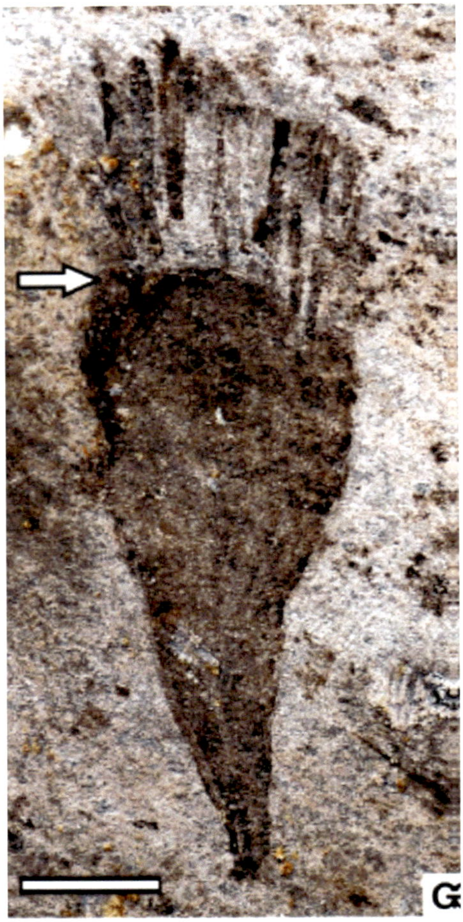


**Figure 8.11** Lantianella. Scale bar is 5 mm

Reproduced with permission from Wan, B., Yuan, X., Chen, Z., Guan, C., Pang, K., Tang, Q. and Xiao, S., 2016. Systematic description of putative animal fossils from the early Ediacaran Lantian Formation of South China. Palaeontology, *59*(4), pp.515–532.

just the few localities mentioned here. They have turned up in western Ukraine, the valleys of the Ural Mountains, the Mackenzie Mountains of north-western Canada, along the Olenek River in northern Siberia, and in many other places. The Dengying Formation in the Yangtze Gorges region of central China has a typical collection of fossils near its base—the Shibantan biota, including *Charnia*, *Pteridium*, *Rangea*, *Aspidella* and so forth—which thin out upwards until they are replaced by *Cloudina* near the top, next to the overlying Cambrian deposits.

Over large areas, the seabed itself was sealed by a thick, tough microbial mat consisting of the glued and felted sheets and strands of bacteria, perhaps intermingled with fungi and algae. Embedded in this mat were discs or seal-shaped organisms like

*Aspidella*. Some of these were the holdfasts of frond-like organisms like *Charnia* and *Rangea*, which formed a weedy sort of lawn on the seabed. Lying free on the surface of the mat were other fronds, such as *Fractofusus*, together with long, apparently segmented organisms like *Dickinsonia* and *Spriggina*. Most communities featured small mounds with a more or less elaborate series of ridges such as *Parvancorina* and *Tribrachidium*. In some, there were bag-like organisms such as *Ernietta*. All were sedentary and probably completely sessile, although some might have been feebly motile. They lived at all depths, from shallow sunlit seas to deep dark water. They spread throughout the world and have been found in dozens of different localities. They were not a short-lived phase, either: essentially similar kinds of organism flourished for at least 35 My before the Cambrian period began.

## The Ediacaran Period

Preston Cloud and Martin Glaessner got together in 1982 to argue that the late PreCambrian fossils were so widespread and distinctive that the period when they occurred deserved a special name. In fact, it already had one, the Vendian, which had been proposed by the Russian geologist Boris Sokolov to cover the period from the beginning of the Marinoan glaciation to the base of the Cambrian. 'Vendian' was even coloured separately on Soviet geological maps for a time, but it never received the imprimatur of the International Commission on Stratigraphy. Instead, the Commission, inevitably, set up a Working Group on the Terminal Proterozoic Period in 1989, which visited all the sites put forward for the honour of hosting the period, including the White Sea coast and the Yangtze Gorges. When the final ballot was held 15 years later, the completely unsurprising choice was a site in the Flinders Ranges, where the classical fossils had been discovered. It is on Enorama Creek, where the gold spike at the base of the cap carbonate now marks the beginning of the Ediacaran Period. It is placed at the end rather than the beginning of the ice age, but otherwise Cloud and Glaessner are in agreement with Sokolov, so if you continue to call this period the Vendian there is little danger of confusion.

## The end of the Ediacarans

The radiation of animals in the early Cambrian tore up the old PreCambrian script that had directed earth systems for the last 2 billion years and replaced it with a new and more complicated set of rules. Large muscular worms could burrow through the microbial mat deep into the sediment beneath. The simplest burrows are called *Skolithos*; they are unbranched vertical burrows which are probably made by a variety of worms, since they are found throughout the Phanerozoic, from the early Cambrian to the present day. There were many variations on this theme: *Treptichnus*, for example, made a more complicated forked burrow by continually probing from side to side.

Once large motile animals had evolved, the microbial mat was doomed. Whether it covered a large proportion of the sea floor or only a few quiet areas, it would soon be cleared away by burrowing worms and scraping arthropods. Dolf Seilacher called this the 'agronomic revolution'. Burrowing broke up the microbial mat and in doing so stirred and aerated the sediment. When arthropods with biting mouthparts appeared they could shred the remains of the mat. The seabed that had previously been covered with a thin tough biofilm became a uniform surface of naked churned mud. Any sessile creature now had to have a substantial holdfast rather than simply anchoring itself to the mat. Motile animals such as priapulids and chaetognaths could also eat smaller animals, setting up longer food chains and more complex food webs than ever before. The new animal body plans that had evolved in the early Cambrian had radically rewritten the ground rules of marine ecology that had lasted for the previous 2 billion years.

The immediate consequence was the demise of the first fauna of fronds and discs, which depended on the mat and disappeared with it. Perhaps there was also a community of animals specialized for life in or on the mat which became extinct at the same time, although no certain trace of them has been found. The animals that flourished would have been those that were at home on the new sea floor as it emerged at the beginning of the Cambrian.

The classical Ediacaran biota lasted from the Gaskiers glaciation to the opening of the Cambrian. The earliest communities of large Ediacaran organisms yet found are in south-eastern Newfoundland; those at Mistaken Point itself are about 565 My old, while the less spectacular fossils found at Pigeon Cove, at the north-western edge of the Reserve, are somewhat older and date from about 578 Mya. The latest are the fronds and bags found in southern Namibia, which flourished in the very latest Ediacaran; those from Farm Swartpunt are 545–542 My old. They are accompanied by plugs of sediment that may be first traces of burrowing worms. The Cambrian period proper, when vertical burrows become abundant, began only a few metres above, but in that brief interval of time the distinctive Ediacaran biota had vanished completely and forever.

It may have been a little less dramatic than that. A few fronds of *Swartpuntia* have been found just above the base of the Cambrian in South Australia and California, for example. The entire Namibian community is markedly less diverse than the rich fauna of the Flinders Ranges and the White Sea coast, perhaps because many types had already become extinct. There are also a few species in Cambrian communities that might have evolved from Ediacaran precursors, since they have a superficial resemblance to these earlier forms. For example, *Thaumaptilon* is a large frond-like form from the Burgess Shale which resembles a sea-pen—or a rangeomorph. It has a long, rather carrot-shaped holdfast bearing a blade with a prominent midline and 20–40 slightly curved branches arranged in pairs opposite one another. The branches have tiny bumps that Simon Conway Morris, who first described the fossil, interpreted as retracted zooids. *Stromatoveris* is a smaller frond-like form found at Chengjiang. It has a stalk that is prolonged into a prominent midline from which about a dozen narrow branches fan out on either side. It has been likened to a rangeomorph, but there is no indication of secondary branching; nor is there any sign of zooids, so it was not a sea-pen either. What kind of animals *Thaumaptilon* and *Stromatoveris* were remains a mystery, but I do not think that they can be convincingly likened to any member of the Ediacaran fauna.

Although a few survivors might have lingered on in some obscure or protected localities, the distinctive Ediacaran community of fronds, discs, buttons and bags that had populated the sea floor for 35 My or more had been swept away in a geological instant, too quickly for the event to register in the rocks. At one horizon there is the typical Ediacaran assemblage; only a little higher it has been replaced by worm burrows and small shellies, with rarely any hint of a gradual transition between the two (Figure 8.12). There were to be several mass extinctions during the Phanerozoic, but perhaps none as complete or as surprising as the mass extinction of the Ediacaran biota that accompanied the initial radiation of animals.

## The cause of the hiatus

What caused the great disappearance at the end of the Ediacaran? The turmoil of the Cryogenian and Ediacaran was over, and the Cambrian was a quieter period, so it seems unlikely to have been caused by purely physical disturbances. There are certainly no signs of a catastrophe; no ash layer deposited by volcanic eruptions nor an iridium layer to reveal the impact of an asteroid. The mudstones at Fortune Head and elsewhere continue smoothly upwards with no break to mark the transition from Ediacaran to Cambrian.

On the other hand, animals were clearly diversifying, becoming larger and beginning to evolve mineralized skeletons at the Ediacaran-Cambrian boundary, and they have generally been held responsible. The combination of geological uniformity with biological revolution is unique in Earth history.

We can rule out the possibility that animals simply ate the Ediacarans, because there is no evidence of bite-marks or other signs of damage. Anyway, the jaws that would enable animals to bite had not yet evolved.

Animals may have been more efficient suspension feeders than the Ediacarans. Filopodia that extend 10–30 μm are effective feeding devices for a single cell feeding on other single-celled organisms, but large multicellular creatures can use them only to capture cells that have approached very close to its surface. They are like flypaper, passively capturing anything that strays close enough. Sponges are much more efficient because they have a pump which creates a stream of water which they can filter the bacteria. Ciliated-tentacle suspension feeders such as bryozoans and brachiopods also have a pump, which they use to direct a stream of water towards their mouth, then swallowing the algae it contains. Once a cilia-powered pump had evolved, the Ediacarans could not compete effectively either for bacteria or for algae, and would eventually be replaced by animals. It might be thought that the process of competitive replacement would take a long time, with many Ediacaran types lingering on deep into the Cambrian. but this underestimates the speed of exponential processes. Suppose that possessing a pump gives animals a modest 1% advantage over Ediacarans. If we start out with a billion Ediacarans and only a single animal, these numbers will be reversed—a billion animals and only a single Ediacaran—just 4,166

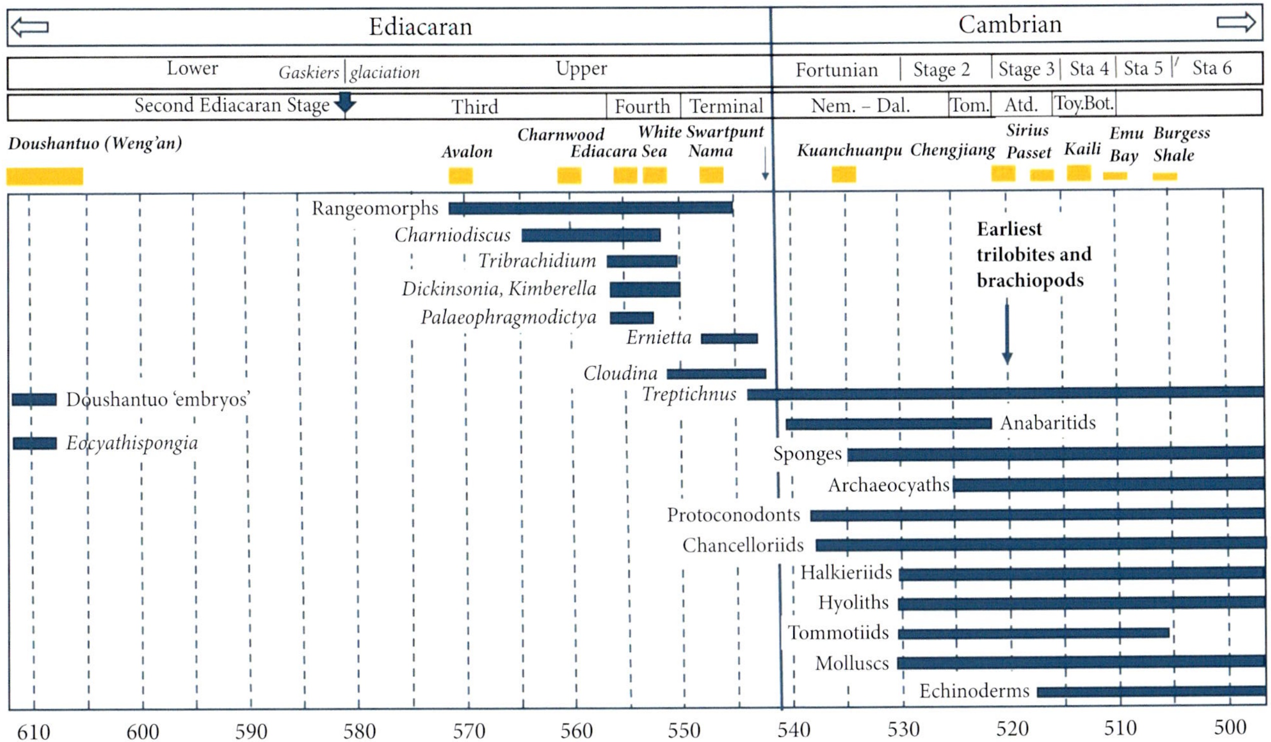


**Figure 8.12** The hiatus

generations later. Even for long-lived animals, this is too short an interval of time to register in the rocks.

The remaining possibility is that the Ediacaran biota was physically rooted in, and completely dependent on, the microbial mat that covered much of the sea floor at all depths. Once this was broken up by burrowing worms such as *Skolithos* and *Treptichnus* the fronds and discs were unable to survive. The force required to penetrate a biofilm no doubt depends on its composition, but the early Cambrian worms made burrows up to a centimetre across, and must have been powerful animals. They would also have needed something to push against; modern worms thicken the hind end of their body, and perhaps their distant ancestors did something similar. I am still not quite sure how a worm could use muscular force to break a microbial mat overlying soft sediment—if you lay a sheet of clingfilm on top of a bowl of porridge, you can apply a great deal of pressure with the tip of a finger without producing more than an indentation. Perhaps there were flaws in the mat, weak spots caused by local dissolution or the grazing of *Kimberella*. However it was managed, the ancient microbial mat had been almost completely cleared as the Cambrian got under way, and the elephant-skin texture it left behind on the surface of the rock was no longer seen.

The new sea floor, cleared of potential competitors and with its covering bulldozed away, was then available for the expansion and radiation of animals. The hiatus marking the transition from Ediacaran to Cambrian was a sea change in biology; a Black Swan event that set evolution on a new trajectory.

## The missing sponges

The seafloor communities of late Ediacaran times are not at all what might have been expected. For one thing, there are few if any sponges. The classical view is that the earliest animal communities consisted of sponges, or rather the ancestors of sponges. Sponges certainly branched off early and fossilize readily (many flint nodules were initially formed by siliceous sponges). Hence, sponge fossils should dominate the earliest stages of animal evolution. Perhaps they do, but the evidence is far from conclusive. There is a recent report of a large stalked organism found in the late Ediacaran of the Dengying Formation which looks something like a sponge. The surface of its body is divided into a nested set of rectangles, which is not at all sponge-like, but it seems to be the best candidate of about the right size and date. It is a unique specimen, however, and there are few other indubitable sponge fossils in the Ediacaran.

Indeed, a group of three very well-qualified British biologists, Jonathan Antcliffe and Marting Brasier of Oxford and Richard Callow of Aberdeen, concluded that there is no convincing evidence for fossil sponges before the Cambrian. Small beads a few millimetres long from Cryogenian deposits in Namibia some 760 My old were supposed to be fossils of the sponge *Otavia*, because they are pitted with small holes which were taken to be pores. But they might be no more interesting than battered grains of sediment. There are high concentrations of a substance produced mainly by sponges (24-isopropyl cholestane, if you're a biochemist) in rocks of Cryogenian or early Ediacaran age, which might indicate the hidden presence of a large community of sponges. But the same substance is produced by some seaweeds, so the evidence is inconclusive. Simple siliceous sponge spicules (linear and pointed at both ends) have been reported from north-western Canada that may date from as early as the Marinoan glaciation, and others have been found in rocks of similar age in South China. Silica can form long narrow crystals by purely physical processes, however, so these structures do not necessarily show that sponges were around so early. I was shown a fossil shaped like a narrow isosceles triangle at Mistaken Point that was said to be the remains of a small sponge called *Thectardis*. But *Thectardis* lacks any definite sponge features, or for that matter features of any kind, so 'sponge' is little more than a guess founded on the assumption that there must have been sponges in the Ediacaran. Some bag-shaped fossils from Russia (*Fedomia*) and Namibia (*Ausia*) might be sponges, but they too have no unmistakeable sponge characteristics. Even *Palaeophragmodictya*, confidently identified as a sponge from the Flinders Ranges, might be simply the attachment disc of an unknown frond. Antcliffe and his co-authors conclude that the oldest fossils that can be confidently identified as coming from a sponge are some star-shaped spicules from the lowermost Cambrian of Iran and Siberia (and of course the abundant body fossils of Cambrian archaeocyaths). Some might find this scepticism exaggerated, but it would be impossible to direct a comparable criticism against any phylum from the Cambrian on.

## Conulariids

There are not many polyps, either, although many Ediacaran fossils were originally said to be medusae and a few have been interpreted as polyps. Much depends on how you place the square, distinctively ribbed tubes of conulariids. These have been associated with a wide range of phyla from tubeworms to molluscs, but the prevailing opinion (as I mentioned in connection with *Olivooides* at Kuanchuanpu) is that they housed the polyps of scyphozoans, probably coronate jellyfish. Coronate polyps live attached to hard surfaces (such as pebbles or fragments of bivalve shell) in the deep sea. They build chitinous tubes several centimetres long which are circular in cross-section, so the whole structure is a long narrow cone. The polyp is predatory; its broad open end bears a crown of many narrow solid tentacles, which capture prey and drag it into the mouth. The tube has prominent rings at intervals of about 0.1 mm and also less prominent longitudinal ridges; there are also tooth-like projections at intervals along the inside of the tube which are handy for identifying species but whose function is unknown. The polyp gives rise to medusae by a process called 'strobilation', which is found only in scyphozoans: the polyp is divided transversely by fission zones into a stack of discs, each of which develops into a miniature medusa, the ephyra, which is eventually liberated, swims away and grows up into the familiar jellyfish.

Conulariids are similar in some respects, for example the prominent thickened rings along the tube. It is conceivable that the ringed tube evolved independently in coronates and conulariids because it gives flexibility. A corrugated plate is stiff in bending in the direction of the corrugations but flexible in the perpendicular direction, which is why corrugated iron rather than sheet metal is preferred for roofing.

A corrugated tube, with the corrugations arranged transversely, can bend in a current which is strong enough to topple a rigid tube. This sounds reasonable for a coronate polyp, but whether it applies to conulariids, which could bend in only four directions (because of their square cross-section), and were at least partly mineralized anyway, is less clear.

The alternative is that the ringed tube implies common ancestry, so that coronates have evolved from conulariids by altering the shape of their tube from a pyramid to a cone and losing mineralization. The conulariid tube is mineralized (with calcium phosphate), has low incomplete partitions but lacks the 'teeth' of coronate tubes, and is square in cross-section, so that, rather than a cone, it looks like an improbably stretched pyramid. The narrow end of the tube bears a small attachment disc, and the broad end is partly closed by mineralized flaps continuous with the wall of the tube. If tentacles or, better still, strobilation were found there would be no doubt what conulariids were; but no tentacles have yet been described (and it is not clear how they would protrude past the flaps), while the evidence for strobilation rests on a single doubtful specimen.

One remaining question is when they first appeared. The oldest reputed fossil is *Vendoconularia*, from the upper Ediacaran of the White Sea coast, but only a single specimen is known, it has a bag-shaped body, and the stratum in which it was found has not yet been dated securely. There are similar fossils of about the same age or a little later from Russia and Namibia which have a similar appearance and may be non-mineralized conulariids, or something related to or even ancestral to conulariids.

*Corumbella* is a different story. Many specimens of this unusual animal have been reported from the cement quarries near Corumbá, in the Mato Grosso do Sul of southern Brazil, near the border with Paraguay. The fossils are found alongside *Cloudina* in shales of the Tamengo Formation, which have been dated to about 543 Mya, so were laid down at about the same time as the Nama Formation in Namibia. Similar fossils have been found in Paraguay and Nevada. *Corumbella* is a calcareous but flexible tube more or less square in cross-section, formed of four L-shaped bars that alternate along the midline of each face, forming prominent rings. Each bar consists of two sclerites which grip the base of the bar above to form a flexible joint. The row of bars produces a square cross-section, but apart from that there is little resemblance to a conulariid, and how *Corumbella* was related to other groups remains a puzzle.

A variety of conulariid-like fossils have been found in early Cambrian deposits, including the Punctatus developmental stage of *Olivooides* from Kuanchuanpu. They became more abundant in the succeeding Ordovician period, but were never very diverse, and eventually became extinct near the close of the Triassic period, about 200 Mya. For a long-lived and fairly abundant group of fairly large animals the conulariids remain annoyingly mysterious. The prevailing opinion is that conulariids and coronates are closely related, but it is only if strobilation can be clearly demonstrated, preferably in several specimens, that we will know for sure. Until it is, other ideas remain possible, such as Dolf Seilacher's notion that conulariids and scyphozoans are separate natural groups that share a common ancestor. They might even be a whole extinct phylum, as others have suggested. Granted that almost everyone thinks that the conulariid tube was built by a polyp, it would still be reassuring to have a few specimens with tentacles, but these are lacking even in the best-preserved fossils.

## The missing polyps

Indeed, PreCambrian body fossils of polyps are seldom found. The star of the Mistaken Point display at The Rooms, for example, is a very strange fossil called *Haootia*, which is imaginatively reconstructed (by a group which included Martin Brasier, who was so dismissive of early sponge fossils) as a stalked jellyfish (Figure 8.13). These can still be found in modern seas, although they are perhaps the least-known class of cnidarians. Prominent strands or filaments which drape the fossil were interpreted as muscle fibres. There is also a recent report of a polyp from Charnwood, happily preserved on a slab that also bears *Charnia* and *Bradgatia*, so it can be directly compared with rangeomorph fronds. The specimen, named *Auroralumina*, is presumably part of a colony because it consists of two cuplike structures connected by a branching stolon (Figure 8.14). The cups are encased in some resistant material and are crowned by numerous strap-like tentacles. A possible medusa named *Bjarmia* has been described from the White Sea region. It is a bowl-shaped structure with ridges on its rim that might be tentacles. *Haootia, Auroralumina* and *Bjarmia* are all large (with dimensions in centimetres) and all are worryingly rare—*Auroralumina* and *Bjarmia* are unique and only two specimens of *Haootia* have been found so far.

Whether or not the tentacles of these animals could catch large prey is not known. All living cnidarians use stinging cells, the cnidocytes, to capture large living prey. The cnidocyte is a very complicated affair, perhaps the most complicated of all animal cells. Its basic machinery consists of a capsule surrounding a central chamber in which is stored a long coiled filament tipped with a sharp, poisoned harpoon. The cell also has a detachable lid and a tuft of cilia that acts as a trigger. The trigger is pulled when a small animal such as a copepod or a cladoceran brushes past (or when you accidentally kick a stranded jellyfish while walking along the beach). This releases a flow of

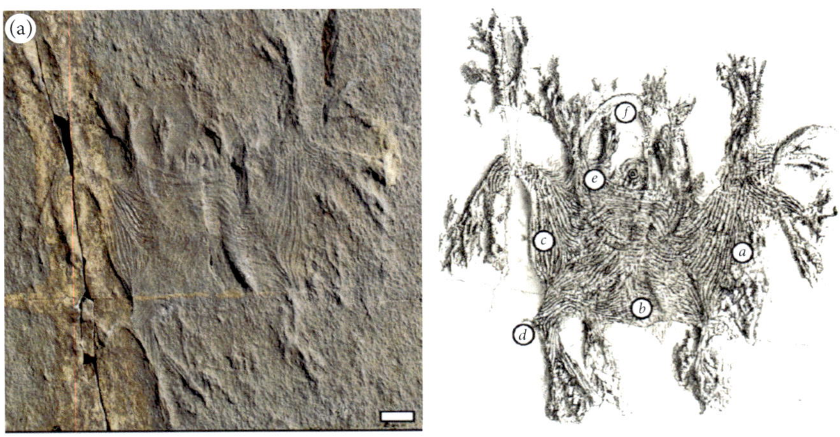


**Figure 8.13** Haootia. Scale bar is 10 mm

Reproduced from Liu, A.G., Matthews, J.J., Menon, L.R., McIlroy, D. and Brasier, M.D., 2014. *Haootia quadriformis* n. gen., n. sp., interpreted as a muscular cnidarian impression from the Late Ediacaran period (approx. 560 Ma). *Proceedings of the Royal Society B: Biological Sciences, 281*(1793), p.20141202.

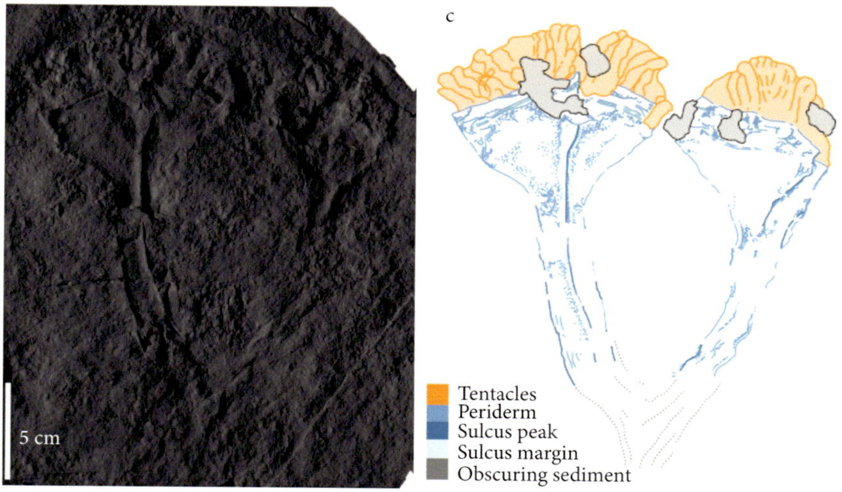


**Figure 8.14** Auroralumina

Reproduced with permission from Dunn, F.S., Kenchington, C.G., Parry, L.A., Clark, J.W., Kendall, R.S. and Wilby, P.R., 2022. A crown-group cnidarian from the Ediacaran of Charnwood Forest, UK. *Nature Ecology & Evolution*, 6(8), pp.1095–1104.

calcium ions from the capsule into the central chamber, causing an osmotic gradient across the cell wall and thereby a rush of water into the cytoplasm. This raises the internal pressure of the cell to such an extent that the lid flips open and the harpoon, tethered by its filament, is discharged at high velocity. The bad news for the copepod is that it will be quickly killed when the harpoons from a battery of cnidocytes blast through its chitinous armour and flood its tissues with nerve agents. The bad news for zoologists is that it is difficult to understand how such a complicated device could have evolved gradually from simpler but functional antecedents. One promising idea is that the cnidocyte evolved first as an anchoring device that had nothing to do with prey capture. The harpoons were shot into the substrate, perhaps a microbial mat, and served to hold the polyp in place. It was only later that it was weaponized to capture prey—not copepods, which had not yet evolved, but small worms, perhaps, wandering about in the mud or through the mat. This notion requires that anchorage cnidocytes evolved in a single lineage only (otherwise several kinds of animal would have cnidocytes), that only a single descendant of this lineage weaponized the cell (otherwise there would still be cnidocytes still used for peaceful purposes) and that the descendants of this lineage replaced all their competitors and have never looked back (otherwise there would be cnidarians that had no cnidocytes). There is nothing implausible about this argument, which is just one example of the difficulty of explaining how complex specialized cells evolve when no intermediates exist.

The formidable team of Leo Buss and Dolf Seilacher were also puzzled by the origin of cnidocytes, but came up with a completely different explanation. Buss and Seilacher suggested that they were originally acquired from microsporidians, a group of intracellular parasites related to fungi, which penetrate a host cell with a structure that resembles the working part of a cnidocyte. Whatever the merits of this scheme (which involves the translocation of the relevant battery of genes from the parasite to its host), it clearly implies that the first polyps lacked cnidocytes. Perhaps *Ernietta* was a polyp that lacked both tentacles and cnidocytes, and simply collected unicellular algae by gravitational settlement in its bag-shaped body.

The absence or rarity of sponges and polyps underlines one of the chief features of the first fauna, that it is entirely unexpected. A PreCambrian fossil record consisting mostly of sponges and polyps would be consistent with comparative morphology and gene-based phylogenetics. It would have welded fossils, body plans and genomes into a unified account of animal origins and early evolution. Instead, we have a very odd assemblage of organisms that palaeontologists have struggled to relate to any living groups. Instead of solving the mystery, the fossils that were eventually found only added to it.

# 9
# Sea Changes

There was a biological sea change at the boundary between the Ediacaran and the Cambrian, as one assemblage gave way to another. It is conceivable that this was the outcome of a purely internal dynamic, the inevitable and more or less predictable unfolding of a scheme already prefigured in the genome. I shall explore this possibility later, in the section about orthogenesis. The other option is that there was a literal sea change during the late Ediacaran, a change in the physical characteristics of the ocean which provoked the rise of the Cambrian fauna and the demise of the Ediacarans.

There was indeed a profound sea change during the Ediacaran. The previous billion years seem to have been a long calm stretch of time, so calm that the period has been called 'the Boring Billion'. The ocean was full of living organisms, but they were all single cells except for a few delicate threads like *Bangiomorpha*. There were no massive multicellular creatures and certainly no large animals. There is little evidence of major geological events such as mountain building or ice ages. There was not even much movement of tectonic plates: most of the land surface was fused into the single huge continent of Rodinia and had been for hundreds of millions of years. It was about 800 Mya, during the NeoProterozoic, that the world began to change.

Rodinia began to break up, pried apart by volcanic eruptions, as magma, forced up from the mantle, found lines of weakness in the old continent. There is still disagreement about the details, but by the early Cambrian, some 200 My later, Rodinia had split apart into the four continents I've already mentioned (Gondwana, Laurentia, Siberia and Baltica) plus some other bits and pieces. These fragments began drifting, and occasionally breaking up or colliding, in the new, tectonically active, post-Cambrian world before reaching their current configuration. One consequence of all this activity was the expansion of shelf seas, the shallow water fringing each of the new landmasses that would later be occupied by animals.

## Ice ages and impacts

Not yet, though. The break-up of Rodinia was followed by two prolonged ice ages: the Sturtian at about 710 Mya, and the Marinoan at about 635 Mya. Together the Sturtian and the Marinoan span the Cryogenian part of the NeoProterozoic, the cold snap immediately before the Ediacaran. Both seem to have lasted about 15 My and left behind clear signals of glaciation. After these two events there was a third glaciation, the Gaskiers glaciation, about 580 Mya. It was much shorter than the previous two, lasting only about 300,000 years, and it was probably not worldwide.

*The Sea-Change*. Graham Bell, Oxford University Press. © Graham Bell (2026).
DOI: 10.1093/9780197797464.003.0011

just before Ediacaran times. Just how much more oxygenated the sea became is even more difficult to say, but most guesses are about 10%, or perhaps a little more, of the current value.

The balance of the evidence, then, suggests that during the Ediacaran the ocean shifted from below to above the critical oxygen concentration defined by the Pasteur point. The Pasteur point is the oxygen concentration above which metabolism switches from anaerobic fermentation to aerobic respiration. It corresponds very roughly to about 1% of the current oxygen partial pressure of the atmosphere. At this level, diffusion alone can provide enough oxygen to support aerobic metabolism—provided that the animal is no more than about 100 μm thick. The earliest animals, before the evolution of specialized respiratory structures, must have been very small.

The ocean appears to have been poorly oxygenated at the beginning of the Ediacaran, and most organisms lived below the Pasteur point. During the Ediacaran the oxygen concentration rose, until by the opening of the Cambrian most organisms were living well above the Pasteur point, and could breathe oxygen if they were small enough.

## Mineralization

Other features of ocean chemistry were changing at the same time. In particular, there seems to have been an unprecedented increase in phosphorus concentrations, which has left its mark in the huge phosphorite deposits now mined in China and elsewhere. Phosphorite is normally formed when cold water rich in phosphate is brought up from the deep sea to the surface, where the phosphate precipitates out as grains of fluorapatite or some similar mineral and sinks back to the deep sea floor. Phosphate-rich rock has been formed throughout Earth history; in North America there are the Permian deposits of Wyoming and Idaho, for example, or the Miocene deposits of Florida. The Ediacaran and early Cambrian, however, stand out as a period in which massive amounts of phosphorus must have suddenly become available in the ocean.

Calcium concentration in the ocean was changing at the same time. The ratio of magnesium to calcium decreased sharply between the late Ediacaran and early Cambrian, suggesting an increase in calcium concentration. The best evidence comes from marine cements, which show a complicated transition from dolomite (calcium magnesium carbonate) in the early Ediacaran to aragonite (a crystal form of calcium carbonate) in the late Ediacaran to calcite (a different crystal form of calcium carbonate) in the Cambrian. This shift would have facilitated biomineralization and the synthesis of hard shells by animals during the early Cambrian.

## Acritarchs

With an ice age and an asteroid impact, the NeoProterozoic must have been an exciting time to be around, particularly in South Australia, and we would expect to find equally strong biological signals as communities all over the world were altered or destroyed. There were as yet no dinosaurs to be killed off, but there were acritarchs and these might have been seriously affected. Acritarchs are the only eukaryotes with a fossil record throughout the NeoProterozoic, and they do indeed fluctuate strongly in diversity and morphology. How these signals should be interpreted, however, is far from clear.

The first problem is whether the Snowball Earth scene should be taken literally. The image of a frozen Earth, a ball of ice, encapsulates the idea. Beneath the ice, the ocean would be dark and stagnant; it would also be anoxic, because all the oxygen would have been used up by respiration, without being restored by photosynthesis. Acritarchs, if they really were phytoplankton, would have become extinct, or at best lingered on in isolated refugia. There would have been plenty of melt ponds on the surface of the ice sheet, as there are in Antarctica today. There would also have been dust wells, small cylinders of water formed when dust blown in from the land settles on the uneven surface of the ice sheet, heats up because it is dark, and melts a hole a foot or so deep into the ice. Dust wells can occupy several percent of the ice surface, so they are not merely curiosities. (They were first described by the Norwegian explorer Adolf Nordenskjöld, who complained that they were so common on the Greenland ice sheet that it was difficult to find a place to lay his sleeping-bag. He called them cryoconite holes and thought they were formed by cosmic dust.) It is an appealing notion (to a lab scientist like me) to view the holes as a vast array of culture tubes, continually renewed, housing a very long-term experiment in which almost anything, including the first animals, might have evolved. They do indeed support a diverse biota of cyanobacteria, green algae and even small animals such as bdelloid rotifers and tardigrades. One difficulty, though, is that lakes, ponds and wells of any size on the surface of an ice sheet are filled with freshwater from melted ice, and are colonized by freshwater organisms. These would have to adapt to seawater once the whole ice sheet had melted into the ocean. This is not necessarily an insuperable obstacle: one of my students, Josianne Lachapelle, took a strain of our standard lab *Chlamydomonas*, normally a soil alga which is poisoned by low concentrations of salt, and by gradually increasing the concentration of salt managed to evolve a strain capable of growing in seawater within a few hundred generations. Still, the notion that modern marine phytoplankton, not to mention animals, originally evolved in freshwater is an unexpected complication for which there is no evidence at all.

The diametrically opposite point of view is that Snowball Earth never existed. The geological evidence for equatorial glaciation is misleading and mostly represents sediments formed by the break-up of Rodinia. It is true that tillite and cap carbonates are more likely to engage geologists than the lay public, but I find the scratches and dropstones difficult to dismiss. So I am not sure that the opponents of Snowball Earth have made a strong case either.

There is a compromise, of course, the Slushball Earth scenario, in which there are prolonged glacial periods with ice covering much but by no means all of the ocean surface, leaving enough free water for the acritarchs (or their planktonic stages) to enjoy. After all, acritarchs did not become entirely extinct, although species diversity seems to have declined markedly during both Sturtian and Marinoan glaciations. Indeed, some individual species are recorded before, during and after both glaciations, which seems inconsistent with a long-lasting complete shell of ice covering the ocean. On the other hand, species of acritarch are distinguished solely by a few simple morphological features, which might have evolved independently in several lineages at different times. The persistence of acritarchs throughout the NeoProterozoic, however, seems difficult to reconcile with Snowball Earth, and I suspect that if a vote were taken most geologists would favour some version of the Slushball.

The morphology of acritarchs, although simple enough, was in constant flux during the NeoProterozoic. The very oldest acritarchs, dating back a billion years or more, have a smooth, more or less spherical envelope. What they looked like inside is unknown, but they may represent the settling-in period, so to speak, when eukaryotes evolved nuclei, linear chromosomes, cell compartments and regular cell division based on microtubules. They acquired endosymbiotic bacteria that were eventually domesticated as mitochondria and plastids. Later on, they evolved a sexual cycle of meiosis, the special division that produces haploid gametes from diploid germ-line cells, followed by pairwise gamete fusion, which restores diploidy in the zygote. Acritarchs are thought to have evolved all these features before the onset of the NeoProterozoic, although this is only a guess. They certainly became more diverse in appearance, with the originally smooth surface of the envelope erupting into spines, spikes, hairs and tufts of various shapes and sizes. What these were for is not entirely clear.

Long, fine hair-like processes might make a large cell sink more slowly through the sea, as a dandelion seed sinks slowly in air. A small spherical algal cell 10 μm in diameter will sink slowly anyway, taking four days to fall 1m through seawater (according to Stokes' Law, which you may dimly remember from elementary physics classes), and will readily remain in the photic zone near the surface of the ocean. A comparable cell 100 μm in diameter will sink 1m in about an hour and will reach the seabed a kilometre down in about a month. It will remain in the plankton near the surface only if it can become buoyant by adjusting its physiology or teasing out its surface into hairs or filaments. The largest acritarchs are about 1 mm in size and must be benthic organisms because they sink too quickly (1 m in 30–40 seconds) to remain near the surface regardless of how hairy they were.

A few stout spines will not have much effect on how fast the cell sinks, but might protect it from being eaten by something with a mouth—a mouth big enough to swallow the envelope without spines, but too small to swallow it with spines. This would imply the existence of animals at least a millimetre or two long in the middle Ediacaran, when large spiny acritarchs became abundant. More plausibly, the spines might have protected acritarchs because they were long enough to baffle the filopodia deployed on the surface of protists such as forams.

A rather simplistic view of acritarchs, then, is that they comprised two main groups. The first group included small plain and large hairy forms which were phytoplankton living in the photic zone. The species in the second group were large plain and spiny forms strewn on the seabed, perhaps resting stages of phytoplankton that would later split open to liberate many smaller cells that would rise or swim towards the surface of the sea. They certainly resemble the asexual cysts and sexual zygotes of modern algae with this sort of life cycle, alternating between an active photosynthetic stage and a larger dormant stage. There are quite a lot of acritarchs, however, which don't fit into such a simple scheme. In particular, some of the largest Ediacaran acritarchs are covered with short curly ribbons which do not obviously serve to float or protect the organism but do bear an astonishing resemblance to crustacean egg cases. If that is what they were, they must have been produced by early or mid-Ediacaran arthropods which have left no other trace of their existence.

## The Ediacaran upheaval

Clearly, there were momentous changes to the whole Earth system in the later part of the NeoProterozoic, in the 100 My or so leading up to the Cambrian. It is difficult to resist the temptation to arrange these changes in a causal chain that neatly links asteroid impacts, ice ages, ocean chemistry and the carbon cycle. It is even more difficult to resist speculating how these changes led to the fronds of Mistaken Point and the other enigmatic but unmistakeable remains of large creatures that appeared and flourished at about the same time. Hardly anyone who has written on the topic has been able to resist one or other of these temptations, and most have fallen for both. I shall, too, but I shall separate them, setting the stage first and dealing with the fossils later.

There are two ground rules that I'll use to discipline what might otherwise become unbridled guesswork. The first is to invoke only those processes that can be demonstrated to occur nowadays or that have left clear physical evidence (such as fossils) in the rocks. Hypothetical events (such as asteroid impacts in deep water) are not allowed. The second is to prefer fat signals, clear indications of worldwide change supported by several independent sources of evidence. A faint signal that might mark a purely regional event, or a beguiling but tenuous interpretation of a signal, will both be ignored.

### Glaciation and deglaciation

We can begin with the series of events that meet both strictures, the extensive and prolonged glacial periods of the Cryogenian. I have no idea why ice ages should suddenly appear after a billion years of mostly fine weather. Perhaps it was connected with the break-up of Rodinia, but even the experts seem at a loss. At all events, it has been demonstrated beyond reasonable doubt that there were two long-lived glaciations during the Cryogenian during which the land and most of the sea (how much is still a matter of debate) was covered with a thick layer of ice. It is equally certain that glaciation was followed by deglaciation. As the world warmed up the ice began to melt, forming cryoconite holes, summer ponds and eventually large lakes on the surface of the ice sheets. The glaciers, too, would begin to shrink. The meltwater does not flow over the surface of a glacier, but rather finds its own level, percolating through cracks and fissures in the ice, and flows beneath. The hydrodynamics of melting glaciers are still not fully understood (partly because the sole of a glacier is an awkward place to reach) but meltwater flow and discharge can be studied by intrepid geographers in Antarctica and Greenland today.

Glaciers first bulldoze away the soil and then grind over the underlying rock, eventually reducing it to a gritty paste of rock flour. At the height of an ice age, they have scooped out the U-shaped valleys characteristic of a glaciated landscape and cut deep trenches into the shoreline. During deglaciation, the meltwater flowing along their soles is injected, not at the surface of the sea, but rather several hundred metres beneath, onto or into the sediment of the continental shelf. The rock flour itself is rich in nutrients and so the meltwater will fertilize the ocean. More importantly, this meltwater is fresh and therefore slightly less dense than seawater, so it must rise to the surface, mobilizing a much larger volume of seawater and muddy sediment as it does so. The buoyancy of freshwater is the engine that drives upwelling.

The sediment was made up of the thin rain of dead phytoplankton sinking from the surface waters, together with the bacteria (and possibly fungi) metabolizing their remains. It would have been rich in light carbon, so when it was upwelled by deglaciation it would have left a pronounced negative $\delta^{13}C$ signature in carbonate rock. The Acraman impact was terrestrial and therefore did not contribute to the upwelling, nor is there any evidence for a large marine impact at this time.

## Acritarchs and oxygenation

The sediment of the Cryogenian seafloor had been accumulating for a billion years, and when lifted to the surface would enrich the photic zone with nutrients, including nitrate and phosphate. This would have two main effects. The first would be to form phosphate rock by precipitation, accounting for the vast phosphorite deposits of South China. The second would be to induce a worldwide phytoplankton bloom. The bloom turned out to be permanent, perhaps because it led to a persistent phosphorus cycle in the ocean, with organic phosphorus compounds formed by phytoplankton remineralized at depth into phosphate, which in turn supports phytoplankton growth when lifted to the surface. Moreover, it was a bloom dominated for the first time by eukaryotes, the acritarchs. Cyanobacteria have an advantage when nutrients are scarce because they are small and therefore have a high ratio of surface area to volume. When nutrients are readily available this advantage is annulled and larger eukaryotic algae have a competitive advantage.

A permanent shift towards a more productive phytoplankton community dominated by acritarchs would have led to increased photosynthesis and thus increased oxygen release. The stores of readily oxidizable minerals had already been exhausted, so the oxygen concentration of the ocean and atmosphere would increase towards present-day levels.

This simple scheme links together glaciation, deglaciation, upwelling, the Shuram $\delta^{13}C$ excursion, phytoplankton and increased oxygenation. It is, of course, a good deal more complicated than that (as I can hear every expert reader exclaiming). I have not mentioned the importance of trace elements, the evidence from strontium isotopes and a dozen other themes that doubtless played a part in the Ediacaran upheaval. It may suffice, however, to set the scene for the main event, the first appearance of large complex organisms soon after the Gaskiers glaciation.

# 10
# The Elusive Ancestor

You now have most of the leading facts about the early history of animals, so far as they are known, and a sense of how they have been handled in the past. It is at this point that I should announce some strikingly original theory that would tie together all these facts and clear up the misunderstandings that have accumulated over the last couple of centuries. Unfortunately, I haven't been able to think of one. The best I can do is to suggest that the answer you come up with will depend on the answers you give to three questions.

## Questions

The first question is: did animals evolve early or late? By 'early' I mean that the molecular clock tells the right time, in which case polyps and worms have a long hidden history with little or no fossil record, originating about the time of the Marinoan glaciation, about 635 Mya. Sponges are even older, going back to the Sturtian glaciation, about 710 Mya. By 'late' I mean that the fossil record should be taken literally, so that the first animals lived at more or less the same time as the first fossils. This would imply that the first animals lived either during or after the Gaskiers glaciation some 580 Mya, if the fossils preserved at Pigeon Cove were animals, or near the base of the Cambrian, about 540 Mya, when the first vertical burrows were made.

The second question follows from the first: were the Ediacaran fossils animals or not? If they were, then the search for an ancestor has ended, at least as far as fossils can help. If not, they might have been any kind of giant multinucleate or multicellular protist, belonging either to some familiar group such as foraminiferans or to some unknown extinct group, and the search for an ancestor continues.

Finally: was the first animal benthic or pelagic? Structures for locomotion and feeding used by sessile or crawling animals are characteristically different from those used by swimming animals, so the appearance of the ancestor would reflect its way of life. The dichotomy is almost (but not exactly) the same as the distinction between the sponge-first and ctenophore-first versions of phylogenetics. Sponges are benthic, of course, and the first ctenophores, notwithstanding the Chengjiang polyps, were presumably pelagic, since the comb rows of fused cilia are an effective way of modifying ciliary propulsion to accelerate a massive body but are unnecessary for suspension feeding.

*The Sea-Change*. Graham Bell, Oxford University Press. © Graham Bell (2026).
DOI: 10.1093/9780197797464.003.0012

## Eight scenarios

These three yes/no questions together define a set of eight possible scenarios for the early evolution of animals. I hesitate to call them theories, but they will at least serve to illustrate the range of published speculations.

### Scenario 1: The Garden of Ediacara

(Late/Animals/Benthic)

This is the current default explanation, I think. The Ediacaran biota were animals, and included the ancestors of modern groups, besides other forms which have since become extinct. Martin Glaessner thought that the fronds were akin to sea pens, a modern group of sessile cnidarians. Mikhail Fedonkin thinks that *Kimberella* was a mollusc, while *Dickinsonia* was a segmented animal in a group he calls the Proarticulate, which included the ancestors of modern arthropods. In this scenario, the 'Cambrian Explosion' was already in train by the Ediacaran, whose biota foreshadowed the subsequent radiation of their Cambrian successors.

The phrase 'the garden of Ediacara' was coined by Mark McMenamin, who meant it literally. He solved the main difficulty of interpreting the Ediacaran biota as animals—no mouth—by supposing that they harboured photosynthetic algae, rather like many modern corals. Provided that they lived in the upper sunlit layer of the sea they could rely on photosynthesis carried out by the algae to fuel their metabolism. This is reasonable for the Flinders Ranges or the White Sea coast, whose sandstones and shales were laid down in a shallow sea, but is questionable for Mistaken Point and Charnwood, which are thought to have been deep-water deposits, formed below the depth to which light can penetrate. McMenamin suggested that the fossils found in such localities had been transported from shallower water, but their intact bodies and common alignment strongly suggest that they lived in place. They might have farmed bacteria rather than algae, of course, but there is no sign of the specialized structures that modern organisms, such as ventworms, use to house their endosymbionts. The prevailing view is that the Ediacaran organisms simply absorbed nutrients dissolved in seawater. I have pointed out that the concentration of nutrients in seawater is exceedingly low, and moreover that fronds and discs are the wrong shape to compete with bacteria. Nevertheless, many living soft-bodied marine invertebrates are able to supply some of their diet by the active transport of nutrients across their body wall, so this explanation cannot be entirely ruled out.

The case for the Ediacarans being proto-sponges has been strongly argued by Thomas Cavalier-Smith, a well-known comparative biologist. A pelagic choanoflagellate colony evolved into a benthic pre-sponge by differentiating a non-ciliated anchor cell that secreted a mixture of mucopolysaccharides and collagen to glue it to a hard surface. It could compete with pelagic colonies because it would be able to filter a larger volume of water. The glue attaching it to a pebble became an extracellular matrix with wandering amoebocytes, the origin of mesenchyme. This little ball of cells evolved into a sponge by moving the ciliated cells to the interior of the colony, ventilating it with ostia and an osculum, and covering it with a protective layer of flattened cells. Rangeomorphs such as *Charnia* were proto-sponges with choanocytes on both sides of the frond; prostrate forms such as *Dickinsonia* had choanocytes only on their upper surface.

## Scenario 2: The Sea of jellyfish

(Late/Animals/Pelagic)

This used to be the default explanation: Reg Sprigg confidently classified *Ediacaria* and other circular or oval impressions in the Pound Quartzite of the Flinders Ranges as jellyfish, stranded on a sandy beach. Martin Glaessner and Mary Wade broadly agreed; Wade even interpreted *Kimberella* as a jellyfish lying on its side, whose gonads and the puckered zones between them appeared as a frilly skirt. This interpretation lasted 50 years, until Jim Gehling and Guy Narbonne showed that the sandstone was deposited below the tidemark and the supposed jellyfish were merely the holdfasts of fronds. The idea has not quite died. Dmitriy Grazhdankin, a Russian geologist based at Novosibirsk, found a fossil in latest Ediacaran rocks of the White Sea coast that he interpreted as the medusa of a coronate jellyfish, buried in the mud because it was weighed down by pebbles. The specimen—it is unique—is a couple of inches across and has a bowl-shaped body with a ruffled margin. The article in which it is described is apologetically titled 'Forbidden Fruits in the Garden of Ediacara', which perhaps best describes its reception.

The problem of what a cnidarian might have eaten is even more acute for a medusa than for a polyp. A cup-shaped or bag-shaped polyp, even if it lacked tentacles or cnidocytes, might enclose a quiet zone where sinking particles could be engulfed by the cells which lined it, this option is not available to a pelagic medusa, which would have to make ends meet by absorbing dissolved nutrients. This is not impossible—some living jellyfish can do it—but, as with polyps, it is disconcerting to have to rely on such a marginal method of feeding to explain how early animals made a living.

If the earliest animal was a medusa, it presumably evolved from a pelagic colonial choanoflagellate. This is an awkward transition. A free-swimming pelagic colony has its flagella on the outside, where they propel the colony by directing the water stream away from it, together with the bacteria it contains. A more efficient arrangement for feeding is to put the flagella on the inside, so the water stream is directed to the interior of the colony where the bacteria can be captured efficiently. Consequently, the arrangement that maximizes propulsion inhibits feeding, whereas the arrangement that maximizes feeding more or less abolishes propulsion. The solution to this problem is to use environmental cues to flip between flagella-out and flagella-in conformations. This may seem onerous, but a team led by Thibaut Brunet and Ben Larson in Nicole King's lab at Berkeley showed that an extraordinary choanoflagellate called *Choanoeca*, which they found in splash pools on the coast of Curaçao, was capable of performing this extraordinary feat. The cells of its cup-shaped colony are yoked together by their collars, which contain contractile fibres of actomyosin. In the light, these fibres are relaxed, the collar is more or less cylindrical, and the flagella point in. In the dark, the fibres contract, the collars become cone-shaped and, because they are connected, force the colony to turn inside-out, with the flagella now pointing out. Consequently, the colony can shift between propulsion and feeding according to the level of illumination.

## Scenario 3: The Vendobionts

(Late/Protist/Benthic)

This is Seilacher's heresy. The Ediacaran fronds and bags were not animals at all, but rather protists with quilted bodies made up of fluid-filled compartments. (I think that the Lilo air mat is a better analogy, but quilts are traditional.) More specifically, they were the ancestors of xenos and komokis, which are still abundant in the deep sea. Neither has any fossil record (although almost any muddy organic blob might be a komoki), but they are probably related to forams, which became calcified in the Carboniferous period and are thereafter abundant as fossils. Non-calcified forams are not often fossilized, but were definitely present in the early Cambrian. Stephen Culver, a geologist trained in Wales who moved to Old Dominion University in Norfolk, Virginia, isolated fossil forams from strata in the Walidiala Valley in southern Senegal, near the border with Guinea. They built a coiled tube from grains of silt bound together by an organic glue. They are quite small, only a fraction of a millimetre across, and are not divided into compartments, but they do prove that forams lived alongside small shellies in the early Cambrian. Moreover, there seems to have been a major radiation of forams at about the time of the Ediacaran/Cambrian transition, according to gene-based phylogenetic trees.

A similar idea was suggested by Hans Pflug, who worked on *Pteridinium* and *Ernietta* from Namibia. He thought that they were colonial organisms, formed of small cup-shaped modules reminiscent of zooids. Each cup had an internal cavity formed by invagination where dissolved nutrients could be assimilated. The cups were arranged in rows separated by a zig-zag suture, something like the segmentation of some animals. The Namibian fossils were not animals, however, but belonged to a quite different group, which Pflug named the Petalonamae, the petal-shaped organisms, without specifying their affiliations. He wrote in a rather dense technical German (which I have struggled with), and perhaps partly because of this his views have never been widely accepted by English-speaking authorities.

Granted that the Ediacaran organisms were forams, or some group akin to forams, then the feeding problem is solved: they used filopodia to capture large prey such as acritarchs. Recumbent discs and seals could even consume the microbial mat itself, while erect fronds could harvest the open water. The organisms were large because rising oxygen levels would allow any organism that used oxygen for respiration to evolve larger size, forams as well as animals. Moreover, there are species living today in the deep sea which prove that large forams are viable. Seilacher's ideas have not been accepted by the authorities in the field, but they remain tenable and have not yet been conclusively refuted.

If the Ediacarans were giant protists, the nature of the earliest animals is still up in the air. The default explanation is that first animal was a benthic colonial choanoflagellate attached to the seabed, or to some hard surface on the seabed. By bringing the choanocytes into the interior of the colony this evolved into sponges with flagellated chambers, whose existence is testified by isolated spicules, biomarkers, and the occasional body fossil such as *Palaeophragmodictya*. Sponges must have been part of the earliest communities, despite being very shy about announcing their presence. When these earliest animals lived is uncertain; I have assigned them a late date, but they might just as well have lived much earlier.

## Scenario 4: Ctenophores first

(Late/Protist/Pelagic)

This scenario is based on phylogenetics rather than fossils. Despite the long history and solid arguments for a sponge-first history of animals, the alternative view that the ctenophore lineage was the first to diverge has gained ground, at least among people who put their trust in DNA. It is always possible, of course, that the earliest ctenophores were benthic rather than pelagic, like the Chengjiang polyps. If the first animals were pelagic, though, the molecular tree suggests that they were ctenophores. Until recently, the fossil record offered no support—PreCambrian ctenophores are even rarer than PreCambrian sponges—but in 2008 some fossils that might well be ctenophores were reported from China.

The fossils are found in a bed of black shale near Wenghui village in central China, about 600 km north-east of Kunming. The shale began life as an organic-rich sediment flooring a shallow sea. It was laid down about 580–560 Mya, at about the same time as the oldest fronds at Mistaken Point, and contains many fossils. Most are seaweeds, but there are also numerous impressions of a very odd organism called *Eoandromeda*. It is a disc about an inch across which bears eight arms arranged in a spiral, with tips pointing clockwise viewed from above, something like *Tribrachidium* but with eight arms rather than three. The arms bear numerous dark transverse bars (unlike *Tribrachidium*, whose arms are bare), and meet at a dome-shaped structure in the centre of the disc. There is no trace of a mouth, gut or any internal structure.

A group of Chinese scientists headed by Feng Tang of the Academy of Geological Sciences at Beijing (and seconded by Stefan Bengtson) have interpreted the dark bars as fused cilia, similar to the comb rows of ctenophores. This leads to a difficulty. If *Eoandromeda* were firmly attached to the seabed (or a microbial mat) the spiral arrangement of the flaps on the arms will induce a spiral swirl, but this will not form a feeding current because there is no exit for the water. If it were planktonic, it would spin round and round like a Catherine wheel without getting anywhere. Feng Tang's team solved the problem by wrapping the arms around a conical body, so the flaps would propel the organism forwards (or in some direction, at least). This is reasonable, but the evidence for a cone-shaped body is slight and locomotion would be inefficient—much of the energy used in propulsion would be wasted in spinning the body round and round on its axis. It would be tempting to dismiss *Eoandromeda* as a curiosity, but there are lots of specimens, and similar fossils have been found in rocks of about the same age in the Yangtze Gorges and South Australia, so it must have been a regular member of the biota of the time.

Most living ctenophores capture large pelagic prey using sticky harpoons consisting of a coiled thread housed in a colloblast cell on one of the two tentacles. This inevitably suggests that the poison harpoons of cnidarians and the sticky harpoons of ctenophores are derived from a common ancestor, but they are encoded by quite different sets of genes, which seems to show conclusively that they evolved independently. In any case, an Ediacaran ctenophore could not have fed like this because there were no suitable prey available at the time. It must have been a suspension feeder, unless after all *Eoandromeda* was a benthic disc similar to *Tribrachidium* but with eight spiral ridges rather than three.

## Scenario 5: The Plakula

(Early/Animals/Benthic)

This scenario interprets the Ediacarans as placozoan-grade animals feeding by external digestion, like Otto Bütschli's Plakula. It has the great merit of requiring no imagination, since *Trichoplax* serves as a credible relic which can be collected routinely (well, fairly routinely; I have never been able to find any) and cultured in the lab. It has the drawback that none of the Ediacarans are clearly placozoans. It does need a little imagination, then, to see how some of the recumbent rangeomorphs or proarticulates might have placozoan-style ways of life.

Erik Sperling and Jakob Vinther of the geology department at Yale have made a determined attempt to portray *Dickinsonia* as a placozoan-grade animal which flopped down on the mat to cover its prey, using external digestion like a placozoan. Its prey might be large acritarchs, then a fairly novel food source, but since it lay on the surface of the microbial mat, one obvious source of nutrition for *Dickinsonia* is the mat itself. The fronds used the mat only as an anchorage, but the mat could also be used as food, at least in principle. Cyanobacteria can be difficult to eat because they have evolved a wide variety of chemical defences against grazing animals. Perhaps these defences were not in place, however, or at any rate were much weaker, before animals became a problem. Being surrounded by potential food certainly presents an opportunity, but one that can be exploited only if you have the means of eating it. You could simply engulf individual bacterial cells like a ciliate or a choanoflagellate, but then the mat does not provide any great advantage. The mat is a good source of food only if you can eat entire chunks of it, but a relatively large multicellular animal (a few millimetres long, say) can only consume whole sheets or clumps of cells if it has some sort of apparatus for processing food in bulk. It might be a flat animal, for example, able to lie on the surface of the mat, secrete digestive enzymes and absorb the resulting vegetable broth. This is an attractive idea because it is how fungi operate, and placozoans show that animals have retained or re-evolved the ability to use external digestion as a feeding device. Moreover, there are 'resting traces' which may be areas where *Dickinsonia* had settled, consumed the mat, or at any rate the top layers, and passed on. It is also possible, though, that these traces are no more than areas of the mat which were smothered by a dead *Dickinsonia*.

Passing on requires motility, of course. A set of similar-sized prints in the same small area, now a rock surface, certainly suggests movement, but no apparatus clearly evolved for locomotion has so far been described in *Dickinsonia* or any of its contemporaries. Perhaps the ventral surface of some Ediacarans bore a field of cilia which enabled them to crawl like a flatworm. No trackways have yet been described, but perhaps none would be made by an animal gliding over the tough surface of the mat. Or perhaps the movement, if there were any, was purely passive; one is free to imagine individuals floating up, being wafted by a weak current to a new area, and then settling down again to feed.

Suzanne Dufour and Duncan McIlroy of Memorial University in Newfoundland have braved the wrath of Thomas Cavalier-Smith by suggesting that early metazoans were adapted either to hard substrates such as rocks or to soft substrates such as mud. Those on hard substrates evolved into sponges, just as Cavalier-Smith described, but those capable of living on soft substrates evolved into polyps and worms via a placozoan-grade intermediate. Dufour and McIlroy favour the rangeomorph *Fractofusus* rather than *Dickinsonia* as the intermediate, consuming individual bacteria by phagocytosis rather than digesting a whole area of mat, but in either case the first animal, or at any rate the first non-sponge animal, was a Plakula.

## Scenario 6: The Parenchymella

(Early/Animals/Pelagic)

The Ediacarans appear abruptly in the fossil record, without any obvious precursors. They left no remains in the Lantian shales (600 Mya), yet a rich and diverse community was found at Mistaken Point (565 Mya). The Pigeon Cove fossils in the reserve are older (578 Mya) but include *Charnia* and *Charniodiscus* among other fronds and discs familiar from Mistaken Point itself. It was at some time during the interval between the Lantian and Mistaken Point formations that the Ediacarans diversified and became large.

What did they evolve from? Since they were animals (in this scenario) their ancestor was probably a small ciliated creature much like Élie Metchnikoff's Parenchymella. The Ediacarans were unknown when Metchnikoff was writing, of course, but his Parenchymella could have evolved into them. This is an attractive idea because the Parenchymella larva of a sponge, or the similar planula larva of a coral or a sea anemone, is everyone's idea of an ancestor (more to the point, it is Libbie Hyman's too). It is quite small (usually less than a millimetre long) and shaped like a rugby football or a stubby cylinder. It is also quite simple, consisting only of ectoderm, an outer layer of ciliated cells which move it through the water, and an inner core of endodermal cells derived from the wall of the blastula. The apparent simplicity of the planula may be misleading, because it often possesses amoebocytes, neurons, secretory cells and cnidocytes, but its remote ancestor may have had none of these.

Many sponges and cnidarians have small ciliated larvae that could pass for a Parenchymella, so there is no shortage of relics. There is even a fossil. A group from Taiwan and China (this is in the 1990s) described a Parenchymella from the Doushantuo nodules, together with (they claimed) body fossils of sponges. This is about the right age, and the association with sponges is suggestive, but the image they show as evidence is little more than a dark cylinder bordered by hair-like processes interpreted as cilia. *Caveasphaera* is also about the right age and had a solid body that developed by ingression, so it might have developed into the sort of creature that Metchnikoff had in mind.

## Scenario 7: The Invisible Animals

(Early/Protist/Benthic)

This is the Meiofaunal hypothesis, the theory that the earliest animals were the minute creatures of the interstitial community, living in the thin film of water surrounding sand and sediment particles in the mud of the sea floor. Perhaps it is misleading to characterize them as benthic, since they do not perceive the film of water that constitutes their world as either pelagic or benthic, but they would have lived on the sea floor and in this sense qualify as benthic.

This scenario was persuasively urged by Pat Boaden, who was Director of the Queen's University Marine Laboratory, at Portaferry in Northern Ireland, for many years. He once wrote an article on the interstitial fauna of flatworms and gastrotrichs living on the surface of seaweed fronds in Strangford Lough. He showed that this was, so to speak, a tertiary community, which became established only after the seaweeds, the primary community, had been colonized by the secondary community, a lawn of hydroids and bryozoans. This was quite interesting, but what made the paper unforgettable was that it was written entirely in rhyming couplets. This is unusual, to put it mildly, for a technical article published in the primary literature. Here is a sample:

> Analysis of data (Table 1) shows epifauna's the need
> If much meiofauna's to be found on the weed,
> So what was important was the mean cover made
> By the epifaunal species (19%) all over the blade (Fig. IA).
> Bryozoans dominated, e.g., *Alcyonidium hirsutum* (Fleming)
> The numbers were such you could not dispute them.
> Hydroids had also found themselves room—
> Young fronds and holdfast letting them bloom (Fig. 1B).

It may not be Pope, but we can all be grateful that the editors of the journal (*Marine Ecology*) allowed it to be printed.

Boaden pointed out (in more conventional articles) that animals with simple bodies, such as acoels and catenulids, are surprisingly abundant in the anoxic layer of black sand that lies a few centimetres below the surface of the seabed. Moreover, these groups were thought to be the first branches to split from the main trunk of the bilaterian tree. Boaden put these two propositions together to speculate that the first bilaterians, or perhaps the first animals, were minute creatures akin to the modern interstitial fauna, living in the poorly oxygenated PreCambrian mud, before the rise in oxygen levels made large animals feasible. This implies, of course, that there was a cryptic radiation of very small animals during the Ediacaran, which subsequently evolved into the burrowing worms and small shellies of the early Cambrian.

Interstitial animals are seldom preserved as fossils because their minute bodies quickly decompose after death. Animals like catenulids and gastrotrichs just seem to melt into the water. They do not usually leave trace fossils because they are too weak to make burrows. If they use muscle power to move, however, they might leave trails that are later preserved in the rock and can be visualized using X-rays. Narrow trails of this sort, a millimetre or less in width, have been found in a mine near Corumbá, in the same formation that yielded *Corumbella*, just below the boundary between Ediacaran and Cambrian. A large team from Brazil, Canada and the United Kingdom used microtomography, a sort of miniaturized version of medical CAT scans, to view slices through the rock that could be assembled into a three-dimensional picture. What they then saw was a maze of more or less horizontal trackways that could have been made by something like a small nematode, moving with muscles but without appendages.

Even body fossils can be preserved, however, if the animal has hard parts or a tough cuticle, especially if these have been phosphatized. It is still quite astonishing that both a priapulid and a kinorhynch, neither much more than a millimetre in length, have been found in the lowest Cambrian limestones of the Kuanchuanpu formation. These remarkable fossils are quite unmistakeable, minute as they are, and there is no doubt about their attribution: their spiny proboscis, with a terminal mouth, is preserved in microscopic detail and looks much like that of their living counterparts, still foraging in marine sands. Scalidophorans are highly derived animals, so their presence in the early Cambrian not only shows that there was a flourishing interstitial community at the time but also that this community extended deep into the Ediacaran.

Even though the Ediacarans themselves were not animals (according to this scenario) they must have had ancestors. Since the fronds at least consisted of filaments felted together to make a three-dimensional body, the obvious ancestor would have been a simpler branching filament. Rather surprisingly, the Pigeon Cove fronds are found together with impressions of branching filaments about a millimetre wide that do not seem to belong to the cyanobacteria of the microbial mat and are absent from later sites such as Mistaken Point itself. There is no other evidence to connect them with a lineage ancestral to the Ediacarans and the modern xenos and komokis, but the coincidence is striking.

## Scenario 8: The Gastraea

(Early/Protist/Pelagic)

This is Haeckel's original Gastraea hypothesis, which has fascinated generations of zoologists. I have already described it at some length. It gives a straightforward account of the evolution of the mouth and gastric cavity as the consequence of gastrulation by invagination. It is also consistent with the early date for the divergence of phyla suggested by the molecular clock, and with gene-based phylogenies which identify ctenophores as the earliest-branching lineage. Gastrulation in ctenophores, however, does not follow exactly the route foreseen by Haeckel. After the initial cleavage divisions, a sheet of small cells spreads over a clump of larger cells, rather like the early stages of embryogenesis in tadpoles. Once the surface layer of cells is complete, an opening appears at the oral pole followed by the formation of an internal cavity. The smaller cells on the surface are ectoderm, the larger cells inside are endoderm, and there is a group of cells around the opening which migrate into the interior and look suspiciously like mesoderm. Whether or not this peculiar pattern of development leads to a Gastraea seems to be a matter of taste.

Libbie Hyman was not a fan. She thought the theory was 'too beautiful to be true', and raised two objections. In the first place, gastrulation usually occurs by the growth or movement of cells into the interior of the blastula rather than by invagination. Secondly, the gastrula itself is never a larva, but rather a stage leading to the eventual development of a larva. The free-living larva of sponges and cnidarians is a planula or Parenchymella, not a Gastraea. Ctenophores do not have a distinct larval stage, but rather develop into a miniature adult complete with comb rows and tentacles. The Gastraea itself, as an autonomous reproductive adult, is still being sought.

## Answers

Deciding which scenario is most likely to be true involves answering the three questions that I have posed, which is a much more controversial undertaking than simply listing the possibilities. I shall have to take one side or the other in vigorous debates currently being conducted by experts who know much more about the field than I do. The best I can do is to offer an outsider's view, unswayed by prior commitments or sentimental attachment to a particular fauna.

### The amazing embryos

There is a fundamental contradiction between the early date required by the molecular clock and the much later date supported by fossils. This gap was narrowed (according to some) by the discovery of the minute phosphatized fossils of the Weng'an biota in the Doushantuo Formation in South China, in Ghizou Province, south of Kuanchuanpu and about 600 km north-east of Kunming. The fossils have not yet been dated to everyone's satisfaction, but most estimates fall between the Marinoan and Gaskiers glaciations and seem to centre on about 600 Mya. The formation is at any rate considerably older than any of the other Ediacaran sites. Some beds bear numerous small round phosphatic beads, roughly a millimetre or less in diameter, which can be collected by macerating rock samples in dilute acid, as at Kuanchuanpu. The beads are minute fossils, or contain fossils, which are preserved in breathtaking detail, right down to individual cells. Other beds consist of more coherent phosphorite rock, which has to be sliced into very thin sections to see the fossils. The fossils are formed when the original soft tissue is replaced by a film of calcium phosphate, as at Kuanchuanpu.

People who work on the Weng'an fossils tend to fall into two categories. On the one hand, there are enthusiasts like Xunlai Yuan and Zongjun Yin, of the Chinese Academy of Sciences at Nanjing, and David Bottjer of the University of Southern California at Los Angeles. The enthusiasts believe that the Weng'an biota includes microscopic animals, or the embryos of animals, from a very early stage of animal evolution. On the other hand are the sceptics, such as Stefan Bengtson of the Swedish Museum of Natural History in Stockholm and Philip Donoghue of Bristol University. The sceptics not only doubt that the fossils are animals, but even interpret some as purely inorganic structures formed by physical and chemical processes after the sediment was deposited. The controversy reminds me of the Eozoon affair, although of course it need not necessarily have the same outcome.

The distinctive fossil of the phosphatic beads is a multicellular structure, formed of any number of cells between two and several hundred, called *Megasphaera*. There is no doubt that it is a multicellular organism and very little doubt that it is a eukaryote, but beyond that there is little agreement. The enthusiasts maintain that is the embryo of an animal, showing characteristic features of early cleavage such as Y-shaped junctions, showing that cells remain attached to one another. Total cell volume remains the same during development, so that the original mass is cleaved into successively smaller cells, as in the formation of a blastula. There are even groups of cells which do not seem to obey this rule and are interpreted as a germ line distinct from the remaining somatic cells. The sceptics point out that this scheme of cell division is found in several other groups of multicellular organisms, which may also separate germ from soma during development. *Megasphaera* is not exactly like the colonial green alga *Volvox*, but there is an undeniable similarity of overall structure. There is bound to be, because the arrangement of cells in early embryos has few features that are unique to animals—until gastrulation.

After their initial publication in 1998, the Doushantuo fossils were widely publicized 2 years later, when a widely-cited article by Jun-Yuan Chen and his colleagues in China and the United States included dramatic pictures of embryos apparently undergoing gastrulation by invagination. But a similar structure (argues the sceptics) might have been produced by phosphatic cements invading a central cavity. The resemblance to a gastrula is certainly striking, but might be due to mere chance, since the fossils were chosen from thousands of objects. This is a difficult objection to refute. I do not know how many different kinds of conformation might be generated by purely physical processes, but it is easy to imagine sorting through hundreds or thousands of uninteresting blobs before seizing on the one exceptional blob that resembles an embryo. I do not assert that this is what happened, but only that biassed choice might, in principle, explain how an inorganic structure resembling an organism would eventually be found by a diligent searcher.

On the other hand, even the sceptics have been convinced by X-ray microtomography images of *Caveasphaera*, a hollow or solid ball of cells less than a millimetre across, which show the central cavity being gradually filled by strands of cells from the periphery. This is equivalent to gastrulation, not by invagination but rather by the movement of cells from the outside to the inside of the embryo.

The evidence from the Weng'an fossils is still far from conclusive, and some specialists who have studied them are not convinced that they are animal embryos at all. On balance, though. there is a strong case for believing that some of the minute fossils found at Weng'an are early stages in the life of some unknown animal.

One might ask, though, if the embryos are so common, where are the adults and what do they look like? The answer was again supplied, only 4 years later, by Jun-Yuan Chen and a group of colleagues, who had found what seemed clearly to be a bilaterian by screening thin sections of the phosphorite. They named it *Vernanimalcula*, the small animal of the Spring. It was indeed small, barely 200 microns in length, but complete with the mouth, gut and paired coelomic cavities, lined with mesoderm, of a worm-like bilaterian. They and other palaeontologists discussed the evolutionary implications of the find at considerable length, emphasizing its crucial place in animal evolution and its relevance to structures such as blood vessels and activities like swimming.

The sceptics had a field day. Perhaps the fossils had originally been organic, but any cells or tissue had long since been replaced by layers of minerals. The elongate shape of the fossil reflected a glancing section through an organism that might well have been spherical, like an acritarch. Any resemblance to a bilaterian could be explained by selection of a few specimens from thousands of objects. The enthusiasts argued back and produced more specimens, but Bengtson, Donoghue and colleagues brought the debate to a close by publishing an article with the uncompromising title 'A Merciful Death for the "Earliest Bilaterian"'. After all, *Vernanimalcula* had turned out to be little more than a mineral. Rather like Eozoon.

The Doushantuo fossils suggest that animals were living about 600 Mya, and perhaps emerged soon after the Marinoan glaciation. They cannot be assigned to any modern group, but this is to be expected of a very early stage in animal evolution. They do not quite close the gap between the molecular clock and the fossil record, but they do narrow it to the point where it is no longer a major problem, given the imprecision of the clock. This gives us one milestone on the road of animal evolution at around 600 Mya, the other being the burrows of Fortune Head at about 540 Mya. These two milestones indicate that the initial radiation of animals took place during the 60 My or so between the end of the Cryogenian and the base of the Cambrian, during the Ediacaran period.

## The Ediacarans

Any narrative of early animal evolution hinges on how the fronds, discs and other structures scattered through the later part of the Ediacaran are interpreted. There are strong and authoritative voices which insist that they were animals and are the key to understanding early animal evolution. Others are more wary but admit that they might be animals. I should say straight out that I do not believe that they were animals and did not evolve into animals. My opinion goes against the conclusions of experienced palaeontologists who are much more familiar with the fossils than I am, and the reader may well accept their conclusions and stop reading this book.

If you are still reading, which obviously you are, I will try to justify my point of view. My first objection, which will be generally admitted, is that the Ediacarans lack all the main attributes of animals. They have no mouth or gut, no internal structures or spaces, and no indication of muscle, nerve, or sense organs. They are certainly not preserved in *Charnia*, for example, which has no visible mouth nor any internal structures. Nor does it seem to consist of zooids which could be presumed to have possessed these features. The lack of detail might be blamed on its poor preservation as a mere impression of the surface of the organism in a coarse matrix, but this gets us no further forwards. We are forced to admit that *Charnia* shows none of the features that collectively diagnose an animal, and can only classify it as a modern zoophyte, which might be an animal, or a plant, or something entirely different from either. If the Ediacarans were animals, they must have been aleph animals which lived before any of these features had evolved. Perhaps they were; I've already pointed out that the ultimate ancestor of animals must have lacked all the features they subsequently evolved. It is still striking that none of the Ediacarans had any of the features of animals, despite having 25 My to evolve them.

The Ediacarans radiated into a variety of forms. There were fronds, which might be erect or recumbent, simple or branched; discs, which might be plain or bear several ridges extending from a central boss; bags, with strong ribs to support them; and many others. None recalls any familiar kind of animal. There are no intermediates; no sequence of types that might show how a frond or disc or bag was transformed into a recognizable animal, nor even the first steps in such a sequence. The simplest explanation of this stubborn refusal to look like animals is that they were not animals at all.

Ediacarans have been compared to sponges. But none of the Ediacarans bears any obvious resemblance to a sponge. They lack osculum, incurrent pores and flagellated chambers. They have definite form, maintained by a quilted architecture, whereas most sponges are shapeless, with a body supported by spicules. They have also been compared to placozoans, the only other animals which have no mouth. Placozoans are irregular in shape, because shape does not matter for an organism feeding by external digestion. Ediacarans which laid flat on the seabed might have dissolved the mat beneath them, but there is no evidence that they did so, and they were disc-shaped or frond-shaped for no apparent reason. An erect frond could not feed using external digestion at all. In short, there is no reason to suppose that the lack of a mouth can be explained by supposing that the Ediacarans were either proto-sponges or proto-placozoans.

Some of the Ediacarans have a branched body that gives the impression of segments, or segmental limbs. They were at first identified as polychaetes, and a whole group of Proarticulata was later created to accommodate them. The body of an

Ediacaran like *Dickinsonia* has pronounced branches or rays which meet in a midline that is zig-zag because they are spaced alternately. No segmented animal has this arrangement. The branches are plain, without joints, and are united one to the next. These are not the characteristics of arthropod limbs. The concept of Proarticulata should be abandoned.

Large animals are often motile, using muscle power, often coupled to appendages, to move around. There is only scanty evidence for motility in Ediacarans, and what little exists is open to doubt. Many animals are sessile as adults, like corals and tunicates, but these have pelagic larvae which metamorphose into the adult form. There is no evidence for larvae or metamorphosis in Ediacarans.

The body plan of Ediacarans has been worked out in detail for rangeomorphs, where it is based on branching tubes which diminish in diameter with branching order. Animals develop in a quite different fashion, from folded sheets of cells. No Ediacaran is known to have this kind of body plan. It is not just that the body plans are different, however, but rather that it is difficult to see how one could evolve from the other. Branching tubes and folded sheets are fundamentally different ways of constructing a large body and are encoded by different sets of genes that cannot be transformed from one to the other by gradual modification.

The alternative interpretation of the Ediacarans is Seilacher's heresy, that they were gigantic protists able to support the metabolism of a large body by a profusion of nuclei. There are many living examples to show that such a body plan is feasible, for example the deep-sea xenophyophores that Seilacher favoured as a modern equivalent. Whether or not this particular identification is sound can be doubted, but there is no doubt that feeding on bacteria and algae using filopodia is a viable way of life. Erect fronds could feed on plankton suspended in the water column. Benthic forams can eat microbial mats, apparently by kicking loose small clumps of cells with their pseudopodia. They can also crawl (slowly) by hooking their pseudopodia around sediment grains and pulling, although whether this is a feasible means of locomotion for something as large as *Dickinsonia* is dubious, I think.

There are some exceptions to a blanket denial of the animal nature of the Ediacarans: some rare polyps, for example, and above all *Kimberella*. There are three ways to interpret *Kimberella*. The obvious way is to treat it as a member of the community with which it is associated: *Charnia*, *Pteridinium*, *Dickinsonia*, *Parvancorina* and so forth. If these are animals, then *Kimberella* is an animal too. I have rejected this interpretation for the reasons I have given above, but it is currently the prevailing view and you may well decide to accept it. If they are not animals, on the other hand, then neither is *Kimberella*. This is a consistent point of view, but makes it difficult to interpret *Kimberichnus*. The third way is to propose that two kinds of large organism evolved in the late Ediacaran. One was a group, or perhaps several groups, of giant protists, the vendobionts of Adolf Seilacher Most of the community belonged to this group. The other consisted of a few animals, in particular the rare polyps and *Kimberella*. Separating supposed animals from supposed protists resolves the *Kimberichnus* dilemma, but it is a cumbersome idea that necessarily implies that animals not only existed but had grown large during the late Ediacaran.

## Room at the bottom

It is not easy to say whether the first animals were pelagic or benthic. There is a sharp dichotomy between theories which invoke either a pelagic or a benthic ancestor, and for large animals using muscles to move around the dichotomy is equally sharp. Pelagic animals swim in the upper waters; benthic animals crawl across the sea floor, or burrow into it, or are attached to it. These are completely different ways of life which are associated with different and incompatible adaptations. But the first animals are unlikely to have been large, and certainly lacked muscles and appendages. For minute animals a few hundred microns in length the distinction between pelagic and benthic may be less important than how they interact with particles. Pelagic animals inhabit a uniform environment in which they encounter few particles other than faecal pellets and dead zooplankton, neither of which existed before the Cambrian. Benthic animals live in the sediment or in the mat, and must continually navigate among clay, silt and sand grains or a tangle of microbial filaments and colonies. In either case, they have an intimate relationship with a complex world of particles and surfaces.

The heterogeneous world of mud or mat invites active exploration, whereas living in the uniform environment of the water column requires little more than a device to move around and harvest prey. The ability to turn, steer and manoeuvre is at a premium in a maze of particles similar in size to the animal itself, whereas it would be of little use to a pelagic animal. Sense organs will help to navigate and to locate prey. Statocysts will measure angular acceleration as well as telling up from down; chemosensory organs which merely tell a pelagic animal whether a particle is edible or not will enable a benthic animal to detect prey at a distance. The heterogeneity of the benthic environment means there are occasional bonanzas, such as an algal colony on the surface of a sand grain, which can be exploited given the means to do so. In short, mud and mat are more complicated and demanding environments which offer more opportunities and rewards than the open water for evolving a range of body plans.

The most conspicuous minute animals of the open sea are the ciliated-band larvae of marine invertebrates, which despite their very different adult phases are similar in structure, locomotion and diet. Parallel evolution to a common way of life has led to convergence. The interstitial fauna of catenulids, rotifers, gastrotrichs and nematodes, on the other hand, has a great diversity of forms and ways of life. The adaptive radiation of animals that occurred during the Ediacaran required a heterogeneous environment to favour divergent specialization, and so probably took place on the sea floor. If this view is correct, the first animals were benthic.

I think that most, if not all, of the published theories of early animal evolution can be fitted into one of the eight scenarios that I have outlined. In principle, we can use the answers to my three questions to narrow down the list of scenarios into a single candidate, Early/Protist/Benthic, which is scenario 7, The Invisible Animals. This is the meiofaunal hypothesis: the first animals were similar to the modern interstitial fauna, the community of minute animals swimming in the thin film of water between silt particles and sand grains in the mud of the sea floor. They would also have lived in the microbial mat, as they do today, and swam among the holdfasts of a forest of fronds like ants scurrying among the roots of a forest of trees.

# 11
# The Ediacaran Thread

I think that scenario 7, The Invisible Animals, gives a reasonable overall picture of early animals which is consistent with most of the evidence. In particular, it finesses the discrepancy between the molecular clock and the fossil record, and explains why the first animals to leave unmistakeable body fossils should be highly derived. It could be falsified by the discovery of phosphatic nodules like those of Doushantuo, but in a formation of White Sea or Flinders Ranges age, about 550 Mya. Such nodules should contain the contemporary meiofauna, fossils of minute animals similar to the scalidophorans of Kuanchuanpu but from an earlier stage in evolution. If they do not, the theory would eventually have to be abandoned.

Even if this convenient deposit is found, however, the meiofaunal hypothesis falls short of providing a complete account of the early history of animals. It provides a general context for envisaging early animals: they were minute, and lived in the mud among sand grains and fragments of detritus. The modern interstitial fauna shows that minute animals are viable, but all of them are highly derived and are unlikely to resemble their remote ancestors very closely. In fact, the meiofaunal hypothesis does not even specify the body plan of the animals living in the shadow world, beyond asserting that they were minute and had the usual features of animals living in a low-Re regime. It certainly does not specify the order of events that took place and the sequence of ancestors that led from a choanoflagellate to *Kunmingella*.

During the Ediacaran we can, in principle, trace a thread which connects these ancestors to form a single lineage. There are, of course, innumerable side-branches that give the entire evolving assemblage the appearance of a bush rather than a thread. This is a commonplace situation in laboratory experiments which challenge a susceptible clone of bacteria with a lethal stress, such as an antibiotic. A culture containing billions of cells can be maintained for a long period of time without the antibiotic, and will form a continually proliferating bush of billions of lineages. When the antibiotic is added, only a few cells of the billions initially present bear a mutation that confers resistance to the stress, and these few cells alone will be the founders of lineages that persist. Eventually, one of these, usually the lineage with the highest fitness, will replace all the others. At this point the entire evolved population has descended from a single cell in the initial population. All the other billions of cells originally present, resistant or susceptible, were the living dead, which have left no descendants.

The situation is not quite the same for Ediacaran animals, because we can recognize five lineages that survived: the lineages leading to modern sponges, placozoans, ctenophores, cnidarians and bilaterians. This implies a sequence of four common ancestors, each of which diverged into sister groups which have survived to the present day. The sequence is in principle given by phylogenetics, using hundreds of genes to specify the branching order of these five clades. In practice, as we have seen, the results have been contentious and the branching order is far from settled. Moreover,

*The Sea-Change*. Graham Bell, Oxford University Press. © Graham Bell (2026).
DOI: 10.1093/9780197797464.003.0013

phylogenetics alone does not explain why particular body plans have been successful. This is why it is necessary to supplement phylogenetics with a narrative that not only describes the order of events but also explains why each successive modification resulted in a viable and successful lineage.

This is now a well-defined problem: to construct a narrative that follows the thread of ancestors through time and identifies the branching points that lead to surviving sister groups. Four of these branching points are the common ancestors of the five extant clades. The fifth is the ultimate ancestor of all of them, the common ancestor of all animals. These nodes are animals that actually existed in the past. The five constitute a chain of ancestors in which a specified modification gives rise to the next animal in the chain. Some transitions, however, require more than one modification, in which case I shall specify an intermediate ancestor to make the transition evolutionarily plausible. The five common ancestors plus the intermediates together mark a lineage which leads to bilaterians. After that, the order of appearance of phyla is so uncertain that all I can do is to identify a few themes that recur during bilaterian evolution and are primarily responsible for the radiation of the group.

To reconstruct the early history of animals, I will have to revisit the scenarios, borrow imaginary animals, such as the Plakula, and even invent quite hypothetical animals when there seems to be no alternative. The purpose is to construct a narrative: to follow the thread of ancestors during the Ediacaran, describing each intermediate type and specifying the modifications which transform one body plan into the next.

## The narrative

I have tried to construct a narrative based on stages and the transitions between them, each stage being the ancestor of the following stage. The speculative literature is full of imaginary reconstructions and statements that such-and-such an ancestor "must have" existed, followed by a description, and often a drawing, of the author's favourite candidate. I shall try to avoid "must have" and instead restrict myself wherever possible to stages based on body plans that are known to be viable because of the existence of a relic, or, better still, of a fossil. The transitions also have to be anatomically or physiologically plausible, by specifying a particular modification with immediate utility that confers a clear selective advantage. Each transition must also involve a single modification so that evolutionary change is gradual, taking one step at a time. Ideally, both stages and transitions should be not only plausible but also falsifiable, with some clearly articulated reason for rejecting it if new contradictory evidence should come to light. These are counsels of perfection, of course, and the imperfection of the geological record (Darwin's excuse) or inadequate knowledge (my excuse) will sometimes force me to abandon them.

I am uneasily conscious that many, many attempts have been made by very knowledgeable zoologists to do what I propose. Whether a plausible, falsifiable narrative describing the whole course of early animal evolution can be built from our current knowledge is doubtful, and I certainly don't claim to have succeeded. The narrative I shall suggest has a dozen stages, all of which are more or less controversial, and

I don't expect it to be widely accepted by the authorities in the field. I only hope that a better narrative will emerge from the disagreements.

## Stage 1. A colonial choanoflagellate: about the time of the Sturtian glaciation, >700 Mya

Multicellular organisms might arise through partition (separate nucleated compartments dividing up the cytoplasm of a large unicellular organism, usually a ciliate), aggregation (isolated cells coming together to form a single compound individual) or coloniality (failure of cells to separate completely after dividing). All three possibilities have been suggested for the origin of animals, but ciliates turn out to be very distant from animals, while slime moulds are the pinnacle of differentiation for aggregated cells. That leaves coloniality.

It is utterly conventional to begin with a colonial choanoflagellate. There are good reasons for this. Comparative morphologists and molecular phylogeneticists agree that choanoflagellates are the sister group of metazoans, and they often form colonies of cells connected to one another and embedded in an extracellular matrix. Manuel Maldonado, who heads the sponge group of the Spanish Council for Scientific Research, is a lonely dissenting voice who inverts the sequence of events to argue that choanoflagellates are a sort of highly simplified sponge, a unicellular metazoan descending (rather literally in this case) from multicellular ancestors. The ultimate reversion of a multicellular lineage to a unicellular state also occurs in myxozoans (and in certain transmissible dog cancers, but that's another story), so it is not absolutely unprecedented. Sponges are, on this view, highly derived animals with a peculiar cell type specialized to capture bacteria, the ancestor of metazoans must be sought elsewhere, and this narrative fails from the start. Maldonado's theory is consistent with the puzzling absence of Ediacaran sponges, but it implies that choanoflagellates are the sister group of sponges alone, far from the main stem of animal evolution, whereas phylogeneticists agree (for once) that they are sister to the whole of the Metazoa. I shall reluctantly follow the herd and nominate a colonial choanoflagellate as the first step along the road leading to animals.

I do not know whether or not any living species shows us what Stage 1 looked like, nor even whether it was benthic or pelagic. There are plenty of sessile colonies, attached to a pebble or a plant stem, and always a branching single-row filament as far as I know, although the cells may be bunched together as in *Codosiga*. One can find rounded, gelatinous colonies growing in the mud, such as *Spongomonas*, but although it is often difficult to be sure which group they belong to, they are never choanoflagellates—in *Spongomonas*, for example, each cell has two flagella, so it must be a quite different kind of eukaryote. The alternative is a pelagic colony such as *Sphaeroeca*, which is a ball of cells about 300 μm across with a gelatinous filling. This is something like most people's idea of the springboard for animal evolution, and has even been named the 'choanoblastea'. Choanoflagellates have no fossil record, however, so we are reduced to asserting that some similar colonial species 'must have' existed before the Marinoan glaciation. Despite the multitude of relics and the unanimity of the phylogenetic evidence, there is no fossil and probably never will be.

Cryogenian communities that have been exceptionally well preserved seem to be exclusively unicellular and similar to previous communities. They include organisms that are probably eukaryotes, because they are too large to be bacteria, but there are no multicellular forms more complex than a simple filament. We can reasonably conclude that animals had not yet evolved at the time of the Sturtian glaciation, about 710 Mya.

Transition 1. The first transition is to reorganize the cells by a complete or partial inversion so that their cilia point inwards. This will be possible only if the cells are connected, as they are in choanoflagellates, and requires a contractile molecule such as actomyosin, which is known to be synthesized by choanoflagellates. The advantage is that it makes feeding more efficient, by increasing water flow, at the expense of reducing locomotion, since the cilia no longer propel the colony. Complete closure to form a sphere with inwardly pointing cilia is clearly unworkable, since creating a flow of water requires at least two openings, one for inlet and the other for outlet. (Helge Thomsen of the Danish Technical University once described a spherical colony of choanoflagellates with their flagella facing inwards, but I cannot explain how such a colony could either swim or feed.) This configuration is unlikely to have appeared all at once, but adopting a bowl-shaped or saucer-shaped form would allow a modest increase in feeding efficiency without completely sacrificing locomotion.

## Stage 2. An inverted colony: between the Sturtian and Marinoan glaciations, c. 700–650 Mya

This stage has a relic, the pelagic colonial choanoflagellate *Choanoeca*, which uses light as a cue to adopt the better configuration for either feeding or locomotion. There is no fossil, but *Choanoeca* shows that inversion is both plausible and viable.

Transition 2. The next step is the division of labour between specialized kinds of cell, either between somatic cells and the germ line, or between different kinds of somatic cell. Either might have come first, but I have imagined that somatic cells became differentiated first in order to adapt to life on the sea bed, where nutrients are generated and bacteria are more abundant. A special kind of somatic cell evolves to attach the colony to some massive object such as a pebble by secreting a collagen-based glue. This is universal in living sponges, and has the immediate advantage of keeping the individual in a region of high food concentration or otherwise favourable conditions. Feeding on bacteria is facilitated by having the cilia pointing in, and for an attached animal being unable to move this is not a drawback. The animal now has an internal flagellated chamber that can be protected by an external sheath of cells. This is likewise universal among sponges, which are covered with flat cells (pinacocytes) like a tiled roof, interrupted by cells shaped like a doughnut (porocytes) whose central apertures provide multiple inlets (ostia) for the water stream to enter. Sponges even contrive an economy by using the protective cells to secrete the attachment, and perhaps their remote ancestor did too. All that is required to provide an outlet (osculum) is that the external sheath does not wrap around completely, which would evolve immediately because a sealed chamber could not feed. The main difficulty is that any such animal would lack any inlets, because specialized porocytes would not evolve immediately. Without a basement membrane, however, any layer of cells would be likely to have gaps through which water could enter the interior of the body.

## Stage 3. An attached colony: between the Sturtian and Marinoan glaciations, c. 700–650 Mya

At this stage the first fossil appears—perhaps. Among the minute phosphatized structures found in the Doushantuo beds is *Eocyathispongia*, a coiled multicellular tube about 1 mm across. It has a wide aperture (perhaps serving as an osculum), a sheath of flattened cells (pinacoderm), gaps in the body wall for water entry (ostia), and even the possible remains of choanocytes. It certainly has sponge-like features, and the alternatives that have been suggested, such as an agglutinated clump of amoebas or a seaweed, do not seem very plausible. On the other hand, it is the wrong shape for a sponge. A coiled tube loses surface area where the coils are appressed, and indisputable Cambrian sponges (such as those from Chengjiang) are erect or globose, never coiled. Moreover, it is unique; only a single specimen has been described. Perhaps the best course is to suspend judgement until more fossils turn up, if they ever do.

Transition 3. The next transition would have been the separation of soma and germ. That convenient creature *Proterospongia* had collar cells for feeding on the outside and amoeboid cells for reproduction on the inside, although the somatic cells were all alike and their cilia pointed outwards. It was collected from Kew by Saville Kent in 1880, but, as I've emphasized, never seen since. (*Proterospongia*, or *Protospongia*, has been reported subsequently, but only as an undifferentiated colony.) I am not sure what to make of singletons like *Proterospongia*, *Magasphaera* and *Eocyathispongia*, which have been both studied and minutely described by experienced zoologists, so their existence is not in doubt (unlike *Salinella*, which was undoubtedly a phantasm of Johannes Frenzel's imagination) although their interpretation is debatable. It seems safest to view this transition as an hypothetical event which 'must have' occurred but for which there is neither fossil evidence nor living equivalent.

The advantage of having distinct soma and germ lies in separating source from sink. The somatic cells act as the source of the fuel for development, derived ultimately from the bacteria that the collar cells consume. They pass it on to the germ cells, where it is used to drive embryogenesis. The obvious living example is *Volvox*, which although it is nowhere near the ancestry of animals has a clear distinction between soma and germ, and develops along a regular route into a colony with a more or less fixed number of cells. The germ cells normally lie inside the colony, where they are supplied with nutrients by the somatic cells on the outside, but they can be cultured in isolation and still develop into juvenile colonies complete with their own germ cells. Vassiliki Koufopanou took advantage of this convenient property while working as a PhD student in my lab at McGill to compare the growth of isolated germ cells with those in normal intact colonies. She found that normal lineages with intact colonies grew faster than isolated germ cells at high nutrient concentrations, which is predicted by the source-sink hypothesis and is consistent with their distribution in nature (look in heavily fertilized farm ponds if you want to find *Volvox*). It is a bit of a stretch to extrapolate from this experiment to the upwelling of nutrients after the Marinoan glaciation, but there is at least a clear physiological mechanism and some experimental support for this transition.

The division of these early events into stages may be artificial, and in reality somatic and sexual differentiation may well have evolved at the same time, or even with modifications of the sexual system preceding the emergence of specialized somatic cells. Since we have no fossil evidence, it is impossible to be sure.

## Stage 4. The proto-sponge, between the Sturtian and Marinoan glaciations, c. 700–650 Mya

Whatever the course of events, at this stage we have a sponge with somatic and sexual differentiation. It begins to constitute an individual rather than a colony, but its individuality is precarious because it is not yet fully protected from selfish somatic mutations. Sponges have no gonads, and their germ line remains wide open; archaeocytes in the mesohyl are transformed into ova and choanocytes from the flagellated chambers turn into sperm. Perhaps this is why sponges have never evolved complex body plans. Body fossils of fragments of sponges have been reported from the Doushantuo nodules, but the claim has been contested.

Although sponges might be doomed by their unconstrained development to remain simple, they could at least grow larger. In fact, they can grow very large; the largest sponges found today can easily accommodate a diver. The advantage of large size is to make feeding more efficient by generating a more powerful water stream with an effective Bernoulli updraft. Oxygen supply is not the constraint that it would be for other kinds of animal, because sponges are effectively all outside, with a constantly renewed supply of circulating seawater. Pre-Marinoan sponges might, in principle, have grown to any reasonable size. A more serious constraint, however, might be the danger of collapse, since there is only a gelatinous layer with little mechanical strength between the outer covering of flattened cells and the flagellated chambers. Modern sponges can grow large because their bodies are stiffened by a framework of interlocking or fused spicules made of calcium carbonate or silica. Before the evolution of spicules, sponges might have been restricted to small size, or thin films, for purely mechanical reasons. This would at least explain the paucity of sponge fossils before the appearance of unequivocal spicules in the early Cambrian.

At this point we encounter a problem. 'It seems impossible', according to Claus Nielsen, 'to derive eumetazoans from an adult sponge'. Most zoologists would agree with him. There are three possible ways of escaping this quandary. One is to scrap the previous stages that I've proposed, and accept instead that ctenophores were the first lineage to diverge from the animal stem. If they evolved from a ciliated ancestor, then muscle and nerve must have evolved independently in ctenophores and in the lineage leading to cnidarians and bilaterians. Alternatively, if the ancestor already possessed muscle and nerve, these were subsequently lost in the evolution of a simplified sponge body plan. I have rejected this possibility because the parallel evolution of muscle and nerve seems unlikely (to me), while an ancestor of minute size would not need muscle power for locomotion. It remains nevertheless a serious alternative to the scheme that I have suggested.

The second option is to deny that there is any problem. Thomas Cavalier-Smith maintains that Nielsen's quandary was an unjustifiable failure of the imagination. It is not at all difficult to evolve a polyp from a sponge: 'On the contrary, to evolve a coelenterate from a stem sponge depended on preexisting epithelial adhaerens junctions, enabling extracellular digestion'. The interior cavity of the sponge is converted into a gut, with the osculum as mouth, and once choanocytes have been lost and cnidocytes gained, a functional feeding polyp is the result. Wnt proteins are known to be localized around the osculum as they are around the mouth, which suggests that the mouth was derived from an osculum. There are some problems with this vision, however. The osculum is an outlet, not an inlet; the internal cavity is lined with flagellated chambers and has little resemblance to a gut; and the adhaerens junctions hold cells together but do not enable or participate in the secretion of digestive enzymes. Moreover, the *Wnt* genes of sponges are quite different from those of cnidarians and bilaterians. It is perfectly true that the problem is solved if the internal cavity of a sponge can evolve into a gut, but I do not see how this can happen.

The final option is to derive a eumetazoan, not from an adult sponge, but from a larval stage in the life cycle which is much simpler and less specialized. This implies the evolution of a complex life cycle, which has some unexpected consequences.

Transition 4. A sessile creature growing in a productive site might be able to make and release thousands of fertilized eggs, but only the few that are swept by currents to similar sites will develop into adults and prosper like their parents. It all depends on chance, like a poplar tree releasing a million balls of fluff so that one or two will be blown to fertile soil. The probability of success would be much higher if the parents invested in a few highly-engineered propagules rather than just hoping for the best. This kind of propagule is a larva. A larva is a multicellular creature, much larger than the smallest egg but necessarily much smaller than the parent, typically a few hundred microns in length. Its size implies that it belongs to the interstitial community and is propelled by cilia, like ciliates and rotifers, which are about as big. Because it lives in different conditions, and has evolved to move rather than to feed, it is likely to have a body plan radically different from the adult. Indeed, if you were shown the larva of a sponge or a starfish or a flatworm it would be impossible to guess what the adult looked like, unless you were a zoologist of course. Most marine animals have a biphasic life cycle like this, alternating between two completely different kinds of individual.

For a lineage of sponges to evolve a larva, however, it must solve an evolutionary paradox. The ciliation constraint implies that cells can divide and differentiate only if they lose their flagella and move to the interior, leaving the exterior cells ciliated to propel the larva. The adult sponge, however, is built in the reverse order: the ciliated cells are on the inside and the non-ciliated cells on the outside. In this respect, a colonial choanoflagellate makes a poor ancestor, and it must have taken a long time to resolve this contradiction. Perhaps the diversity of larvae found among sponges—making a larva with both ciliated and non-ciliated cells on the outside, developing from a solid blastula and so forth—reflects the ancient range of attempts to cope with this difficulty. It was still necessary, however, to permit cells to transform readily from one somatic cell type to another. This is a makeshift solution because it makes it impossible to segregate the germ line in order to prevent the incursion of selfish cell lineages.

The larval stage of the life cycle is often said to achieve dispersal, which it certainly will if the larva is ejected from a sessile parent into the pelagic zone above the seafloor. It does not seem reasonable, however, to suppose that a swimming pelagic larva evolves because it disperses more actively than a passively drifting egg. Even a slow current runs at 1–10 cm/second, and will carry an animal 100 μm long 100–1,000 times faster than it can swim. The advantage of the larva is rather that it can manoeuvre to discover the best site for settlement, which an egg cannot do at all. The propagule might just be dropped into the mud, of course, in which case a larva might well be able to swim a few centimetres away from its parent, whereas an egg would be unable to move at all.

The larvae of modern animals, even sponges, are rather complex creatures with an array of different cell types. Any larva must reorganize its entire body in order to develop into a quite different kind of individual, the adult. This requires a specific process of metamorphosis, which in many cases involves discarding most of the larval tissues while rededicating the rest to develop into the adult. The evolution of metamorphosis does not seem to have investigated in much detail, but perhaps the earliest larvae were little more than ciliated embryos which followed the normal course of development into the adult without any unusual change of form.

The evolution of a larva has an unexpected consequence for the germ line. The fuel needed for locomotion and development must be supplied by the yolk of the egg, since it would be unreasonable to suppose that the original larva came fully equipped with mouth, gut and digestion. All modern sponges still have yolk-fuelled larvae, in fact. A large yolky egg could be produced either by the fusion of two medium-sized gametes or by the fusion of a large gamete with a very small one, to take the extreme cases. But if a moderately sized egg would produce a viable, if somewhat inferior, larva, then a population that produced moderately-sized eggs could be invaded by a type which produced very small gametes, because it could produce proportionately more of them. The population could not be taken over completely by this type, however, because fusion of two very small gametes would produce a very small egg incapable of developing into a viable larva. The consequence of evolving a larva, then, is to favour two strongly contrasted kinds of gamete, large immotile eggs and small motile sperm.

Moreover, eggs and sperm have different gender, so that either fuses only with the other. It is not the only possibility: some kinds of organism have a similar arrangement, ferns and *Volvox* for example; in others it is modified because the conditions are different, such as the ovules and pollen of land plants; while seaweeds may have morphologically identical gametes and fungi may have hundreds of genders. Choanoflagellates are sexual, but their gametes are very elusive and are certainly not morphologically distinct. Eggs and sperm are a shared derived character of all animals (except those that have lost sex altogether), including sponges, so we can infer that this sharp distinction had already evolved in their common ancestor, probably as the consequence of producing large yolky eggs that developed into a larva.

## Stage 5. The animal, after the Marinoan glaciation, c. 600 Mya

This transition yields an animal: a sponge with biphasic life cycle in which the fertilized egg develops into a larva consisting of a ciliated epithelium and an internal mass of amoeboid cells. Such a larva cannot evolve into a polyp or any other eumetazoan, however, because it cannot feed and therefore cannot reproduce. Further evolution along this route depends on the larva being able to feed.

Transition 5. Several routes leading to a feeding larva might be imagined. The simplest is to retain choanocytes and continue to feed on bacteria, but this means a compromise between the demands of locomotion and feeding, and in any case yields only another kind of sponge. Many larvae harbour endosymbiotic algae, which contribute organic carbon to the energy budget, but it is stretching credulity to suppose that an early planula would have come fully equipped with a population of domesticated algae.

Another option is to absorb external nutrients. This is a very general attribute of animals that lack a cuticle or some other barrier that blocks uptake. Sponges, in particular, seem competent to take up dissolved organic carbon substrates, even at low concentration. The planula larvae of cnidarians can do the same; I have not come across any conclusive evidence that sponge larvae take up external substrates, but it would be odd if they cannot, given that adult sponges and cnidarian larvae both can. Whether or not even a small animal can live entirely on dissolved organic matter is debatable, but seems beside the point: even a rudimentary ability to take up dissolved organic matter would provide the essential preadaptation for supplementing this source of nutrition by secreting digestive enzymes and absorbing the products.

A feeding larva could secrete digestive enzymes onto its food and assimilate the substances released by the lysed cells. This might be done inside the animal, given that a mouth and gastric cavity had been formed after gastrulation by invagination. This is the Gastraea hypothesis, which requires that the larva could take up particulate food and that the endoderm lining the gastric cavity immediately began to secrete digestive enzymes. This is a double coincidence, which seems too much of a coincidence to me. The remaining possibility is that in one lineage the epithelium of the larva began to secrete the digestive enzymes that had previously been sequestered inside vacuoles in the cytoplasm, and passively absorbed the products of digestion. This could evolve from modest beginnings, with some degree of external digestion merely subsidizing a pre-existing ability to take up dissolved organic matter and making possible a somewhat longer lifespan for the larval stage, enabling broader dispersal or better choice of settlement site. The strategy is unlikely to succeed in the open water, where the enzymes and any products of digestion would quickly diffuse away. It seems more likely to evolve among benthic animals living in the sediment, where the concentration of dissolved organic matter is an order of magnitude greater than in the open water and where the organism is in intimate contact with a food source.

This is not obviously improbable; it is how fungi feed, although I cannot find any evidence that planula larvae actually release digestive enzymes. Living in a microbial mat would reduce the need to recognize specific food sources. There was in any case no shortage of edible objects: acritarchs littered the surface of the microbial mat and provided an attractive and hitherto unexploited source of food for any small animal capable of locating and digesting them. Natural selection would then improve the efficiency of feeding, for example, by secreting from a defined area of epithelium only when it came into contact with an edible object.

## Stage 6. The parenchymella, c. 600 Mya

The outcome of this would have been a sponge-like animal with a long-lived benthic larva feeding on algal cells which it lyses by releasing digestive enzymes. Modern sponges have all kinds of larvae that correspond more or less to this description, a sort of benthic parenchymella, similar to Elie Metchnikoff's imaginary animal that I described in Scenario 6. It is similar to a simplified planula, one without all the bells and whistles that have evolved since. The only possible fossil is the parenchymella found in the Doushantuo nodules, but as I have said, this claim has been contested.

There are clear counter-examples to a mouthless absorptive larva, however, such as the planula larvae of sea anemones, which have a mouth that can swallow particulate food and a gastric cavity that can digest it. What I cannot swallow is that this arrangement should have appeared all at once, so I have been driven to imagine a planula larva or parenchymella capable of external digestion. Claus Nielsen does not find it surprising that early sponges should have had a larva with a mouth, and since he is very familiar with sponge larvae (and has been seconded by a group from the European Molecular Biology Laboratory in Heidelberg) it is possible that I am entirely mistaken and this stage is unnecessary.

Transition 6. No matter how the larva fed, it would have eventually developed into just another sponge. The crucial step that led to animals with a more ambitious body plan was to allow a feeding larva to reproduce, short-circuiting the life cycle and basing subsequent evolution on the modification of the sponge larva rather than the adult. The question asked by Nielsen was, 'are we derived sponge larvae?', and the answer he gave was, 'yes'. I agree, although I have chosen a different route from sponge to eumetazoan.

The precocious maturation of germ cells in juvenile individuals is not unusual. Neoteny, the ability of a juvenile stage to reproduce, has been reported from many species from arthropods to chordates. When I was a graduate student I investigated the life history of the smooth newt, a common amphibian in Britain. I trapped about 5,000 adult newts from ponds near Oxford and York, and found a single neotenous individual with the external gills of the larva but the size and form of an adult. In some species, such as the blind cave-dwelling axolotl, the entire species is neotenic. If a neotenic species radiates, it may give rise to a whole group in which all species reproduce as juveniles, such as the Larvacea, which are neotenic tunicates.

## Stage 7. The Plakula, c. 600 Mya

I have imagined a minute interstitial animal resembling a benthic planula larva, covered with cilia that it uses to creep on or in the microbial mat covering the sea floor, feeding on the algae of the mat by secreting digestive enzymes, and reproducing by eggs and sperm. This is close to Otto Büschli's Plakula, which, he said, evolved without much change into the Placozoa. The Plakula could detect and digest algae more easily by evolving from a more or less oval shape into a thin flat plate bounded by a single layer of cells. The lower cells, apposed to the substrate, were ciliated, enabling the animal to crawl. They could also release digestive enzymes, once prey had been detected and the cilia temporarily halted. In placozoans, a special kind of cell releases granules, presumably containing digestive enzymes, which lyse algal cells. This stage corresponds to *Trichoplax*, which makes an adequate relic. A similar division of labour between locomotion and digestion might or might not have been present in the Plakula. The upper cell layer seems less important, but necessarily establishes a distinction between dorsal and ventral surfaces, which foreshadows the distinction between ectoderm and endoderm.

Transition 7. The ventral sole of placozoans forms a dome over its prey which is enclosed on all sides by the flat edge of the body. This makes assimilation more efficient by blocking the lysed cell contents from diffusing wastefully into the surrounding seawater. It could be made more efficient still by forming a more spacious pocket over the prey, because more cells could participate in secreting enzymes and taking up the products of digestion. It might be possible for a placozoan to form a pocket like this. The cells of the sole are tacked together with adhaerens junctions that link inside each cell to the actin fibrils of the cytoplasm. The cytoskeleton itself, then, provides a motor capable of changing the shape of the cells so that they form a pocket. The result would be a sort of temporary gastric cavity as a space in which digestion occurs outside the cells but inside the animal. Placozoans are said to be able do this, and (rather surprisingly) several individuals can cooperate to attack a field of algae, together forming a spacious pocket which presumably provides a benefit to everyone.

A permanent pocket formed during the course of development would be better still because it would save the work of continually re-creating a temporary structure. It would also evade the ciliation constraint, because the internalized cells of the pocket no longer need cilia for locomotion and are free to divide and develop.

Forming a permanent pocket would be much easier if the sole were a real epithelium, a sheet of cells united and oriented by a basement membrane, that could be moulded into shape by the coordinated movement of the whole sheet. Unfortunately, the basement membrane has a serious drawback: cells cannot pass through it. It might be thin, but it is very tough, formed of cross-linked protein molecules arranged like chicken wire. This prevents cells from moving from outside to inside during gastrulation. It can be circumvented, of course: cnidarian development is based on cells doing exactly that, presumably by local dissolution of the membrane. But this necessitates a further modification, which is very unlikely to have evolved simultaneously with the first appearance of a basement membrane. This common means of gastrulation therefore seems unlikely to be primitive, so we must look for alternatives. A sheet of smaller cells might be able to spread over larger cells, relegating them to the interior, and this often happens, in frogs and chicks for example. But it requires the prior evolution of an unequal cell division in early cleavage, which is likewise unlikely to have evolved at the same time as gastrulation. That leaves invagination among the common types of cell movement leading to the formation of a gastrula. Cells do not need to pass through the basement membrane at all when gastrulation is by invagination because the apical ends of the epithelial cells point outwards in the ectoderm and inwards in the endoderm. Perhaps Haeckel was right after all.

There is a simpler alternative, however: the Plakula might just bend to form a shallow bowl, then a cup, and finally a hollow flattened ball with an opening that could function as a mouth and an internal cavity lined with digestive epithelium from the sole. It is invagination by stealth, so to speak. This is a rare mechanism of gastrulation nowadays—it is found only in a few species of coral—but it shows how a placozoan body plan, with digestion in an external pocket, could evolve gradually into a cnidarian or bilaterian body plan, with digestion in an internal cavity.

Placozoans do not have epithelia with basement membranes. Nor do sponges, of course, except for one group of obscure sponges with a correspondingly long name, the Homoscleromorpha. These are soft sponges which have tiny spicules or lack them altogether, and were first found in the darkness of Mediterranean caves. Their larvae develop inside the mother and are released as hollow balls consisting of a single layer of cells complete with basement membrane. They subsequently settle and curl up so as to put the ciliated cells inside, rather like my stage 2, where they form the flagellated chambers of the adult. The basement membrane suggests that homoscleromorph sponges might be close to the ancestry of the eumetazoans. Whether or not this is the case, their larvae certainly show that a small simple animal can be bounded by an epithelium.

## Stage 8. The individual, c. 600 Mya

This transition results in an animal with a ventral mouth leading to a gastric cavity in which algae and bacteria were lysed and digested. The individual, whose constant form and definite limits develop by a regular succession of somatic cell lineages, is

A muscle fibre must be told to contract by a nerve cell; muscle and nerve go together, so the initial function of muscle requires the prior evolution of nerve cells. A nerve cell secretes a substance that changes the behaviour of an individual, or part of an individual, when a condition is met. All nerve cells are on-off switches, basically simple devices that can nevertheless be arranged in very complex arrays, culminating in the central nervous systems and nerve cords of bilaterians.

The origin and early evolution of nervous systems have been debated for more than a century. One influential idea is that to begin with isolated muscle cells were embedded in an epithelium. Sensory cells then evolved from the epithelial cells next to the muscle cells, and eventually the two were connected by nerve cells. The original nerve channel thus consisted of three elements: the sensory cell, the nerve cell and the muscle cell. This is a sensible theory that summarizes a good deal of physiology and is backed up by comparative evidence—simple chains of two or three elements do indeed occur in cnidarians and ctenophores. However, it does not explain why an epithelium should have muscle cells in the first place. Nerve cells need to have had an indispensable function of some kind before the evolution of muscle.

The most likely guess, I think, was made by Gaspar Jékely, a developmental biologist at the Max Planck Institute in Tubingen, who suggested that the earliest nerve cells evolved to control the beating of cilia in minute animals. Single-celled creatures such as ciliates can regulate how their cilia beat even without a nervous system, so the equivalent of a Plakula or bilaterogastraea need only to transfer this pre-existing system to a multicellular platform by evolving a division of labour between specialized cell types. Sponge larvae have both sensory cells and ciliated cells, and can use chemical cues to modify locomotion. Perhaps the function of the earliest approach to a nervous system was as simple as secreting a peptide that inhibited ciliary beat when a chemical cue indicated that a suitable site had been reached, causing the larva to stop moving and settle there.

The range and reliability of such a primitive system will be limited by diffusion, and both will be improved by a further division of labour separating signal reception from transmission. This leads to nerve cells specialized for transmitting the signal to an effector. The signal which is transmitted need bear no relation to the cue which initially triggered the sensory cell. It can be quite arbitrary; the same signal could instruct a ciliated cell to wave or a muscle cell to contract. If specialized nerve cells were already in place when muscle cells first evolved they could readily be co-opted to control muscle contraction.

## Stage 10. The proto-polyp, 580–560 Mya

We have now reached the common ancestor of cnidarians and bilaterians. It is imagined (again, there is no fossil or relic) to be a small elongate animal formed by folding epithelia to make a two-layered body consisting of ectoderm and endoderm. Once the nerve cell had evolved as a specialized conditional switch it could evolve into the diffuse nerve net of polyps, a system of interconnected nerve cells ramifying throughout both ectoderm and endoderm that cause muscle fibres, which likewise form a network, to contract, closing the mouth or squeezing the gut or bending the body. The nerve cells of polyps are neurons, in which the signal is conducted electrically via a long extension of the cell and transmitted to a muscle cell, or to the next neuron in the chain, by a synapse. The neuron evolved in the common ancestor of cnidarians and bilaterians because it has two immediate advantages. In the first place, the distance that a signal has to travel by diffusion is limited to the gap between cells. Secondly, the signal can be effectively amplified because the strength of the response is not necessarily proportional to the strength of the signal; a very weak signal can elicit a very vigorous response. The coordination of muscle and nerve makes the animal far more flexible. A further improvement is to arrange the muscle fibres as a grid with circular and longitudinal strands, which makes it possible to change the shape of the body by shortening it, lengthening it, or bending it in any direction.

The proto-polyp had a mouth leading to a pocket lined with cells able to secrete digestive enzymes, which equipped it to take to crime as the first predator, feeding on other small animals. There were probably enough of them, since sandy beaches and microbial mats today support about 1–10 metazoans per cc, or about 5–50 small animals per teaspoon. Most of them eat bacteria or algae (or detritus, which amounts to the same thing), but some, such as nemerteans and polychaetes, attack other small animals. Recall *Mesostoma*, the predatory flatworm which everts its pharynx and pours digestive enzymes over its prey. The animals that lived in the PreCambrian mats and sediments had not yet evolved a pharynx, but scarcely needed to: spitting digestive enzymes onto a feebly motile prey would have immobilized it and created a soup of partly digested tissue that could be scooped up by the mouth. Most lineages no doubt continued to lead blameless lives eating microbes, but they would themselves have been a novel and abundant source of food that would sooner or later be exploited.

I imagine (there is no fossil) one lineage to have lived for a time as a small motile animal, before attaching itself to a sand grain and lying in wait for prey to pass by. This lineage would eventually evolve into cnidarians. There is no reason to suppose that motile and sessile phases were at first very different from one another, or that the switch from one to the other took place at some strictly determined age. The ways of life pursued by a motile forager and a sessile ambush predator are different in many respects, however, which is likely to favour the gradual evolution of divergent adaptations. These will eventually lead to pronounced differences in the form of the two phases, with a steadily more dramatic process of metamorphosis linking the two. This would constitute a life cycle with a larva, the motile phase, and an adult, the sessile phase.

The advantage of becoming an ambush predator is that it avoids the metabolic cost of movement. This will not affect the rate at which a predator encounters randomly moving prey very much, if at all. Consequently, an ambush predator will succeed when it first appears, provided that the cost of movement is depressed more than the marginal benefit is reduced by encountering fewer prey. Its overall advantage will diminish as it spreads, however, because ambush predators, if they stick to their posts, will never meet. The most likely outcome of this evolutionary dance is a mixture of ambush and roving predators, in proportions that depend on how these strategies affect metabolic costs and encounter rates. An ambush predator cannot sweep the board, but must instead co-exist with roving predators. The lineages leading to cnidarians (ambush predators) and bilaterians (roving predators), along with many other lineages that subsequently became extinct, would have been balanced in this way.

A benthic ambush is best set by choosing a feature likely to attract or concentrate prey, such as a channel in the sediment or mat, and staying there with one end of the body attached to the substrate. The mouth will tend to move to the other end of the body because in this position it is free to grasp prey without the body wall getting in the way. This would be a proto-polyp whose oral-aboral axis corresponds to the anterior-posterior axis of its ancestors. It would have looked something like *Protohydra*, the small polyp that lives in the sediment of saltmarshes, except that it would have been a much feebler predator because it lacked the cnidocytes that *Protohydra* uses to catch its prey.

Since all cnidarians have cnidocytes, whereas no other animal has them, their common ancestor presumably did too. The molecular-clock estimate of the divergence of anthozoans and medusozoans gives us a date for the evolution of cnidocytes, although (like most such estimates) it has very wide confidence intervals. To be on the safe side (the upper 95% confidence interval), cnidocytes had evolved (or had been acquired from microsporidians) about 668 Mya, which is well before the Marinoan glaciation and fully 100 My before the Ediacarans of Mistaken Point. Large polyps have been reported from both Mistaken Point and Charnwood, but the very long period of time during which there is no fossil evidence of polyps underscores the incongruity of the molecular clock with the fossil record.

The opposite of a benthic ambush predator is a roving pelagic predator, and the pelagic zone is so large (by far the largest habitable space on Earth) and so full of potential food (several billion tons of bacteria and algae) that it must have been invaded many times. The problem that any invader has to solve is how to remain afloat. Organic carbon is slightly denser than seawater, so organisms tend to sink. The problem is worse for larger organisms because they have less surface area relative to volume: bacteria will sink only very slowly (their own body length in about 2 minutes), whereas algae will sink faster (their own body length in a few seconds) and an animal, even a small animal only a millimetre long, will sink much faster (several body lengths per second). Animals must be able to paddle fast in order to stay afloat.

They must also be able to eat. Anything larger than a colonial choanoflagellate needs a mouth able to swallow particulate food and a gastric cavity able to digest it. All pelagic organisms larger than a few hundred microns are animals with a mouth. If the comparative evidence can be trusted, animals occupied the pelagic zone only after the evolution of the mouth.

One solution to the sinking problem is to retain the cilia of the benthic ancestor (a bilaterogastarea with muscle, according to this narrative) but glue them together as rows of little paddles. An animal with comb rows and a mouth is a ctenophore, able to move and eat in the open water despite being several centimetres or more in length. Ctenophores lack any sign of a distinctive larval phase; or perhaps they should be viewed as gigantic neotenic larvae which have descended from much smaller ancestors via the evolution of comb rows as a device that enables cilia to propel a massive body. Presumably they evolved large body size—there are no minute ctenophores nowadays—when the ocean became more oxygenated after the Gaskiers glaciation. They already had (according to this narrative) a mouth and digestion inside a gastric cavity, and would have been able to feed on pelagic acritarchs. But when did the lineage leading to modern ctenophores become separated from all others?

The radical proposal coming out of genomics claims that ctenophores branched off very early indeed, before sponges, so that they are the sister group of all other animals. This is Scenario 4, Ctenophores first. The molecular clock dates this separation deep into the NeoProterozoic, before any of the ice ages, perhaps as long as 800 Mya. The scenario, and its associated date, is supported by gene-based phylogenetics but implies that not only muscle and nerve but also mouth, gastric cavity and external digestion evolved independently in ctenophores and in all other animals. Many zoologists find it difficult to believe this degree of parallel evolution, but others, equally expert, do not think it incredible. It is, of course, also possible that the common ancestor of animals evolved muscle and nerve, which has been retained by ctenophores and by the sister groups of cnidarians and bilaterians, but lost by sponges, along with mouth and gastric cavity.

A morphologically more reasonable option is that ctenophores diverged before the last common ancestor of cnidarians and bilaterians but after sponges. If *Eoandromeda* was a large pelagic ctenophore then we can date this divergence some time before the Wenghui fauna of 580–560 Mya, conveniently around the time of the Gaskiers glaciation. It is always possible, of course, that the divergence took place later, in which case ctenophores are the sister group of either the cnidarians (justifying the conventional clade Coelenterata) or the bilaterians (justifying another clade so unfashionable it has no name as far as I know). If I had to choose, I would plump for the earlier possibility, even though this is inconsistent with the genomic dating and implies that ctenophores diverged after both sponges and placozoans.

The other common group of pelagic animals with a gelatinous body are the medusae, the jellyfish proper, which are able to swim by jet propulsion: the circular muscles of the bell contract, forcing a stream of water out and propelling the animal forwards by reaction, and are then re-extended by the springy mesoglea. There are many claims for Ediacaran medusae, dating back to the original descriptions of the Flinders Ranges fauna, but most of these are dubious. The strongest cases can be made for *Haootia* (Mistaken Point) and the presumed medusoid stage of conulariids (latest Ediacaran) and *Olivooides* (Kuanchuanpu). Otherwise, the earliest unequivocal jellyfish are from the mid-Cambrian Burgess Shale. Jellyfish soon decompose and are seldom fossilized; nevertheless, their scanty fossil record suggests that they evolved quite late in the Cambrian, after other kinds of animal had already entered the pelagic zone.

The other main device for living in the pelagic zone is to use fins or limbs to transduce muscular contraction into locomotion. The evolution of appendages, however, could take place only after a long sequence of events leading to the radiation of bilaterians. This began with an inconspicuous lineage of the proto-polyp's prey in which the whole course of development had been turned upside-down.

Transition 10. I have already mentioned that the initial asymmetry of the egg cytoplasm governs the subsequent course of development, so that the animal pole of the egg becomes the oral pole of the polyp. The crucial event in development is gastrulation, the movement of cells from the outside layer to the interior of the animal. This usually occurs through invagination in sea-anemones, corals and jellyfish, in which the site of gastrulation is the blastopore, marking the entrance to the interior pocket lined with endoderm. In hydroids, gastrulation usually just involves cells moving into the interior, so there is no blastopore, but the point on the wall of the blastula where the cells first invade can be recognised as the site of gastrulation. In either case, the site of gastrulation corresponds to the animal pole of the egg, and to the oral pole of the polyp. In one egregious lineage this rule no longer held: instead, the site of gastrulation shifted to the opposite side of the egg, its vegetal pole, but the mouth continued to be formed at the animal pole. This is what separates cnidarians such as corals and jellyfish from their sister group, the bilaterians, the whole host of animals from starfish to butterflies that has filled sea, lake, land and air with a wide variety of forms. One wonders why an apparently inconsequential detail of development should lead to such an unparalleled adaptive radiation.

The best answer I know was given by Mark Martindale. The cnidarian system makes use of an inevitable asymmetry in the cytoplasm of the egg to create gradients of morphogens able to pattern the developing embryo by a complex network of signals between cells that switch genes on or off. This works very well, but entails a grave disadvantage: as the network grows more complicated, the signals increasingly interfere with one another. Separating the mouth (and other structures formed from ectoderm) from the site of gastrulation (where the endoderm begins to form) is a division of labour which reduces signal interference and permits a greater degree of specialization among structures formed both from ectoderm and endoderm.

Once the endoderm has become disentangled from the signal pathways of the ectoderm, a further division of labour is possible. The endoderm of cnidarians contains two kinds of cell, digestive cells and muscle cells. Separating the two permits a further level of specialization. Moreover, it leads to a different kind of tissue. Cnidarians develop as folded sheets of epithelia, whereas bilaterians have a second broad category of cells, the mesenchyme. Epithelial cells can be transformed into mesenchyme cells by losing their attachment to one another and hauling themselves through the basement membrane into the extracellular matrix beneath. Here they live like fish in a rather viscous pond, oriented front-to-rear rather than base-to-apex, and capable of slow amoeboid movement. They can differentiate into any kind of somatic tissue, including muscle, blood and internal organs such as liver and kidney.

Mesenchyme is not an exclusively bilaterian tissue. Placozoans lack both basement membrane and extracellular matrix, but they do have a sort of mesenchyme, in the form of interconnected 'fibre cells' whose function is far from clear, although they may just scavenge debris and intruding bacteria. Cnidarians may also have wandering cells in their mesogloea. Nevertheless, the extensive development and differentiation of mesenchyme is a bilaterian innovation. This new tissue is derived from a new germ layer, the mesoderm, which forms most of the structures between body wall and gut wall. The ultimate consequence of the switch of the site of gastrulation from animal to vegetal pole is thus the addition of a third germ layer to the ectoderm and endoderm of cnidarians. This is what underlies the great bilaterian radiation.

The initial tissue formed from mesoderm was muscle, not as the isolated cells or strands of cnidarians but as solid blocks capable of exerting considerable force. It still cannot be used for propulsion (because the animals are still too small) but it can be used for steering, by changing the shape of the body. Contracting the longitudinal muscle on one side of the body would turn it in that direction, or even reverse itself completely. This may sound trivial, but for an animal threading its way through the sand and silt of the sea floor, or through the clumps and strands of algae in the mat, it would be a major accomplishment.

(If you have access to a low-power microscope and take a sample of detritus from a local pond you will usually see minute elongate animals twisting and turning rapidly among the fragments of plant stems. These are gastrotrichs. They don't really belong here, being much more complex than an acoel, but they serve to illustrate the rapid flexible directional movement that muscle strands confer, even in a small animal using cilia for propulsion.)

From a formless blob crawling slowly over the sea floor in no particular direction one lineage of animals had evolved into a nimble worm exploring the labyrinth of sediment or mat. A Martian which alighted from its spacecraft and looked closely enough to see it might not be very impressed, though. It was, after all, just a small strange creature dwarfed by the fronds that towered high above it.

## Stage 11. The proto-acoel: c. 570 Mya

Nevertheless, it was equipped with muscle and nerve and would be capable of excavating trails or horizontal burrows like those of *Helminthoidichnites*, a common trace fossil in the Flinders Ranges. What kind of animal made them could only be guessed at before Scott Evans, then a graduate student in Mary Droser's lab at Riverside, noticed some tiny indentations on the base of sandstone beds which also bore *Helminthoidichnites*. These were the impressions made by a cigar-shaped animal about the size of a grain of rice. Nothing is known of the anatomy of *Ikaria* beyond its size and shape, but if it really did make the *Helminthoidichnites* burrows it must have had muscle in its body wall. At all events, it is exactly the sort of small benthic animal that is anticipated by the narrative at this point.

The animal is now a proto-acoel with the first hint of cephalization: the nervous system of acoels is a nerve net which shows little advance on that of a polyp, except that it is concentrated at the forward end and in some species foreshadows a brain. Moreover, it is hooked up to a statocyst that will tell up from down and presumably might also detect acceleration. Modern acoels have acquired a third *Hox* gene to complete the basic set of three *Hox* genes—anterior, central and posterior—which suffice to guide the development of a simple worm.

Perhaps acoels are not the most appropriate choice for a relic, though, since they seem to have given up external digestion and instead have a gut crammed with phagocytic cells. A better choice might be *Nemertoderma*, an equally simple (and even more obscure) marine worm which at least has a gut with a lumen. Like acoels, *Nemertoderma* has no circulatory or excretory systems, nor even discrete gonads, although it does have both circular and longitudinal muscle fibres. It is about the size and shape of a grain of rice, like *Ikaria*. Gene-based trees often show acoels and nemertodermatids as the sister group to all other bilaterians, branching very early from the main stem, so they make a good relic to show what the earliest worms might have looked like.

It seems likely that the bloom of acritarchs during the Ediacaran drove an increase in ocean oxygenation after the Gaskiers glaciation, about 580 Mya. This was a challenge—oxygen is highly toxic—but it was also an opportunity that would have enabled any lineage using aerobic metabolism to evolve larger size, whether or not it was an animal. There is a snag, of course: simply increasing body volume will not do, because diffusion is limited by surface area and no tissues specialized to harvest oxygen (such as gills) yet existed. The problem can be circumvented by making a body from finer and finer tubes in order to maintain surface area, as I've suggested the fronds of first fauna did. It is only a minor problem for sponges and polyps, whose bodies are continually washed by a fresh supply of seawater. It is a serious problem for worms, however, because a worm cannot become large and bulky unless it first evolves some organ that allows them to breathe more easily.

Transition 11. The simplest way of doing this is to replace most of the cells that fill the inside of the proto-acoel with seawater. Carbon dioxide can then be exchanged for a fresh supply of oxygen by diffusion through the body wall. The fluid-filled internal space is a body cavity whose original function was to act as a respiratory organ. How it is formed doesn't affect its function, as long as the body wall remains thin; the most economical way is just to retain the original lumen of the blastula, enclosed by a single-layer epithelium, but if this has been obliterated during development a secondary body cavity must be formed. This secondary body cavity is the coelom.

A body cavity will support respiration only in small animals, however, because of the ineluctable relationship between the volume of the internal reservoir and the surface area available for diffusion. In a large animal, this reservoir would soon become stale as the end-products of metabolism accumulated. In its original state, the coelom would be unnecessary in minute animals (roughly, less than a millimetre in size) and dysfunctional in large animals (roughly, more than a centimetre). It would permit only a modest increase in body size.

The internal reservoir could be refreshed by piercing the body wall with pores through which a stream of seawater could be driven by cilia, but this is a clumsy expedient that would leak manufactured goods such as proteins as well as admitting unwelcome guests such as bacteria. A much better solution would be to modify the ciliated cells of the ectoderm as filters. Fluid from the interior would pass through a membrane (rather like the collar of choanocytes) that allowed small molecules (such as urea) to pass through while retaining large molecules (such as proteins). The filtrate is then discharged to the outside. This is how a flame cell operates (it is named from the flickering of the cilia, which under the microscope looks something like the flickering of a candle flame). An animal with flame cells has an excretory system capable of purifying the fluid contained in its secondary body cavity without wasting resources or risking infection. It can be further improved by grouping the flame cells together to form a primitive kidney, the protonephridium, and draining the filtrate through a duct to the outside. Acoels and nemertodermatids do not have an excretory system, but all other free-living bilaterians have some kind of excretory system, even if it is only a protonephridium, and can proceed immediately to evolve larger bodies.

## Stage 12. The protodeuterostome: c. 560 Mya

The stem groups of modern phyla, according to the narrative I have proposed, were small animals that lived during the Ediacaran, perhaps between 570 Mya (the proto-acoel, common ancestor of bilaterians) and 560 Mya (the protodeuterostome, common ancestor of protostomes and deuterostomes). The characteristic features that separate protostomes and deuterostomes evolved after increased oxygenation made the evolution of large size possible, because the body plan of deuterostomes doesn't work in a minute animal. The apparatus of pharynx + atrium + outlet requires some roomy internal spaces, so the basic deuterostome body plan can evolve only in a fairly large animal.

Although these are just guesses, I think they are likely to be close to the truth. The existence of a recognizable group at a given time necessarily implies the previous existence of all its ancestors. The crown groups already present in the early Cambrian, especially the priapulids and kinorhynchs of Kuanchuanpu, prove that the proto-acoel and the protodeuterostome must already have lived (and died). *Ikaria* makes a good proto-acoel and lived at the right time. By the early Cambrian, the spiralian and ecdysozoan lineages must already have separated; and even within ecdysozoans the scalidophoran (priapulids) and arthropod (lobopods) lineages must have separated too. All this took time. Exactly how long it took is uncertain, but 10 My to evolve the basic body plans of bilaterians does not seem excessive, particularly if they evolved exceptionally fast.

The situation is not without precedent; indeed, there is an example much closer to our own immediate ancestry. During most of the Mesozoic, mammals were small creatures living in the trees, or on the ground in the forest, under the feet of dinosaurs. Once the giant sauropods had been wiped out by the Chicxulub asteroid, the surviving mammals soon radiated into a great diversity of ways of life, some calling for animals almost as large as dinosaurs, such as elephants, or even larger, the whales. Several major groups evolved and flourished for a time, only to die out with no descendants—multituberculates, uintatheres, creodonts. The question is, whether all these multifarious mammals descended from a single post-apocalypse lineage, or whether there was a cryptic radiation, perhaps extending far back into the Mesozoic, of lineages which, although sharing few or none of the distinctive features of later groups, were nevertheless their ancestors. Without a label, it is difficult to be sure, even though there are (in this case) plenty of fossils. With far fewer fossils, it is not surprising that the early history of animals should be so contentious.

Large bodies evolved after the oceans became better oxygenated. Many of the shared derived characters of bilaterian phyla—number of coelomic compartments, radula, appendages, hydraulic system and so forth—are features of large animals that would not be adaptive at minute size. If the split of protostomes from deuterostomes occurred when all animals were minute it might not be detectable, but instead would be manifest only once they had evolved large size.

The first large body fossil (if you accept my conclusion that the vendozoans were not animals) is *Kimberella*, which is quite numerous at White Sea coast and Flinders Ranges localities. The *Kimberichnus* scratches it made strongly suggest that it was a coelomate bilaterian. It raises a crucial question: if most animals were minute, why did only some of them become much larger? The answer, I suspect, is that large size will evolve more readily and more rapidly in animals whose body plan works reasonably well at any scale.

Some body plans work only at small scale and are found only in small or minute animals such as rotifers and gastrotrichs, which have muscle and nerve but use cilia for propulsion. None of these have a distinct larval phase.

Other body plans work only at large scale. For example, the longitudinal fibres in one transverse strip of an elongate body can be used re-extend the fibres in the adjacent strip. This is how molluscs move, using transverse ripples in a solid muscular foot. As one strip is lifted from the substrate, its longitudinal fibres contract, moving the rear of the body forwards, and the strip is then lowered again. This does not require a coelom (even turbellarians can manage it), although it does require some means of preventing slippage, usually a weakly adhesive mucus. It also requires a fairly massive block of muscle, and even so snails can creep only very slowly. Other body plans work only at large scale because they incorporate an element that is difficult or unprofitable to miniaturize, such as the hydraulic system of echinoderms or the segmentation of many annelids. Animals such as molluscs, echinoderms and annelids require a radical metamorphosis to bridge the gap between a minute larva and a large adult.

Only a few body plans work at both scales without reorganization, so that evolving a large body does not require a distinctive larval phase. For example, priapulids, known to be abundant in the early Cambrian, might have already evolved. The priapulid plan of a large open coelom with an eversible proboscis and the muscles to extend and retract it allows movement by rope-and-grapnel haulage. This works well, if rather slowly and jerkily, over a broad range of body size. Modern priapulids range over two orders of magnitude in length, from 1 or 2 mm for interstitial species to 10 cm or more for species that burrow in the mud, while retaining the same basic body plan. Gehling rejects this possibility because *Kimberella* did not have the tubular appearance of a priapulid, but it must nevertheless have had a coelom in order to extend its proboscis. The burrows of *Treptichnus* may have been made by a priapulid; minute priapulids are certainly present at Kuanchuanpu, and large priapulids are common members of the seafloor communities of Chengjiang and Field. All of these are interstitial or burrowing animals, but *Kimberella* shows how priapulids might have evolved from a simpler predecessor with a different lifestyle. Unlike marine molluscs, echinoderms and annelids, modern priapulids have no distinctive larval phase; a 'larva' is sometimes described, but it is essentially just a smaller version of the adult.

Whether *Kimberella* was close to the ancestry of any modern group is still not clear. It might have been a proto-annelid; unsegmented annelids such as sipunculids have an eversible proboscis, but (according to gene-based phylogenies) descend from segmented ancestors without a proboscis. It seems more likely to have been a proto-mollusc, creeping along on a broad foot, but the proboscis seems very un-molluscan. Perhaps it was a proto-priapulid, since priapulids do have an eversible armed proboscis and were around shortly afterwards. If any of these options is eventually verified *Kimberella* could still not have been the protodeuterostome, since it lived, not only after the divergence of protostomes from deuterostomes, but after the separation of spiralian and ecdysozoan protostomes. In turn, this would imply that the protodeuterostome lived much earlier than White Sea times 555 Mya.

Or it might not have been a proto-anything, but rather a distinct group far from the ancestral lineage of bilaterians. *Kimberella* is a single name that conceals a great deal of morphological variation, from broadly oval to teardrop-shaped to highly elongate forms, suggesting a modest adaptive radiation. It was widely distributed and was abundant for about 5 My before completely disappearing along with the rest of the Ediacaran biota. It left no descendants; there is no further radiation of *Kimberella*-like forms in the early Cambrian. We shall solve the mystery of *Kimberella* only if a Chengjiang-type Lagerstätte of White Sea age is discovered in some formation that the geologists have not yet investigated.

The long thread of ancestors comes to an end with the protodeuterostome, the last common ancestor of all the familiar bilaterian animals—snails, crabs, frogs and more or less everything you are likely to run into that isn't a sponge, a coral or a jellyfish (Figure 11.1). From this point on, the thread frays out into a bush, and it is no longer possible to trace any single line of descent leading to an orderly succession of phyla.

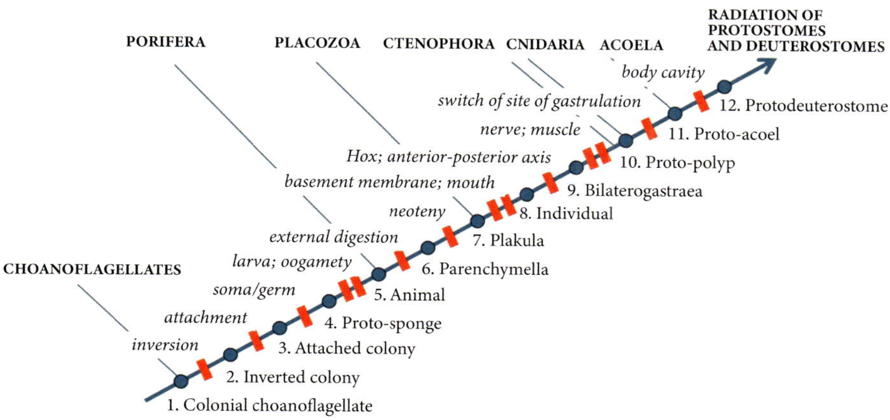


**Figure 11.1** The narrative

# 12
# The Cambrian Tapestry

The bush is, of course, the great diversification of animals during the early Cambrian, the 'Cambrian explosion', which soon led to the Chengjiang and Burgess Shale communities. There are two points of view about what happened during the first 20 My of the Cambrian. The first is that there was a real burst of animal diversity at this time, which is the straightforward interpretation of the fossil record. The second point of view is there was only a burst of fossils, left behind by animals which were for the first time able to make mineralized skeletons because of a change in ocean chemistry. The reason for this dichotomy is the disagreement between fossils and phylogenomics, which is made worse if the Ediacarans are not accepted as animals. This creates a gap of almost 200 My between the first fossils and the inferred origin of animals around the time of the Sturtian glaciation, and a second gap of about 100 My between the first fossils and the inferred split of cnidarian and bilaterian lineages around the time of the Marinoan glaciation. A little inconsistency between such very different kinds of evidence might be expected, but these are very wide gaps indeed.

## The cryptic history of animals

If they are real, the gaps imply that there was a long cryptic history of animals, extending deep into the Ediacaran, during the stages vaguely dated as 'c. 600 Mya', around the time of the Doushantuo nodules, when animals diversified into a variety of clades which became apparent only once they became visible as fossils. I do not doubt that the gaps are real, because coelomate worms had already evolved by the base of the Cambrian (*Treptichnus*) and soon became very abundant (*Skolithos*); predators with mouth spines comparable to chaetognaths are found in the Fortunian (*Protohertzina*); lobopods with segmentation and appendages are part of the small shelly fauna (*Microdictyon*); and unmistakeable scalidophorans have been found at Kuanchuanpu (*Eopriapulites*). The fossil record shows that highly derived bilaterians had already evolved by the first 5 or 6 million years of the Cambrian. They could not have appeared abruptly out of nowhere, so the fossils found in the earliest Cambrian clearly demonstrate that there must have been a long Ediacaran history of animals.

There must also have been a long prior history of the Ediacarans themselves, if you accept that they were not animals. There were then two groups able to evolve large bodies when oxygen became more freely available after the Gaskiers glaciation. One was derived from branching filaments. Some of these were able to grew upwards to form an erect three-dimensional body by felting the filaments together to form a frond. These evolved into the range of body plans, including the sessile discs, bowls and fronds, that constitute the Ediacarans. They fed by catching algae and protists

*The Sea-Change*. Graham Bell, Oxford University Press. © Graham Bell (2026).
DOI: 10.1093/9780197797464.003.0014

with filopodia, and were similar to the xenos and komokis of the deep sea floor. The other included sheets of cells that could be folded to form a three-dimensional body that evolved first into sessile sponge-like creatures and subsequently into small motile animals that included the ancestors of placozoans, ctenophores, cnidarians and bilaterians.

One leading feature of the phylogenomic trees intended to show the order in which these groups of animals evolved is the length of the basal branches. The branch leading to metazoans from choanoflagellates is relatively long. I have assigned this to scenario 3 ('Vendobionts'), leading from colonial choanoflagellates to sponges. To make it consistent with the clock, stages 1 to 4 must be very early and occupy the first gap, between the Sturtian glaciation about 730–705 Mya and the Marinoan glaciation about 650–635 Mya. There is no way to be sure, because there are no fossils.

The branches separating placozoans, cnidarians and bilaterians are much shorter and indicate rapid evolution. Both morphological and molecular trees agree that cnidarians and bilaterians are sister taxa, although the position of placozoans and ctenophores remains uncertain. I have imagined a sequence of minute ciliated animals from stages 5 to 10 that occupies most of the second gap, after the Marinoan glaciation until late in the Ediacaran. This gap includes the Doushantuo phosphate nodules, about 600 Mya, which may represent this early radiation, although the stages I have suggested have not been preserved and remain purely imaginary.

Finally, the branches leading to the bilaterian phyla are very short indeed, indicating very rapid evolution. These correspond to stages 11 and 12, the bilaterians that evolved after the switch of the site of gastrulation, the assembly of the *Hox* cluster and the origin of mesoderm. All of the stages from 5 to 12 correspond to scenario 7 ('The Invisible Animals'), although I have borrowed some terms, such as 'parenchymella' and 'Plakula', from other scenarios when they portray the appropriate kind of body plan. The history of early animals is cryptic because they were small and soft-bodied and have left few fossils. They correspond roughly to the modern interstitial fauna of minute animals crawling or swimming in the mud.

The best candidate for an early bilaterian is *Ikaria*, but it took the keen eyes of a graduate student to notice these tiny fossils, even though the sandstone blocks were from a well-known locality and were scribbled over with the trace fossil *Helminthoidichnites*. These minute animals evolved larger bodies and the apparatus needed to support and propel them, such as coelom and muscle, after stage 11. This occurred after the seas became better oxygenated at about the time of the Gaskiers glaciation, about 580 Mya, because of a prolonged acritarch bloom that led to the Shuram excursion of light carbon. Animals did not immediately become bigger because the apparatus needed for large bodies, and the genes required to encode it, took a long time to evolve. The earliest fossils are the rare polyps found at Charnwood and Mistaken Point, about 565–560 Mya. Large bilaterians first appear as the body fossil and feeding traces of *Kimberella* in the Flinders Ranges and White Sea coast deposits about 555–550 Mya. *Yilingia* was a segmented worm-like animal which lived about 550 Mya. Large coelomate worms made the probing traces and vertical burrows that first appear at the base of the Cambrian, about 540 Mya. These dates are the rather shaky palaeontological pegs on which the closing sections of the narrative are hung.

## Stem groups

The order in which stem groups appear in the fossil record should correspond to the branching order of the phylogenetic tree. The stem groups of modern phyla, according to the narrative I have laid out, were minute animals living in the Ediacaran. If all kinds of animal were readily fossilizable and came with labels indicating their stem group, even if they had no shared derived characters, then the inferred phylogeny could be compared directly with the observed fossils. Unfortunately, neither of these conditions is met. Minute soft-bodied animals are fossilized only in exceptional circumstances, and unlabelled stem groups can be difficult to recognize.

The fossil evidence for stem groups in the Ediacaran is scanty, apart from the Doushantuo nodules. These have yielded the diverse Weng'an biota, but the fossils are difficult to place. If *Caveasphaera* is the embryo of an animal, the adult is unknown; the presumptive gastrulas have been challenged; and *Vernanimalcula* is a mineral concretion. Otherwise, the fossil record includes only large animals. There is the dubious proto-ctenophore *Eoandromeda* and the even more dubious proto-arthropod *Keretsa*; a few rare cnidarians; and of course *Kimberella*. The trace fossils are more convincing. Horizontal burrows like *Oldhamia* show that small worms lived in or under the microbial mat, although which modern groups they gave rise to, if any, remains unknown.

Even in the early Cambrian, it is seldom obvious that a fossil belongs to the stem group of a particular phylum. The literature is full of claims and counter-claims because many of the shared derived characters that build a phylum consist of soft tissue that is not readily fossilized, and if it is has been reduced to vague lines and smudges open to competing interpretations. A strong case can be made out for *Wiwaxia* as a stem mollusc and lobopods such as *Microdictyon* as stem arthropods. The Chengjiang arthropods are scattered along the stems of modern arthropod classes, before their characteristic patterns of segment identity had evolved. Whether tommotiids such as *Lapworthella* are stem brachiopods, or *Yunnanozoon* is a stem deuterostome is less certain, while interpreting *Dinomischus* as a stem ctenophore is, as I've said, a daring guess.

All sorts of other animals appear early in the Cambrian and flourish for a while before disappearing for ever. Archaeocyaths, chancelloriids, hyoliths and conulariids were all very abundant at one time but are difficult to place in any modern group. Whether they should be placed in the stem of a modern phylum, and if so which, remains debatable, despite confident claims from time to time that their true nature has been discovered at last. Many more kinds of animal may well have existed but left little or nothing behind, the dark matter of zoology that may include the imaginary stages in the chain of ancestors that I have proposed.

The cast of characters was well established by the time that the Chengjiang and Sirius Passet communities flourished. The order of appearance of the phyla is much more uncertain because of the inconclusive fossil record and the short-branch problem that continues to plague phylogenetics. There are several themes, though, which crop up repeatedly in the early Cambrian radiation of body plans. Most of them are about eating, or about moving around in order to eat, because getting food dominates how the body is constructed.

## Insides

The radiation of large animals in the early Cambrian followed the evolution of the inside. The first lineage with an inside was not a sponge, because sponges are all outside: whatever is sucked in through the ostia can go migrate to any part of the sponge body—consequently a rich and diverse community of bacteria flourishes in the extracellular matrix, the collagenous ground substance in which the sponge cells are embedded. Polyps have an inside of sorts because they have a mouth leading to a cavity in which the food captured by the tentacles is digested. The sheet of cells that covers the outside of the body continues as the lining of the digestive cavity, however, so the cavity itself is on the outside of the animal. The real inside is rather notional, consisting merely of the gelatinous layer between the body covering and the cavity lining, which is usually devoid of cells, although it may harbour a few bacteria and the wandering amoeboid cells that mop them up.

A burrowing worm, on the other hand, has a real inside, a closed fluid-filled space between the body wall and the gut wall. This body cavity can be used as a hydrostatic skeleton so that the circular and longitudinal muscles of the body wall can oppose one another, making it feasible to burrow through viscous sediment. Making a horizontal burrow does not absolutely need a coelom, but it is much easier when the force of muscle contraction is instantly transmitted without loss to all parts of the body, or at least to all the parts bordering the coelom. The worm-like animals that shuffled slowly and laboriously through the top layer of sediment have left the remains of their burrow systems as horizontal trace fossils. It is unlikely that cilia would have provided enough force to make them. Instead, the animal making them was probably a coelomate worm moving by peristalsis. The oldest trace fossils produced by animals are the unbranched, shallow, horizontal burrows reported from the White Sea and Flinders Ranges assemblages. They become more common close to the base of the Cambrian; the branched tunnels of *Oldhamia* are early Cambrian. Some of the Cambrian burrows seem to have been actively backfilled, which would imply that the animal that made them had a through-gut as well as a coelom.

The abundance and variety of burrows increased sharply during the earliest part of the Cambrian. In particular, vertical burrows such as *Treptichnus* had appeared by the beginning of the Cambrian period, 541 Mya. These could have been constructed by proboscis extension: the circular muscles are used to evert a proboscis rapidly enough to make a hole in the sediment, into which the animal can drag itself by contracting a set of longitudinal muscles while using the proboscis as an anchor. In this way, the combination of circular and longitudinal muscles with the coelom as a hydrostatic skeleton generates sufficient force to permit vertical burrowing. A variety of simple worms, such as nemerteans (acoelomate, but with an eversible proboscis housed in a fluid-filled cavity at the front end), sipunculids and priapulids use a method like this for constructing their burrows. *Treptichnus* itself may have been made by a priapulid. The characteristic side-to-side pattern of the burrow would have been produced by probing the sediment to find food, the mouth being at the tip of the proboscis.

The inside of the worm is closed to the environment and must be kept rigorously sterile, lest any invading bacteria set up a fatal infection. A fluid-filled cavity within the body creates a health hazard even if it can be kept sterile, however, because it will inevitably accumulate waste products, so it must be scrubbed regularly. An open cavity such as an atrium can simply be flushed out with seawater, but a completely closed cavity like a coelom necessarily implies that even the most rudimentary excretory system would have an immediate advantage. This need consist of little more than a primitive kind of kidney such as a protonephridium as a filtering system and a pore or duct to expel the filtrate. At the same time the coelom offers a space to store eggs and sperm, aggregated into discrete gonads to wall off the germ line from interlopers. These could share the kidney duct, since its purpose is to ship things from inside to outside. The protodeuterostome had some such arrangement, and all its descendants—all bilaterians except acoels and nemertodermatids—have inherited it.

The coelom and its contents, together with the gut, comprise the visceral being. The stem of lower-ranked (more inclusive) groups of animals, including the chain of ancestors that I have proposed, are predominantly visceral, with only rudiments of the somatic structures that evolve later.

## Segments

Worms with an undivided coelom find it difficult to burrow, because the force of muscle contraction is transmitted instantly to all parts of the body. It is by no means impossible—this is how large priapulids burrow—but it is slow and inefficient, proceeding in a series of jerks as the worm hauls itself through the sediment. Using this probe-and-pull method of digging a hole and then crawling into it is made much easier if the coelom is divided into two compartments by a rigid watertight membrane, the septum. Lengthening and shortening can then be conducted independently, because a change in pressure is felt only in one of the two regions rather than in the whole body. In practice, septa are neither perfectly watertight nor rigid, but any degree of separation will increase the efficiency of burrowing.

Most animals with a coelom divide it into a few (two or three) compartments or many segments. It is unlikely that compartments were the first step towards segmentation, however tempting the idea might be; they seem rather to be the end-points of adaptation to different ways of life. The converse is much more likely: some kinds of worm whose body plan is based on compartments rather than segments, such as sipunculids and echiurans, have probably evolved by simplifying a fully segmented ancestor and are now classified as annelids.

Rapid continuous burrowing is made possible only if a series of peristaltic waves, involving the temporary thickening of part of the body, can be passed along the length of the worm, driving it smoothly forwards. Dividing the coelom into a series of more or less watertight compartments, each with its own set of muscles, allows local control of muscle contraction and thereby greatly facilitates peristaltic burrowing. The result is a segmented worm.

Making a vertical burrow requires a coelom to transmit the force generated by body wall muscles and is much easier for a segmented animal. Trace fossils such as *Skolithos* suggest that segmented worms were fairly abundant in the early Cambrian, and are large enough—*Skolithos* burrows can be an inch wide or even more—that they could only have been made by animals using muscle power. They are just traces, though, without the body fossil that would provide hard evidence for the interpretation. Both trace and body fossils were at last found together by a team from Nanjing in the Denying formation of the Yangtze Gorges area in South China. These rocks were laid down in the last 10 My of the Ediacaran. There is no doubt that body and trace belong together; in one remarkable case the body fossil was found at the end of the horizontal trail it had been making on the surface of the sediment. The animal was named *Yilingia* and is a segmented worm. It is gratifyingly large, up to a foot long and an inch wide, and consists of about 50 segments, all alike and without any obvious modification at the head end. The segments themselves are rather odd, though. Each consists of three lobes, a central rhomboid-shaped lobe flanked by a more or less triangular lobe on either side. These lateral lobes point backwards and might have been appendages, although there is no sign that they could be flexed. *Yilingia* is another unexpected find—not so much for the segmentation as for the fancy work on each segment.

Segmentation is not without its drawbacks. The major blood vessels and nerve tracts must pass through each segment across gaps in the septum which must then be made watertight. Large worms are viable only when these gaps have been created and sealed, which might require a long process of gradual modification. Dividing up the body into a series of more or less independent units involves supplying each with its own circulatory, nervous and excretory systems, which prevents them from condensing into specialized organs. This limits the evolvability of a simple segmented worm (although it would not prevent it from evolving in the first place).

## Tentacles

The sea is full of single-celled organisms which can be harvested by animals using the right kind of apparatus. Some sessile animals can pump a stream of water through their body and filter the bacteria and algae from it. The pump is always powered by cilia, but the filter has been modified in many ways by different groups: amphioxus and tunicates filter the stream through slits in the pharynx, for example, whereas clams strain it though pores in their gills. The alternative is to move a sheet of water over the surface of the body, again by cilia, and collect the plankton by a filter, mucus groove or inertial trap. Animals that feed like this have a ring of tentacles around their mouth, and it is worth asking why this distinctive feature should have evolved independently in animals as different as crinoids, tubeworms and brachiopods.

The tentacle physically connects the benthic and pelagic zones of the sea by enabling a sessile benthic animal to catch pelagic prey. The food-collecting ability of a suspension feeder will be proportional to the area of the external surface of the body wall available for inserting cilia. One way of increasing this would be to extend the body wall by a certain amount, giving a longer animal with more cilia. A better solution is to replace the body wall with a ring of tentacles, whose combined surface area may greatly exceed that of a simple extension of the body wall. Assuming that both body and tentacles are cylindrical, a ring of six stout tentacles has three times the surface area produced by just increasing the length of the body by the same amount.

This advantage increases with the number of tentacles, but with diminishing returns, so that a large number of fine tentacles has a surface area exceeding a simple extension of the body wall by a factor approaching π. The same argument applies to the armed tentacle, since the number of cnidocytes will also be proportional to surface area.

Hence, prolonging the body with tentacles offers a large immediate advantage in food-collecting ability for both suspension feeders and predators. This must be paid for, of course, since a greater surface area implies a greater volume of tissue, which is expensive to produce. If the tentacle wall is the same thickness as the body wall then cost and benefit are the same, and tentacles are an expensive investment that might pay off if prey are scarce. It seems reasonable to suppose that the tentacle wall can be made thinner than the body wall, in which case tentacles will be profitable, although not as profitable as a simple comparison of surface area would suggest.

Both *Haootia* (Mistaken Point) and *Auroralumina* (Charnwood) seem to have tentacles, but what they caught with their tentacles is difficult to say. If they were suspension feeders then they could have captured acritarchs, which were numerous at the time. Modern sea anemones, on the other hand, catch zooplankton, mainly copepods and the larvae of crabs, shrimps and molluscs. None of these existed in the PreCambrian, as far as we know. It might be more reasonable to imagine armed tentacles evolving in the early Cambrian, when other kinds of predator (such as chaetognaths) were already feeding on the first zooplankton (such as bradoriids).

## Jaws

The unarmed mouth is useful for collecting the phytoplankton harvested by tentacles, but it cannot attack anything larger or more active without being modified. Polyps modified another part of the body, the armed tentacle, to capture and kill prey that could then be swallowed. Priapulids improved their method of clambering in the sand by equipping their proboscis with spines that occasionally snagged another worm, and in this way became predators. Flatworms and starfish are able to evert their pharynx and convert the tissues of their prey into soup by pouring digestive juices onto them. Other bilaterians evolved a mouth with a hard rim that can be closed forcibly, and for the first time were able to bite.

There is an odd clade of animals which seems to diverge early from the main stem leading to annelids and molluscs despite having rather complex jaws. Some are pelagic: the rotifers of freshwater, which have pharyngeal jaws that crush the algae they feed on, and the marine chaetognaths, which capture large prey with external jaws equipped with long pointed spines. Others are benthic and deeply obscure (they were not even described until the 1950s), although they can be abundant in marine sands. Gnathostomulids are very small unsegmented worms with little in the way of a coelom—they don't need one—and a simple body plan, but what they do have is a pair of jaws, made of rods of chitin and closed by strong muscles. They use these jaws to scrape algae and fungi from the sand grains amongst which they live. There are no fossil gnathostomulids, but the living animals show how jaws can evolve to aid feeding even in minute benthic worms.

## The mouth

All these themes point directly or indirectly to the mouth, as the entrance to a gastric cavity or gut, as the major potentiating innovation responsible for the radiation of animals. The mouth and how it is connected to the food-collecting apparatus together characterize the stem groups of modern phyla. The importance of the mouth and its associated apparatus as a food-collecting device for animals eating other organisms is obvious and need not be emphasized. It belongs in the same category as appendages, body cavities and muscle layers. Unlike these other features, however, the mouth fundamentally alters the relationship between an organism and its environment by transforming a public good into a private good.

Public goods are resources produced by some individuals that can be consumed by all. They may be depletable or not; roads are depletable, because traffic jams result when everyone tries to use them at once, whereas street lighting is not, because the amount of illumination is the same no matter how many people are out at night. Fungi and bacteria secrete digestive enzymes to break down large molecules outside themselves into smaller molecules that they can absorb and use as substrates for metabolism. This creates a depletable public good that can be consumed by a neighbour that needs substrates but does not secrete enzymes. Cheating like this is more profitable than honest work, but the profit depends on a plentiful supply of honest neighbours, and as this dwindles the profit dwindles too, until at the limit a population consisting entirely of cheats cannot obtain substrates at all and starves to death. Before this happens, though, the population will reach an equilibrium at which the cheats prosper because it is just—only just—worthwhile for their honest neighbours to manufacture and secrete the enzymes that provide everyone with substrates. This elegant hypothesis has been tested by Duncan Greig and Mike Travisano, two of the most talented experimental evolutionary biologists, by using two strains of yeast, one (the honest type) able to secrete the invertase used to hydrolyse sucrose outside the cell, and the other (the cheat) not. The outcome of manipulating the initial density of cells was exactly as predicted: in dense populations cheating was profitable, but the profit grew less as the density declined, until cheats were actually less fit than their honest neighbours when the population was sparse. External digestion, then, creates depletable public goods that can be exploited by cheats.

Cheats can be frustrated by converting a public good into a private good. A private good is one that is enjoyed by a single individual, who is able to exclude other individuals from consuming it. There are two ways of converting food from a public good to a private good. One is phagocytosis, or intracellular digestion. This is perfectly effective, and it is the only option for single cells eating bacteria—or for sponges, using batteries of choanocytes. The other option is the mouth.

The mouth combines external digestion with a gastric cavity to convert food into a private good. It is the equivalent in a multicellular animal to phagocytosis in a single-celled protist. It is more versatile, though, because the food can be any size, from the phytoplankton screened out by the lophophore of a bryozoan to the seal caught by the jaw of a great white shark. Angler fish can even swallow prey larger than their own bodies. The versatility of the mouth leads to a vast range of diets and the associated ways of life: all the manifold variations on suspension feeding, deposit feeding, scraping, scavenging, herbivory and predation. Large or small, once the food has been swallowed it becomes private property that cannot be exploited by others.

When food is a public good, regular development is hindered because loss-of-function mutants can make a living. Converting it into a private good, conversely, promotes regular development and thereby the evolution of the individual. The mouth, with external digestion in a gastric cavity, provides the Darwinian rationale for the evolution of individuality because it is only the individual which profits from the capture of prey. Once the mouth has appeared, the poorly-defined and poorly-defended individuality of sponges and placozoans can evolve through maternal predisposition and a separate germ line into the regular development of individuals. Only then can the body plans of animals diversify, beginning with the stem groups of the modern phyla.

## Crown groups

When animals became larger, muscle power was applied to locomotion and yet more ways of life were opened up. The stem groups of modern phyla evolved during the Ediacaran and early Cambrian; by the end of the Cambrian, or at the latest by the middle of the Ordovician, the crown groups of all fossilizable phyla had appeared. The order in which they appear is still in dispute, but the range of body plans was constrained by the structures and arrangements that had already evolved in the stem groups.

### Outsides

The gradual shift from a purely visceral body to one that also has somatic elements marks the emergence of crown groups, which are mostly (although not exclusively) defined by features borne on the outside, such as the radula of molluscs or the compound eyes of arthropods or the hollow tentacles of a lophophore.

In particular, the segmental body plan provides a substrate for the evolution of a corresponding series of appendages as outgrowths from the body wall. In annelids these are fairly simple. Earthworms merely have bristles which facilitate burrowing by increasing friction with the soil. Polychaetes can bear more elaborate structures, the parapodia, which aid serpentine swimming or crawling. The segmental appendages of arthropods evolved into even more complicated structures, the jointed limbs, whose pattern of elaboration is a major feature of the body plan of crustaceans, barnacles, centipedes, insects, spiders, and all their relatives. It is the elaboration of appendages in segmented worm-like animals that enables them to crawl and swim.

The most anterior segments are the first to encounter new opportunities, risks and challenges in a crawling animal. There is an obvious benefit in placing eyes and antennae at the front, where they can sense any change in conditions and relay the information to a nervous system capable of instructing the appendages borne by more posterior segments to take appropriate action. Because of this necessary asymmetry between front and rear, the front few segments tend to bear sense organs and brain, and evolve into a head. This reflects the gradual separation of somatic from visceral features which is most marked in arthropods and chordates. This process began in the early Cambrian, was well under way by Chengjiang times, and was more or less complete by the end of the period, when the crown groups of most modern phyla can be recognized. Like all biological trends, the shift to a more somatic animal was not uniform and could even be put in reverse, as in the great echinoderm renunciation of eyes, brain, gill slits, segmental muscles and fins. Not all the crown groups originated during the Cambrian, either; bryozoans, for example, waited until later to make their appearance (at least as fossils), and minute soft-bodied animals like gastrotrichs have no fossil record at all.

## The arms race

The evolution of muscle-powered motility, appendages and jaws sparked an arms race between predators and prey in the early Cambrian, 540–530 Mya. The small shellies, wearing swords or shields on the outside, document the struggle between predators and prey which began in the early Cambrian and has gone on ever since. Predators like *Protohertzina* evolved sharp tusks to stab soft-bodied prey and make it more difficult for the prey to escape. Prey like *Anabarites* responded by secreting shells, helped by changes in ocean chemistry that made it easier to precipitate calcium and phosphorus. The race has no definite finish line; predators will continue to evolve stronger jaws and sharper tusks, and prey will respond by growing thicker shells. The outcome was the small shelly fauna, which included whole organisms protected by calcium carbonate or calcium phosphate armour, such as molluscs and hyoliths, as well as larger animals such as *Microdictyon*, with shields slung along the length of its body.

From time to time a predator invents some novel kind of weapon, such as the cnidocytes of cnidarians. Once these had evolved (or had been acquired from microsporidians) they could be used to capture small living prey. The hunter became the hunted when animals with jaws had become large enough to graze on soft-bodied prey. Cnidocytes can be used for defence as well as attack, however: sea anemones use them for protection against grazing nudibranch molluscs—'sea slugs'—as well as for capturing prey such as small shrimps. The nudibranchs respond by secreting a mucus that specifically inhibits cnidocyte discharge, so they can eat the anemone—and steal its cnidocytes, which they relocate to their gill tufts and use to protect themselves in turn.

The arms race stimulates both quantitative and qualitative improvements in weaponry and armour. Since ecological as well as dynamic principles have not changed since the Cambrian, no doubt such chain reactions of adaptations and counter-adaptations were set in motion when the earliest predators began to attack defenceless prey. The arms race was accelerated by the evolution of the head and new ways of seeking out prey and avoiding predators. Predators evolved complex image-forming eyes and became capable of discerning dilute chemical cues; prey responded by building hard or tough envelopes, or simply fled when they detected a predator nearby. Opportunists turned the table and evolved into parasites attached to the outside or invading the inside of larger animals that would otherwise eat them. The coevolutionary struggle that began in the Cambrian has continued ever since everywhere.

## Starfish from Mars

Despite the arms race, few if any novel kinds of large robust animals have appeared since the Cambrian, and one might well ask why. Animals are exceedingly diverse, but whether or not they embody all possible designs of a multicellular heterotrophic organism is difficult to say. I suspect they do not, although I can't prove it, and I suspect that my impression is based on some notably bizarre body plans. It does not seem very likely, for example, that the tube feet of echinoderms are the inevitable outcome of a radiation that has encompassed every viable combination of characters. If a future expedition to Mars returns with a fossil segmented worm I shall understand it as an example of parallel evolution, but if it returns with a fossil starfish I shall suspect a hoax. If animals came in a predictable range of body plans, we should not find it so difficult to interpret *Vetulicola*, *Dinomischus* or *Eldonia*.

This argument implies that there are kinds of animal that have not evolved, and imposes a duty to specify them. Imagining body plans and ways of life that have not evolved is difficult; we prefer the easier task of explaining those that have. Animal body plans use three of the six classical simple machines: lever, wedge and inclined plane. All are incorporated in devices that use linear or reciprocating motion, such as limbs, teeth and fins. There are also many examples of pulleys, for example to change the angle at which a tendon pulls, but the block is always a groove lined with a slippery substance such as cartilage, rather than a wheel mounted on an axle. There are no examples of wheels, in fact, nor of screws. Rotatory motion is presumably ruled out because it is incompatible with a tissue supplied with blood vessels.

There may be a variety of reasons for the failure of animals to use all possible devices. In the first place, some missing faculties may reflect constraints common to all organisms. Animals cannot detect or emit radio waves, for example, but no other kind of organism can either. Other absences might be due to narrower ancestral constraints. No animals use cellulose in their body wall (except tunicates, who borrowed it from bacteria) and few can even digest it without the aid of symbiotic bacteria, despite its abundance and availability as the main structural carbohydrate of seaweeds and land plants. Yet others may not have evolved because a way of life is inviable. The lack of aerial zooplankton, buoyed by secreting hydrogen into a bladder, might be because there is not enough to eat in the clouds.

Such instances apart, however, there seem to be some obvious gaps in animal design. Some specify structures or arrangements for locomotion that have rarely or never evolved. Sails are used as means of propulsion only by a few jellyfish, despite making long-distance travel or dispersal cheap. There are no double-headed animals able to move and feed in both directions without having to turn around. There are a few animals with an undivided coelom, many with a coelom divided into two or three compartments, and lots of fully segmented animals with many compartments. There are very few with several compartments, say five or six. Tardigrades have four segments, plus a head, and might show that a body plan with several segments works best in small animals.

Other gaps are directly concerned with feeding. For example, no animal has yet rigged up a muscle-powered pump for suspension feeding. Animals use harpoons, but none uses a hook and line to catch its prey (angler fish use lures, but the prey is captured by suction). It is not strictly necessary to make a recurved hook with a sharp point, nor to knot the hook to the line, which most animals would find difficult; indeed, a short straight piece of bone or chitin, glued in the middle to a silk thread, would serve. Animals do not use a noose to trap their prey either, nor visit traplines, although several can spin webs.

There might even be kinds of animal whose whole body is organized in a way completely different to any other, although it is difficult to imagine what they might be. A body plan based on electric motors, for example, might be very successful if it could evolve because it enables rotatory motion that could be used to spin a propeller or, through a gear chain, used to move or manipulate objects. We make electric motors out of metal, but animals could use chemical means of transmitting a current. The air gap between rotor and stator solves the blood vessel problem, because the rotor and shaft could be made of some rigid substance such as chitin or apatite which does not require a blood supply. A motor does require a battery and a magnet, however, and neither is part of the body of most animals. Batteries do exist in electric fish, where they are used for navigation and prey capture. These could be used to create a magnetic field, or magnetite crystals assembled to make a sizeable permanent magnet, although neither possibility has been realized in animals. Moreover, there is the problem of constructing a commutator that reverses the current every time the rotor turns. And the air gap really does have to be air; seawater doesn't work because it conducts electricity and would cause a short circuit. None of these problems is insurmountable, at least in principle, but the difficulty of solving them, with intermediate stages having no clear advantage, perhaps explains why electric motors have not evolved.

Despite the difficulty of imagining non-existent animals, I am left with a strong impression that a range of viable body plans have simply not evolved. Certainly, almost all groups of animals that have a fossil record at all date from the Cambrian, and the failure of a completely novel kind of animal to evolve since then remains a puzzle. Leo Buss suggested that the window for heritable somatic mutations, between the end of the period of maternal determination and the segregation of the germ line, was closed early in the Cambrian, after which novel body plans were unlikely to evolve. This is not a completely satisfactory explanation, because it ignores germline mutations such as the duplication of *Hox* genes, and besides the window is not quite closed in many groups of animals, but it is the best guess that has been made so far.

## Orthogenesis

What caused the unique radiation of animal body plans in the Cambrian? Was it simply natural selection, or must we invoke a different kind of process that guides evolution along a particular track?

One kind of theory finds the source of evolutionary change in the genome itself. Organisms have an innate tendency to evolve in a particular direction, it is said, so the course of evolution reflects the unfolding of a potential embedded within each individual. Ideas of this kind, which can be broadly termed 'orthogenesis', were popularized in the late 19th century by biologists who accepted evolution but were repelled by natural selection. It seemed more plausible to regard clearly progressive change from simpler and looser to more complex and highly integrated (such as the elaboration of the eye from a mere pigmented pit to a precisely engineered camera) as the outcome of a built-in tendency rather than random change. Orthogenesis was popular in the 1920s, when several versions appeared. The best-known is perhaps the 'nomogenesis' of Lev Berg, who invoked a special category of mutation that affected all members of a species simultaneously, in such a way that their offspring had a more advanced body plan. The French philosopher Henri Bergson, advancing his theory of 'creative evolution', thought that the whole species was liable to change from time to time. The American palaeontologist Henry Fairfield Osborn likewise supposed that a distinct category of mutations in 'aristogenes' drove lineages towards a more advanced state.

Both Berg and Osborn came close to the concept of homeotic mutations. A century later, it has been established that the genes responsible for animal body plans do indeed accumulate from more basal to more derived clades. Sponges have no *Hox* genes; cnidarians have one or two; acoels have three; and bilaterians have the full set, and in some cases (chordates like ourselves, for instance) the set has been expanded by repeated re-duplication to an array of 30 or 40 genes. It is difficult to resist the conclusion that once the original *Hox* gene became implicated in development the subsequent evolution of animals was pre-ordained. It follows that the course of evolution could be predicted given the initial state of a lineage, provided that the mechanism of transmission is understood.

The modern version of orthogenesis is the theory of 'molecular drive' put forward by Gabby Dover, a geneticist who spent most of his career at Cambridge. (He became head of the Genetics Department at the University of Leicester in 1992, a few years too late to have Trevor Ford as a colleague.) He specialized in the evolution of gene families (such as the *Hox* genes), whose transmission is affected by events such as unequal crossing-over, which sometimes results in the simultaneous spread of new variants. He was led to propose molecular drive, rather than natural selection, to explain why some gene families, such as ribosomal RNA genes or genes involved in the immune response, are strikingly similar among members of the same species while differing between related species. Like previous directed-evolution theories, it was put forward because natural selection acting on random mutations could not (in the opinion of the proponents) be responsible for rapid directional change. The pace of events in the late Ediacaran and early Cambrian is consistent with the idea that some process of orthogenesis must have been at work.

## Natural selection

The alternative point of view is Darwinism: the diversification of body plans was driven by imperfect competition. Different ways of surviving, moving, feeding and reproducing require different and often antagonistic adaptations. A body plan which is efficient for burrowing is unlikely to be equally efficient in crawling or swimming; ciliated tentacles evolve for suspension feeding but are unable to catch large prey; sessile animals are able to monopolize productive patches of the sea floor whereas motile animals can move on when a patch has become exhausted. Specialization is the outcome of competition, which actively maintains a wide variety of alternative ways of life. The developmental pathways that lead to different body plans, underlain by *Hox* and other kinds of gene, evolved in the early Cambrian to exploit this new range of opportunities. Selection was pushed by the evolution of larger bodies requiring internal spaces and using muscle for propulsion, following the enhanced oxygenation of the ocean at some time after the Gaskiers glaciation.

As for the objection that selection can act only very slowly, both field and lab studies have shown that selection is often strong and is readily capable of driving rapid change, even in the fundamental ecological attributes of organisms. Besides, although the late Ediacaran and early Cambrian might be brief in geological terms, they span several million generations in biological terms, which is plenty of time for the whole parade of early animals to pass.

The balance between internal and external factors is the balance between ancestral constraint and adaptive change. Orthogenesis and natural selection may seem to be starkly opposed explanations of evolution, but nevertheless they can be reconciled. Regular multicellular development requires three conditions: signals that cause a cell to develop in a particular way; gradients in signals so that cells become different; and above all the heritability of the signal gradients. Several kinds of eukaryotes have independently evolved multicellular bodies with somatic differentiation, but the gene families that underlie differentiation and development are different in each case. Animals use *Hox*, *ParaHox* and *Wnt* genes to pattern their bodies, and can take advantage of new opportunities only by modifying these genes. There is no need to invoke a novel directional force to explain animal evolution; natural selection is quite capable of doing the job. Nevertheless, natural selection can act only on variants that arise by mutation, and variation is constrained by the kind of DNA-binding genes that are initially responsible for somatic differentiation.

By the end of the Cambrian the ancestors of modern flatworms, annelids, molluscs, arthropods, echinoderms, hemichordates and chordates have emerged from the murk and can be recognized for what they were. This implies that the *Hox* cluster had by then evolved a particular configuration that governed the characteristic body plan of each phylum. The stems of most or all of these groups were already present in the communities of the Cambrian shore, living in, on or above the muddy bed of a shallow sea 520–500 Mya. This is the end of the line as far as we are concerned; from obscure beginnings the radiation of animals was well under way and the common ancestors of most of the many different kinds of animal we can find nowadays on the sea floor or in the waters above, the crown groups, had already appeared.

Mark Martindale was the last speaker, and proposed a completely different scenario. The debate, after all, was about the structure of the phylogenetic tree, and specifically about the order in which different groups branched from the main stem. The most recent sampling of whole genomes strongly suggested that ctenophores, rather than sponges, were the earliest branch. (In the audience, Libbie Hyman shook her head at this.) He was prepared to admit that acoels were the earliest bilaterians, but the account of early animal evolution must be radically revised to take account of the unique features of ctenophores.

As soon as Martindale had sat down, everyone turned to Darwin for his verdict. Not that anyone was prepared to bow to his authority and change their opinion, but it would be interesting to hear what the old sage thought. Even sages need time to think, however, and it was some time—not that the assembled wraiths took much notice of time—before Darwin had put his notes in order and cleared his throat. I think he would have gone on to say something like this.

'I'm a little surprised that one of the most serious objections to my theory has still not been resolved to everyone's satisfaction after more than 150 years. The geological record must be even more imperfect than I had imagined. It is true that some very remarkable instances of fossilization have been discovered where soft tissues that would normally decompose quickly have been preserved as impressions in the mud. In other cases, the sculpture of the body surface has been preserved by a sheath of calcium phosphate. These exceptional deposits, for which Dr. Seilacher has coined the term 'Lagerstätten', offer us detailed catalogues of a long-vanished world. I had inferred from my theory that the PreCambrian seas must have been full of living creatures, and so they were. The Lagerstätten of Chengjiang, Mistaken Point and other sites have made it clear that pre-trilobite communities were abundant and diverse. But although each provides a detailed picture, such sites are few and far between. It is like trying to reconstruct the plot of a novel from a few scattered pages. No matter how ingenious you are, the connecting thread that ties together these pages might be impossible to discern. The connecting thread of evolution is the innumerable intermediate forms that must have existed but that have left no trace in the fossil record. I do not mean, of course, that we should expect to find a form directly intermediate between, say, a trilobite and a snail, having a mixture of the characters of each, but rather that forms intermediate between either of them and a common but unknown progenitor must have existed in the past. I think that this point is now well understood. Nevertheless, we might well be unable to recognize the parent-form of any two species unless we had a nearly perfect chain of the intermediate links. This we are manifestly unable to do because of the imperfection of the geological record.

I do not find the complete, or nearly complete, replacement of a distinctive Ediacaran community by an equally distinctive Cambrian community at all surprising. Scarcely any palaeontological discovery is more striking than the fact that the forms of life change almost simultaneously throughout the world. The reason for this is that new and improved groups will spread because of their superiority, while old groups will at the same time disappear, leading to a succession of forms that will everywhere tend to correspond.

From its inception, the Ediacaran community consisted of large complex forms which must have had smaller and simpler ancestors. The alternative is that these large forms sprang suddenly into being, but 'Natura non facit saltum'. On my theory of natural selection, we can clearly understand why she should not; for natural selection can act only by taking advantage of slight successive variations; it can never take a leap, but must act by the shortest and slowest steps. The recognition of a distinctive PreCambrian fauna of fronds and other forms, therefore, does not solve the difficulty at all, but merely pushes the search for an ancestor further back in time.

Some possible configurations for this small and simple ancestor are specified by the first three of the talks we have heard today: the Gastraea, the Plakula, and the parenchymella. All are minute, and recall stages in the early embryonic development of animals. The Gastraea theory, indeed, came directly from the observation that the earliest stages of development seem to be very similar in many groups of marine animals whose adults are quite different. There seems to be little doubt that the embryo seems to be left as a sort of picture, preserved by nature, of the ancient and less modified condition of each animal. In the fourth talk, Dr. Boaden conjured up a vision of how these minute forms were the only animals that existed for millions of years, while the defining characteristics of animals gradually evolved and the major groups began to diverge.

My theory requires that smaller and simpler forms should precede those that are larger and more complicated, and I have little doubt that one or another of these scenarios outlined by the first four speakers is close to the truth. Whether or not the large Ediacaran animals were the direct ancestors of modern groups, or were some distance from the direct line of ancestry, or were not animals at all but some quite different kind of organism, was the question discussed by the next three speakers. It seems to me quite crucial to clear up this question unambiguously. It is not one that would have been seen to be vital when I first published my theory. Foraminiferans were simpler than animals and so made perfectly good ancestors; it was not until much later that it was understood that many kinds of organisms had already diverged before the first animals appeared. If the large Ediacaran fossils were gigantic foraminiferans, or some other kind of essentially single-celled organism, they could not have been the ancestors of animals. On the other hand, if they were early sponges or acalephs, then they were close to the line of descent leading to animals. The prevailing view seems to be that they are veritable animals, albeit of uncertain allegiance. I am not convinced, though, that the question is settled beyond dispute.

On the one hand, they appear at exactly the right time, just as I had predicted. This is, I feel, the strongest point in their favour. On the other hand, they lack all the usual attributes of animals, and there are none of the intermediate forms, possessing some attributes but not others, that I would expect to find. I must therefore suspend judgement on this crucial question. Until it is cleared up, our knowledge of the early history of animals must remain incomplete and insecure.

If the first four talks described plausible theories about the structure of the ancestor, without any supporting fossil evidence, while the next three were replete with fossils but lacked any plausible scheme of ancestry, then the final talk lacked both theories and fossils. It is a purely phylogenetic hypothesis, asserting that ctenophores, or rather the ancestors of ctenophores, diverged from the main line of animal evolution before any of the other groups were formed. Speaking as a morphologist, it seems to me a most unlikely speculation, but lacking any knowledge of genomics I am simply not competent to judge its merit.

I am sorry that I have nothing more positive to contribute. The current state of knowledge seems to me too fragile to justify any definitive conclusion. On purely general grounds, for I am not an expert palaeontologist by any means, I can see no reason to dismiss any of the theories that have been advanced today. Perhaps all of them embody some fragment of the truth. I am surprised, as I say, that the difficulty should not have been resolved after so much time and effort. We must wait for fresh evidence before any final conclusion can be reached'.

When Darwin had finished speaking the wraiths began to disappear one by one, melting away like early morning mist, until only Darwin himself was left, a puzzled frown on his face. Finally he too blurred and vanished, leaving the study as it was before the colloquium. Perhaps you are as puzzled as he was, and prefer like him to be agnostic. Or perhaps you have been persuaded by one or another of the schemes that I have described. I have given my own narrative of how events might have unfolded, but you may not agree with all, or any, of the stages that I have suggested. To repeat what I said in the Foreword, it is up to you to make up your mind about one of the most puzzling and absorbing mysteries in the whole of science.

# Bibliography

There are several recent and not-so-recent books that give general accounts of Cambrian or Ediacaran fossils. The most compendious is:

Schopf, J.W. and Klein, C. (eds.), 1992. *The Proterozoic Biosphere.* Cambridge University Press, Cambridge.

This includes chapters by dozens of authors and provides a complete account of what was known at the time about Ediacaran and early Cambrian fossils. However, it runs to 1,300 pages of small type, weighs several kilos, and may be a little technical for most readers. More approachable accounts include:

Brasier, M., 2009. *Darwin's Lost World: the Hidden History of Animal Life.* Oxford University Press, Oxford.

Briggs, D.E.G., Erwin, D.H. and Collier, F.J., 1994. *The Fossils of the Burgess Shale.* Smithsonian Institute Press, Washington, DC, USA.

Conway Morris, S., 1998. *The Crucible of Creation: the Burgess Shale and the Rise of Animals.* Oxford University Press, Oxford.

Erwin, D.E. and Valentine, J.W., 2013. *The Cambrian Explosion: the Construction of Animal Biodiversity.* Roberts and Co, Greenwood Village, CO, USA.

Fedonkin, M.A., Gehling, J.G., Grey, K., Narbonne, G.M. and Vickers-Rich, P., 2007. *The Rise of Animals: Evolution and Diversification of the Kingdom Animalia.* Johns Hopkins University Press, Baltimore, USA.

Gee, H., 1996. *Before the Backbone: Views on the Origin of the Vertebrates.* Chapman & Hall, London.

Glaessner, M.F., 1984. *The Dawn of Animal Life: a Biohistorical Study.* Cambridge University Press, Cambridge.

Gould, S.J., 1989. *Wonderful Life: the Burgess Shale and the Nature of History.* W.W. Norton & Company, New York.

Hou, X.-G., Siveter, D.J., Siveter, D.J., Aldridge, R.J., Cong, P.-., Gabbott, S.E., Ma, X.-Y., Purnell, M.A. and Williams, M., 2017. *The Cambrian Fossils of Chengjiang, China: the Flowering of Early Animal Life.* John Wiley & Sons, New York.

Lipps, J.H. and Signor, P.W. (eds.)., 1992. *Origin and Early Evolution of the Metazoa.* Plenum Press, New York, USA.

Simonetta, A.M. and Conway Morris, S. (eds.)., 1991. *The Early Evolution of Metazoa and the Significance of Problematic Taxa.* Cambridge University Press, Cambridge.

Whittington, H.B., 1985. *The Burgess Shale.* Yale University Press, New Haven, CT, USA.

There are also many review articles. Here is a partial list.

Budd, G.E., 2008. The earliest fossil record of the animals and its significance. *Philosophical Transactions of the Royal Society B: Biological Sciences, 363*(1496), pp.1425–1434.

Budd, G.E. and Jensen, S., 2000. A critical reappraisal of the fossil record of the bilaterian phyla. *Biological Reviews, 75*(2), pp.253–295.

Budd, G.E. and Jensen, S., 2017. The origin of the animals and a 'Savannah' hypothesis for early bilaterian evolution. *Biological Reviews, 92*(1), pp.446–473.

Butterfield, N.J., 2015. The Neoproterozoic. *Current Biology, 25*(19), pp.R859–R863.

Carbone, C. and Narbonne, G.M., 2014. When life got smart: the evolution of behavioral complexity through the Ediacaran and early Cambrian of NW Canada. *Journal of Paleontology, 88*(2), pp.309–330.

Collins, A.G. and Valentine, J.W., 2001. Defining phyla: evolutionary pathways to metazoan body plans. *Evolution & Development*, *3*(6), pp.432–442.

Cong, P., 2023. The early animal radiation: insights from interpreting the Cambrian problematic fossils. *Frontiers in Earth Science*, *11*, p.1120118.

Donoghue, P.C. and Purnell, M.A., 2009. Distinguishing heat from light in debate over controversial fossils. *BioEssays*, *31*(2), pp.178–189.

Droser, M.L. and Gehling, J.G., 2015. The advent of animals: the view from the Ediacaran. *Proceedings of the National Academy of Sciences*, *112*(16), pp.4865–4870.

Erwin, D.H., 2009. Early origin of the bilaterian developmental toolkit. *Philosophical Transactions of the Royal Society B: Biological Sciences*, *364*(1527), pp.2253–2261.

Erwin, D.H., 2015. Early metazoan life: divergence, environment and ecology. *Philosophical Transactions of the Royal Society B: Biological Sciences*, *370*(1684), p.20150036.

Erwin, D.H., 2020. The origin of animal body plans: a view from fossil evidence and the regulatory genome. *Development*, *147*(4), p.dev182899.

Erwin, D.H., 2021. Developmental capacity and the early evolution of animals. *Journal of the Geological Society*, *178*(5), pp.2020–2245.

Erwin, D.H., Laflamme, M., Tweedt, S.M., Sperling, E.A., Pisani, D. and Peterson, K.J., 2011. The Cambrian conundrum: early divergence and later ecological success in the early history of animals. *Science*, *334*(6059), pp.1091–1097.

Fedonkin, M.A., 2003. The origin of the Metazoa in the light of the Proterozoic fossil record. *Paleontological Research*, *7*(1), pp.9–41.

Galis, F., Metz, J.A. and van Alphen, J.J., 2018. Development and evolutionary constraints in animals. *Annual Review of Ecology, Evolution, and Systematics*, *49*(1), pp.499–522.

King, N. and Rokas, A., 2017. Embracing uncertainty in reconstructing early animal evolution. *Current Biology*, *27*(19), pp.R1081–R1088.

MacGabhann, B.A., Schiffbauer, J.D., Hagadorn, J.W., Van Roy, P., Lynch, E.P., Morrison, L. and Murray, J., 2019. Resolution of the earliest metazoan record: differential taphonomy of Ediacaran and Paleozoic fossil molds and casts. *Palaeogeography, Palaeoclimatology, Palaeoecology*, *513*, pp.146–165.

McMenamin, M.A., 1986. The garden of Ediacara. *Palaios*, *1*, pp.178–182.

McMenamin, M., 2023. The Cambrian explosion: macroevolution and biomineralization. *Academia Biology*, *1*(2023), pp.1–7.

Mills, D.B. and Canfield, D.E., 2017. A trophic framework for animal origins. *Geobiology*, *15*(2), pp.197–210.

Morris, S.C., 1987. Cambrian enigmas. *Geology Today*, *3*(3–4), pp.88–92.

Ros-Rocher, N., Pérez-Posada, A., Leger, M.M. and Ruiz-Trillo, I., 2021. The origin of animals: an ancestral reconstruction of the unicellular-to-multicellular transition. *Open Biology*, *11*(2), p.200359.

Schopf, J.W., 2000. Solution to Darwin's dilemma: discovery of the missing Precambrian record of life. *Proceedings of the National Academy of Sciences*, *97*(13), pp.6947–6953.

Schopf, J.W., 2009. The hunt for Precambrian fossils: an abbreviated genealogy of the science. *Precambrian Research*, *173*(1–4), pp.4–9.

Shu, D., Isozaki, Y., Zhang, X., Han, J. and Maruyama, S., 2014. Birth and early evolution of metazoans. *Gondwana Research*, *25*(3), pp.884–895.

Slater, B.J. and Bohlin, M.S., 2022. Animal origins: the record from organic microfossils. *Earth-Science Reviews*, *232*, p.104107.

Waggoner, B., 2003. The Ediacaran biotas in space and time. *Integrative and Comparative Biology*, *43*(1), pp.104–113.

Wood, R., Donoghue, P.C., Lenton, T.M., Liu, A.G. and Poulton, S.W., 2020. The origin and rise of complex life: progress requires interdisciplinary integration and hypothesis testing. *Interface Focus*, *10*(4), p.20200024.

Xiao, S. and Laflamme, M., 2009. On the eve of animal radiation: phylogeny, ecology and evolution of the Ediacara biota. *Trends in Ecology & Evolution*, *24*(1), pp.31–40.

Zhuravlev, A.Y., 2015. The early history of the Metazoa—a paleontologist's viewpoint. *Biology Bulletin Reviews*, *5*, pp.415–461. [English translation of Russian original]

Apart from these general reviews, I have listed my principal sources by chapter and section. This is not an exhaustive list; I have consulted more than 1700 articles, only about 500 of which are listed here.,

## Introduction: Darwin's Dilemma

The quotation in this section is from Chapter 10 of Darwin, C., 1859. *On the Origin of Species by means of Natural Selection, or the Preservation of Favoured Races in the Struggle for Life.* John Murray, London.

### A false start


Adelman, J., 2007. *Eozoön*: debunking the dawn animal. *Endeavour, 31*(3), pp.94–98.

Carpenter, W.B., 1864. IV. On the structure and affinities of Eozoon Canadense. *Proceedings of the Royal Society of London*, 13, pp.545–549.

Dawson, J.W., 1875. *Dawn of Life: Being the History of the Oldest Known Fossil Remains, and Their Relation to Geological Time and to the Development of the Animal Kingdom.* Hodder & Stoughton, London.

Dolan, J.R., 2023. The saga of the false fossil foram *Eozoon. European Journal of Protistology*, 87, p.125955.

Hahn, O., 1876. Is there such a thing as *Eozoon canadense*? A microgeological investigation. *Journal of Natural History, 17*(100), pp.265–282.

King, W. and Rowney, T.H., 1876. Remarks on 'The Dawn of Life,' by Dr. Dawson; to which is added a supplementary note. *Annals and Magazine of Natural History, 17*(101), pp.360–377.

Mitchell, M.E. and O'Brien, C.F., 1971. On *Eozoön canadense. Isis, 62*(3), pp.381–383.

O'Brien, C.F., 1970. *Eozoön canadens:* the dawn animal of Canada. *Isis, 61*(2), pp.206–223.


## Chapter 1. The Golden Spike

### The great Cambrian controversy


Secord, J.A., 1990. *Controversy in Victorian Geology: the Cambrian-Silurian Dispute.* Princeton University Press, Princeton, USA.


### The base of the Cambrian


Landing, E., 1994. Precambrian-Cambrian boundary global stratotype ratified and a new perspective of Cambrian time. *Geology, 22*(2), pp.179–182.

Narbonne, G.M., Myrow, P.M., Landing, E. and Anderson, M.M., 1987. A candidate stratotype for the Precambrian-Cambrian boundary, Fortune Head, Burin Peninsula, southeastern Newfoundland. *Canadian Journal of Earth Sciences*, 24, pp.1277–1293.


## Chapter 2. Mud


Haldane, J. B. S., 1949. Suggestions as to the quantitative measurement of rates of evolution. *Evolution*, 3, pp.51–56.

Weismann, A., 1909. The selection theory. In A. C. Seward (ed.), *Darwin and Modern Science* (pp.18–65). Cambridge University Press, Cambridge.


### Cells

Old cells


Schopf, J.W. (ed.)., 2002. *Life's Origin: the Beginnings of Biological Evolution.* University of California Press, Oakland, CA.

Schopf, J.W., 2021. Precambrian paleobiology: precedents, progress, and prospects. *Frontiers in Ecology and Evolution*, 9, p.707072.

A new kind of cell

Agić, H., 2016. Fossil focus: acritarchs. *Palaeontology Online*, *6*(11), pp.1–13.

Butterfield, N.J., 1997. Plankton ecology and the Proterozoic–Phanerozoic transition. *Paleobiology*, *23*(2), pp.247–262.

Cohen, P.A. and Kodner, R.B., 2022. The earliest history of eukaryotic life: uncovering an evolutionary story through the integration of biological and geological data. *Trends in Ecology & Evolution*, *37*(3), pp.246–256.

Eme, L., Sharpe, S.C., Brown, M.W. and Roger, A.J., 2014. On the age of eukaryotes: evaluating evidence from fossils and molecular clocks. *Cold Spring Harbor Perspectives in Biology*, 6(8), p.a016139.

Knoll, A.H., 2014. Paleobiological perspectives on early eukaryotic evolution. *Cold Spring Harbor Perspectives in Biology*, 6(1), p.a016121.

Martin, F., 1993. Acritarchs: a review. *Biological Reviews*, 68, pp.475–538.

**Seaweed**

Microbial mats

Gehling, J.G., 1999. Microbial mats in terminal Proterozoic siliciclastics; Ediacaran death masks. *Palaios*, *14*(1), pp.40–57.

Prieto-Barajas, C.M., Valencia-Cantero, E. and Santoyo, G., 2018. Microbial mat ecosystems: structure types, functional diversity, and biotechnological application. *Electronic Journal of Biotechnology*, *31*, pp.48–56.

Seilacher, A., 1999. Biomat-related lifestyles in the Precambrian. *Palaios*, *14*(1), pp.86–93.

Tarhan, L.G., Planavsky, N.J., Laumer, C.E., Stolz, J.F. and Reid, R.P., 2013. Microbial mat controls on infaunal abundance and diversity in modern marine microbialites. *Geobiology*, *11*(5), pp.485–497.

The first filament

Bengtson, S., Sallstedt, T., Belivanova, V. and Whitehouse, M., 2017. Three-dimensional preservation of cellular and subcellular structures suggests 1.6 billion-year-old crown-group red algae. *PLoS Biology*, *15*(3), p.e2000735.

Butterfield, N.J., 2000. *Bangiomorpha pubescens* n. gen., n. sp.: implications for the evolution of sex, multicellularity, and the Mesoproterozoic/Neoproterozoic radiation of eukaryotes *Paleobiology*, *26*(3), pp.386–404.

Miao, L., Yin, Z., Knoll, A.H., Qu, Y. and Zhu, M., 2024. 1.63-billion-year-old multicellular eukaryotes from the Chuanlinggou Formation in North China. *Science Advances*, *10*(4), p.eadk3208.

Tang, Q., Pang, K., Yuan, X. and Xiao, S., 2020. A one-billion-year-old multicellular chlorophyte. *Nature Ecology & Evolution*, *4*(4), pp.543–549.

**The shadow world**

Scale

Bardell, D., 2004. The invention of the microscope. *Bios*, *75*(2), pp.78–84.

The interstitial animals

Pennak, R.W., 1968. Historical origins and ramifications of interstitial investigations. *Transactions of the American Microscopical Society*, *87*, pp.214–218.

Rundell, R.J. and Leander, B.S., 2010. Masters of miniaturization: convergent evolution among interstitial eukaryotes. *Bioessays*, *32*(5), pp.430–437.

**Burrows**

Worms

Clark, R.B., 1981. Locomotion and the phylogeny of the Metazoa. *Italian Journal of Zoology*, *48*(1), pp.11–28.

Raff, R.A. and Raff, E.C., 1970. Respiratory mechanisms and the metazoan fossil record. *Nature*, *228*(5275), pp.1003–1005.

Reynolds Number

Purcell, E.M., 1977. Life at low Reynolds number. *American Journal of Physics*, 45, pp.3–11.

Treptichnus

Buatois, L.A., 2018. *Treptichnus pedum* and the Ediacaran–Cambrian boundary: significance and caveats. *Geological Magazine*, *155*(1), pp.174–180.

Mángano, M.G. and Buatois, L.A., 2020. The rise and early evolution of animals: where do we stand from a trace-fossil perspective? *Interface Focus*, *10*(4), p.20190103.

## Chapter 3. Animals and Individuals

### Animals

Mesostoma, Hydra, Spongilla

Hyman, L.H., 1940. *The Invertebrates: Protozoa through Ctenophora*. McGraw-Hill Book Company, New York and London.

Hyman, L.H., 1951. *The Invertebrates: Platyhelminthes and Rhynchocoela. The acoelomate Bilateria*. McGraw-Hill Book Company, New York and London.

Zoophytes

Gibson, S., 2012. On being an animal, or, the eighteenth-century zoophyte controversy in Britain. *History of Science*, *50*(4), pp.453–476.

### Individuals

Eggs and individuals

Extavour, C.G. and Akam, M., 2003. Mechanisms of germ cell specification across the metazoans: epigenesis and preformation. *Development*, 130 (24), pp.5869–5884.

Van Speybroeck, L., De Waele, D. and Van De Vijver, G., 2002. Theories in early embryology: close connections between epigenesis, preformationism, and self-organization. *Annals of the New York Academy of Sciences*, *981*(1), pp.7–49.

Selfish cell lineages

Li, S.I. and Purugganan, M.D., 2011. The cooperative amoeba: *Dictyostelium* as a model for social evolution. *Trends in Genetics*, *27*(2), pp.48–54.

Defending development

Bell, G. and Mooers, A.O., 1997. Size and complexity among multicellular organisms. *Biological Journal of the Linnean Society*, *60*(3), pp.345–363.

Buss, L.W., 1987. *The Evolution of Individuality*. Princeton University Press, Princeton, USA.

## Chapter 4. Ancestors and Embryos

### Ancestors

The phylogenetic tree

Pietsch. T.W., 2012. *Trees of Life: a Visual History of Evolution*. Johns Hopkins University Press, Baltimore, USA.

The new systematics

Schmitt, M., 2003. Willi Hennig and the rise of cladistics. In A. Legakis, S. Sfenthourakis, R. Polymeni, and M. Thessalou-Legaki (eds.) *Proceedings of the 18th International Congress of Zoology*. Pensoft Publishers, Sofia and Moscow.

Are animals a natural group?

Brunet, T. and King, N., 2017. The origin of animal multicellularity and cell differentiation. *Developmental Cell*, *43*(2), pp.124–140.

Dunn, C.W., Giribet, G., Edgecombe, G.D. and Hejnol, A., 2014. Animal phylogeny and its evolutionary implications. *Annual Review of Ecology, Evolution, and Systematics*, *45*(1), pp.371–395.

James-Clark, H., 1867. Conclusive proofs of the animality of the ciliate sponges, and of their affinities with the Infusoria flagellata. *Annals and Magazine of Natural History*, *19*(109), pp.13–18.

The elusive ancestor

Jenner, R.A., 2022. *Ancestors in Evolutionary Biology: Linear Thinking about Branching Trees.* Cambridge University Press, Cambridge.

Stem and crown

Meier, R. and Richter, S., 1992. Suggestions for a more precise usage of proper names of taxa: ambiguities related to the stem lineage concept. *Journal of Zoological Systematics and Evolutionary Research*, *30*(2), pp.81–88.

**Embryos**

Haeckel's universal Darwinism

The three books by Haeckel cited in this section are:

Haeckel, E., 1876. *The History of Creation.* Trans. E. Ray Lankester. Henry King, London.

Haeckel, Ernst., 1897. *The Evolution of Man.* Trans. Anon. D. Appleton, New York.

Haeckel, Ernst., 1900. *The Riddle of the Universe.* Trans. J. McCabe. Watts & Co, London.

Recapitulation

von Baer, K.E., 1828. *Uber Entwickelungsgeschichte der Thiere.* Börntrager, Königsberg.

Cuvier, G., 1834. *Histoire Naturelle des Zoophytes.* De Roret, Paris.

Huxley, T.H., 1849. XXII. On the anatomy and the affinities of the family of the Medusae. *Philosophical Transactions of the Royal Society of London*, 139, pp.413–434.

The two books of illustrations by Haeckel from his monograph on radiolarians have been reprinted as:

Haeckel, E., 2005. *Art Forms from the Ocean: the Radiolarian Atlas of 1862.* Prestel, Munich.

Haeckel, E., 2015. *Art Forms from the Abyss: the Radiolarian Atlas of 1862.* Prestel, Munich.

The drawings by Haeckel of a conserved stage in the development of vertebrates are from *The Evolution of Man*, originally published as:

Haeckel, E., 1874. *Anthropogenie oder Entwickelungsgeschichte des Menschen.* Engelmann, Leipzig.

The recent articles criticizing it are, in chronological order:

Pennisi, E., 1997. Haeckel's embryos: fraud rediscovered. *Science*, 277, pp.1435.

Richardson, M.K., Hanken, J., Gooneratne, M.L., Pieau, C., Raynaud, A., Selwood, L. and Wright, G.M., 1997. There is no highly conserved embryonic stage in the vertebrates: implications for current theories of evolution and development. *Anatomy and Embryology*, 196, pp.91–106.

Hanken, J. and Richardson, M.K., 1998. Haeckel's embryos. *Science*, 279, 1283.

Richardson, M.K., Hanken, J., Selwood, L., Wright, G.M., Richards, R.J., Pieau, C. and Raynaud, A., 1998. Haeckel, embryos, and evolution. *Science*, 280, pp.983.

Watts, E., Levit, G.S. and Hossfeld, U., 2019. Ernst Haeckel's contribution to Evo-Devo and scientific debate: a re-evaluation of Haeckel's controversial illustrations in US textbooks in response to creationist accusations. *Theory in Biosciences*, *138*, pp.9–29.

The Gastraea

Grell, K.G., 1979. Die Gastraea-Theorie. *Medizinhistorisches Journal*, 14, pp.275–291.

Haeckel, E., 1874. The Gastraea-theory, the phylogenetic classification of the animal kingdom and the homology of the germ-lamellae. Trans. E. Perceval Wright. *Journal of Cell Science*, *2*(54), pp.142–165.

Levit, G.S., Hoßfeld, U., Naumann, B., Lukas, P. and Olsson, L., 2020. The biogenetic law and the Gastraea theory: from Ernst Haeckel's discoveries to contemporary views. *Journal of Experimental Zoology*, *338*, pp.13–27.

Leys, S.P. and Eerkes-Medrano, D., 2005. Gastrulation in calcareous sponges: in search of Haeckel's gastraea. *Integrative and Comparative Biology*, *45*(2), pp.342–351.

Nielsen, C., 1985. Animal phylogeny in the light of the trochaea theory. *Biological Journal of the Linnean Society*, *25*(3), pp.243–299.

Nielsen, C., 2013. Life cycle evolution: was the eumetazoan ancestor a holopelagic, planktotrophic gastraea? *BMC Evolutionary Biology*, *13*, pp.1–18.

The ciliation constraint

Buss, L.W., 1987. *The Evolution of Individuality*. Princeton University Press, Princeton, USA.

King, N., 2004. The unicellular ancestry of animal development. *Developmental Cell*, *7*(3), pp.313–325.

The Plakula

Bütschli, O., 1921. *Vorlesungen über Vergleichende Anatomie*. Springer, Berlin.

Metchnikoff, E., 1882. Vergleichend-embryologische Studien. 3. Ueber die Gastrula einiger Metazoen. *Zeitschrift für Wissenschaftliche Zoologie 37*, pp.286–313.

von Graff, L., 1891. *Die Organisation der Turbellaria Acoela*. Engelmann, Leipzig.

## Chapter 5. The Grand Design

### Morphologists

Huxley, J., 1942. *Evolution, the Modern Synthesis*. George Allen & Unwin, London.

Hyman, L.H., 1940–1967. *The Invertebrates* (6 vols.). McGraw-Hill, New York, USA.

### The choice of characters

Martindale, M.Q., 2005. The evolution of metazoan axial properties. *Nature Reviews Genetics*, *6*(12), pp.917–927.

### The phylum

Budd, G.E. and Jensen, S., 2000. A critical reappraisal of the fossil record of the bilaterian phyla. *Biological Reviews*, *75*(2), pp.253–295.

Valentine, J.W., 2006. *On the Origin of Phyla*. Chicago University Press, Chicago, USA.

### The great debate

Blair, S.S., 2008. Segmentation in animals. *Current Biology*, *18*(21), pp.R991–R995.

Chipman, A.D., 2021. The evolution and development of segmented body plans. In N. Nuna de la Rosa and G.B. Müller (eds.), *Evolutionary Developmental Biology* (pp.545–554). Springer International Publishing, Switzerland.

Clark, R.B., 1964. *Dynamics in Metazoan Evolution: the Origin of the Coelom and Segments.* Clarendon Press, Oxford.

Davis, G.K. and Patel, N.H., 1999. The origin and evolution of segmentation. *Trends in Genetics, 15*, pp.M68–M72.

Sarvaas, A.E. du M., 1933. *La Théorie du Coelome.* PhD thesis, University of Utrecht.

Sedgwick, A., 1884. On the origin of metameric segmentation and some other morphological questions. *Journal of Cell Science, 2*(93), pp.43–82.

Tautz, D., 2004. Segmentation. *Developmental cell, 7*(3), pp.301–312.

### The progressive theory of animal evolution

Halanych, K.M. and Passamaneck, Y., 2001. A brief review of metazoan phylogeny and future prospects in Hox-research. *American Zoologist, 41*(3), pp.629–639.

### Relics

The obvious ancestor

Saville-Kent, W., 1880–1882. *A Manual of the Infusoria: Including a Description of All Known Flagellate, Ciliate, and Tentaculiferous Protozoa, British and Foreign, and an Account of the Organization and the Affinities of the Sponges* (3 vols.). David Bogue, London.

The magician's ball

Haeckel E., 1871. Die Catallacten, eine neue Protisten-Gruppe. *Jenaische Zeitschrift für Medicin und Naturwissenschaft, 6*, pp.1–22.

Reynolds, A. and Hülsmann, N., 2008. Ernst Haeckel's discovery of "*Magosphaera planula*": a vestige of metazoan origins? *History and Philosophy of the Life Sciences, 30*, pp.339–386.

Simplified bodies

Adriano, E.A., Zatti, S.A. and Okamura, B., 2022. How to build single-celled cnidarians with worm-like motility: Lessons from Myxozoa. *Journal of Anatomy, 240*(3), pp.475–488.

Chang, E.S., Neuhof, M., Rubinstein, N.D., Diamant, A., Philippe, H., Huchon, D. and Cartwright, P., 2015. Genomic insights into the evolutionary origin of Myxozoa within Cnidaria. *Proceedings of the National Academy of Sciences, 112*(48), pp.14912–14917.

Schiffer, P.H., Robertson, H.E. and Telford, M.J., 2018. Orthonectids are highly degenerate annelid worms. *Current Biology, 28*(12), pp.1970–1974.

Stunkard, H.W., 1954. The life-history and systematic relations of the Mesozoa. *Quarterly Review of Biology, 29*(3), pp.230–244.

The animated tube

Lebour, M.V., 1930. *Protohydra*, a very simple animal (with original sketches from life). *Science Progress in the Twentieth Century (1919–1933), 25*(97), pp.64–69.

Muus, K., 1966. Notes on the biology of *Protohydra leuckarti* Greef (Hydroidea, Protohydridae). *Ophelia, 3*(1), pp.141–150.

The salt animal

Frenzel, J., 1891. *Untersuchungen über die Mikroskopische Fauna Argentiniens.* Stricker, Berlin.

Frenzel, J., 1892. The Mesozoon *Salinella. Journal of Natural History, 9*(49), pp.79–84.

Tihelka, E. and Cai, C., 2021. *Salinella. Current Biology, 31*(18), pp.R1070–R1071.

A ciliated sack

Nakano, H., 2015. What is *Xenoturbella*? *Zoological Letters, 1*, pp.1–8.

Telford, M.J. and Copley, R.R., 2016. Zoology: war of the worms. *Current Biology, 26*(8), pp.R335–R337.

The sticky hairy plate

Miller, D.J. and Ball, E.E., 2008. Animal evolution: *Trichoplax*, trees, and taxonomic turmoil. *Current Biology, 18*(21), pp.R1003–R1005.

Schierwater, B., Osigus, H.J., Bergmann, T., Blackstone, N.W., Hadrys, H., Hauslage, J., Humbert, P.O., Kamm, K., Kvansakul, M., Wysocki, K. and DeSalle, R., 2021. The enigmatic Placozoa part 1: exploring evolutionary controversies and poor ecological knowledge. *BioEssays, 43*(10), p.2100080.

Schulze, F.E., 1883. *Trichoplax adhaerens*, nov. gen., nov. spec. *Zoologisher Anzeiger, 6*, pp.92–97.

## Chapter 6. The Magic Molecule

### Gene-based trees

Field, K.G., Olsen, G.J., Lane, D.J., Giovannoni, S.J., Ghiselin, M.T., Raff, E.C., Pace, N.R. and Raff, R.A., 1988. Molecular phylogeny of the animal kingdom. *Science, 239*(4841), pp.748–753.

Zuckerkandl, E., Jones, R.T. and Pauling, L., 1960. A comparison of animal hemoglobins by tryptic peptide pattern analysis. *Proceedings of the National Academy of Sciences, 46*(10), pp.1349–1360.

Zuckerkandl, E. and Pauling, L., 1965. Molecules as documents of evolutionary history. *Journal of Theoretical Biology, 8*(2), pp.357–366.

### The new animal phylogeny

Adoutte, A., Balavoine, G., Lartillot, N., Lespinet, O., Prud'Homme, B. and De Rosa, R., 2000. The new animal phylogeny: reliability and implications. *Proceedings of the National Academy of Sciences, 97*(9), pp.4453–4456.

Aguinaldo, A.M.A., Turbeville, J.M., Linford, L.S., Rivera, M.C., Garey, J.R., Raff, R.A. and Lake, J.A., 1997. Evidence for a clade of nematodes, arthropods and other moulting animals. *Nature, 387*(6632), pp.489–493.

Hejnol, A., Obst, M., Stamatakis, A., Ott, M., Rouse, G.W., Edgecombe, G.D., Martinez, P., Baguñà, J., Bailly, X., Jondelius, U. and Wiens, M., 2009. Assessing the root of bilaterian animals with scalable phylogenomic methods. *Proceedings of the Royal Society B: Biological Sciences, 276*(1677), pp.4261–4270.

### Phylogenomics

Dunn, C.W., Hejnol, A., Matus, D.Q., Pang, K., Browne, W.E., Smith, S.A., Seaver, E., Rouse, G.W., Obst, M., Edgecombe, G.D. Sørensen, M.V., Haddock, S.H.D., Schmidt-Rhaesa, A., Okusu, A., Kristensen, R.M., Wheeler, W.C., Martindale, M.Q. and Giribet, G., 2008. Broad phylogenomic sampling improves resolution of the animal tree of life. *Nature, 452*(7188), pp.745–749.

Hotaling, S., Kelley, J.L. and Frandsen, P.B., 2021. Toward a genome sequence for every animal: where are we now? *Proceedings of the National Academy of Sciences, 118*(52), p.e2109019118.

Nielsen, C., 2019. Early animal evolution: a morphologist's view. *Royal Society Open Science, 6*(7), p.190638.

Philippe, H., Derelle, R., Lopez, P., Pick, K., Borchiellini, C., Boury-Esnault, N., Vacelet, J., Renard, E., Houliston, E., Quéinnec,E., Da Silva, C., Wincker, P., Le Guyader, H., Leys, S., Jackson, D.J., Schreiber, F., Erpenbeck, D., Morgenstern, B., Wörheide, G. and Manuel, M., 2009. Phylogenomics revives traditional views on deep animal relationships. *Current Biology, 19*(8), pp.706–712.

Ryan, J.F., Pang, K., Schnitzler, C.E., Nguyen, A.D., Moreland, R.T., Simmons, D.K., Koch, B.J., Francis, W.R., Havlak, P., Smith, S.A., Putnam, N.H., Haddock, S.H.D., Dunn, C.W., Wolfsberg, T.G., Mullikin, J.C., Martindale, M.Q. and Baxevanis, A.D., 2013. The genome of the ctenophore *Mnemiopsis leidyi* and its implications for cell type evolution. *Science, 342*(6164), p.1242592.

### The problem of short branches

Laumer, C.E., Fernández, R., Lemer, S., Combosch, D., Kocot, K.M., Riesgo, A. Andrade, S.C., Sterrer, W., Sørensen, M.V. and Giribet, G., 2019. Revisiting metazoan phylogeny with genomic sampling of all phyla. *Proceedings of the Royal Society B, 286*(1906), p.20190831.

Janvier, P., 1997. *Les Chiens Abouent* (book review). 'In retrospect' chosen by Philippe Janvier. *Nature*, *389*, pp.688.

Water filters

Botting, J.P. and Muir, L.A., 2018. Early sponge evolution: a review and phylogenetic framework. *Palaeoworld*, *27*(1), pp.1–29.

Chang, S., Zhang, L., Clausen, S., Bottjer, D.J. and Feng, Q., 2019. The Ediacaran-Cambrian rise of siliceous sponges and development of modern oceanic ecosystems. *Precambrian Research*, *333*, p.105438.

Kloss, T.J., Dornbos, S.Q. and Chen, J.Y., 2009. Paleoecology and taphonomy of the Early Cambrian Maotianshan Shale biota chancelloriid *Allonnia junyuani*: Adaptation to nonactualistic Cambrian substrates. *Palaeogeography, Palaeoclimatology, Palaeoecology*, *277*(3–4), pp.149–157.

Rigby, J.K. and Xian-Guang, H., 1995. Lower Cambrian demosponges and hexactinellid sponges from Yunnan, China. *Journal of Paleontology*, *69*(6), pp.1009–1019.

Moore, J.L., Li, G. and Porter, S.M., 2014. Chancelloriid sclerites from the Lower Cambrian (Meishucunian) of eastern Yunnan, China, and the early history of the group. *Palaeontology*, *57*(4), pp.833–878.

Müller, W.E., 2003. The origin of metazoan complexity: Porifera as integrated animals. *Integrative and Comparative Biology*, *43*(1), pp.3–10.

Wang, X., Hu, S., Gan, L., Wiens, M. and Müller, W.E., 2010. Sponges (Porifera) as living metazoan witnesses from the Neoproterozoic: biomineralization and the concept of their evolutionary success. *Terra Nova*, *22*(1), pp.1–11.

Yuan, X., Xiao, S., Parsley, R.L., Zhou, C., Chen, Z. and Hu, J., 2002. Towering sponges in an Early Cambrian Lagerstatte: disparity between nonbilaterian and bilaterian epifaunal tierers at the Neoproterozoic-Cambrian transition. *Geology*, *30*(4), pp.363–366.

Yun, H., Zhang, X. and Li, L., 2018. Chancelloriid *Allonnia erjiensis* sp. nov. from the Chengjiang Lagerstätte of South China. *Journal of Systematic Palaeontology*, *16*(5), pp.435–444.

Polyps

Han, J., Hu, S., Cartwright, P., Zhao, F., Ou, Q., Kubota, S., Wang, X. and Yang, X., 2016. The earliest pelagic jellyfish with rhopalia from Cambrian Chengjiang Lagerstätte. *Palaeogeography, Palaeoclimatology, Palaeoecology*, *449*, pp.166–173.

Ou, Q., Han, J., Zhang, Z., Shu, D., Sun, G. and Mayer, G., 2017. Three Cambrian fossils assembled into an extinct body plan of cnidarian affinity. *Proceedings of the National Academy of Sciences*, *114*(33), pp.8835–8840.

Zhao, Y., Hou, X.G. and Cong, P.Y., 2023. Tentacular nature of the 'column' of the Cambrian diploblastic *Xianguangia sinica*. *Journal of Systematic Palaeontology*, *21*(1), p.2215787.

Zhao, Y., Parry, L.A., Vinther, J., Dunn, F.S., Li, Y.J., Wei, F., Hou, X.G. and Cong, P.Y., 2021. An early Cambrian polyp reveals an anemone-like ancestor for medusozoan cnidarians. *bioRxiv*, *2021*, https://doi.org/10.1101/2021.12.24.474121.

Zhao, Y., Vinther, J., Parry, L.A., Wei, F., Green, E., Pisani, D., Hou, X., Edgecombe, G.D. and Cong, P., 2019. Cambrian sessile, suspension feeding stem-group ctenophores and evolution of the comb jelly body plan. *Current Biology*, *29*(7), pp.1112–1125.

The lophophore

Harper, D.A., Popov, L.E. and Holmer, L.E., 2017. Brachiopods: origin and early history. *Palaeontology*, *60*(5), pp.609–631.

Liu, F., Skovsted, C.B., Topper, T.P., Zhang, Z. and Shu, D., 2020. Are hyoliths Palaeozoic lophophorates? *National Science Review*, *7*(2), pp.453–469.

Moysiuk, J., Smith, M.R. and Caron, J.B., 2017. Hyoliths are Palaeozoic lophophorates. *Nature*, *541*(7637), pp.394–397.

Zhang, Z., Li, G., Emig, C.C., Han, J., Holmer, L.E. and Shu, D., 2009. Architecture and function of the lophophore in the problematic brachiopod *Heliomedusa orienta* (Early Cambrian, South China). *Geobios*, *42*(5), pp.649–661.

Legs

Boxshall, G.A., 2004. The evolution of arthropod limbs. *Biological Reviews, 79*(2), pp.253–300.

Brandt, D.S., 2007. Multiple-*Rusophycus* (arthropod ichnofossil) assemblages and their significance. *Ichnos, 15*(1), pp.28–43.

Fu, D., Legg, D.A., Daley, A.C., Budd, G.E., Wu, Y. and Zhang, X., 2022. The evolution of biramous appendages revealed by a carapace-bearing Cambrian arthropod. *Philosophical Transactions of the Royal Society B, 377*(1847), p.20210034.

Laere, G. van, 2021. *The Earliest Arthropod Trace Fossils and Their Significance*. PhD thesis, Uppsala University.

Liu, J., Steiner, M., Dunlop, J.A., Keupp, H., Shu, D., Ou, Q., Han, J., Zhang, Z. and Zhang, X., 2011. An armoured Cambrian lobopodian from China with arthropod-like appendages. *Nature, 470*(7335), pp.526–530.

Taylor, D. and Dirks, J.H., 2012. Shape optimization in exoskeletons and endoskeletons. *Journal of the Royal Society Interface, 9*, 3480–3489.

The Chengjiang arthropods

Aria, C., 2022. The origin and early evolution of arthropods. *Biological Reviews, 97*(5), pp.1786–1809.

Daley, A.C., Antcliffe, J.B., Drage, H.B. and Pates, S., 2018. Early fossil record of Euarthropoda and the Cambrian Explosion. *Proceedings of the National Academy of Sciences, 115*(21), pp.5323–5331.

Haug, J.T., Briggs, D.E. and Haug, C., 2012. Morphology and function in the Cambrian Burgess Shale megacheiran arthropod *Leanchoilia superlata* and the application of a descriptive matrix. *BMC Evolutionary Biology, 12*, pp.1–20.

Hou, X., Williams, M., Siveter, D.J., Siveter, D.J., Aldridge, R.J. and Sansom, R.S., 2010. Soft-part anatomy of the Early Cambrian bivalved arthropods *Kunyangella* and *Kunmingella*: significance for the phylogenetic relationships of Bradoriida. *Proceedings of the Royal Society B: Biological Sciences, 277*(1689), pp.1835–1841.

Liu, Y., Hou, X.G. and Bergström, J., 2007. Chengjiang arthropod *Leanchoilia illecebrosa* (Hou, 1987) reconsidered. *GFF, 129*(3), pp.263–272.

McMenamin, M.A., 2020. Bradoriids (Arthropoda) and the Cambrian diversification. *Geosciences, 10*(4), p.119.

Ortega-Hernández, J., Yang, J. and Zhang, X.G., 2018. Fuxianhuiids. *Current Biology, 28*(13), pp.R724–R725.

Perrier, V., Williams, M. and Siveter, D.J., 2015. The fossil record and palaeoenvironmental significance of marine arthropod zooplankton. *Earth-Science Reviews, 146*, pp.146–162.

Zhai, D., Williams, M., Siveter, D.J., Harvey, T.H., Sansom, R.S., Gabbott, S.E., Siveter, D.J., Ma, X., Zhou, R., Liu, Y. and Hou, X., 2019. Variation in appendages in early Cambrian bradoriids reveals a wide range of body plans in stem-euarthropods. *Communications Biology, 2*(1), p.329.

Heads

Holmes, J.D. and Budd, G.E., 2022. Reassessing a cryptic history of early trilobite evolution. *Communications Biology, 5*(1), p.1177.

Vannier, J., Abe, K. and Ikuta, K., 1998. Feeding in myodocopid ostracods: functional morphology and laboratory observations from videos. *Marine Biology, 132*, pp.391–408.

Eyes

Fernald, R.D., 2006. Casting a genetic light on the evolution of eyes. *Science, 313*(5795), pp.1914–1918.

Ma, X., Hou,X., Aldridge, R.J., Siveter, D.J., Siveter, D.J., Gabbott, S.E., Purnell, M.A., Parker, A.R. and Edgecombe, G.D., 2012. Morphology of Cambrian lobopodian eyes from the Chengjiang Lagerstätte and their evolutionary significance. *Arthropod Structure & Development, 41*(5), pp.495–504.

Martin, C., Gross, V., Hering, L., Tepper, B., Jahn, H., de Sena Oliveira, I., Stevenson, P.A. and Mayer, G., 2017. The nervous and visual systems of onychophorans and tardigrades: learning about arthropod evolution from their closest relatives. *Journal of Comparative Physiology A, 203*, pp.565–590.

Nilsson, D.E., 2013. Eye evolution and its functional basis. *Visual Neuroscience, 30*(1-2), pp.5–20.

Zhao, F., Bottjer, D.J., Hu, S., Yin, Z. and Zhu, M., 2013. Complexity and diversity of eyes in Early Cambrian ecosystems. *Scientific Reports, 3*(1), p.2751.

Fins

Van Roy, P., Daley, A.C. and Briggs, D.E., 2015. Anomalocaridid trunk limb homology revealed by a giant filter-feeder with paired flaps. *Nature, 522*(7554), pp.77–80.

Gill slits

Aldridge, R.J., Xian-Guang, H.O.U., Siveter, D.J., Siveter D.J. and Gabbott, S.E., 2007. The systematics and phylogenetic relationships of vetulicolians. *Palaeontology, 50*(1), pp.131–168.

Gee, H., 2001. On being vetulicolian. *Nature, 414*(6862), pp.407–409.

Graham, A. and Richardson, J., 2012. Developmental and evolutionary origins of the pharyngeal apparatus. *EvoDevo, 3*, pp.1–8.

Lacalli, T.C., 2002. Vetulicolians—are they deuterostomes? Chordates? *BioEssays, 24*(3), pp.208–211.

Ou, Q., Morris, S.C., Han, J., Zhang, Z., Liu, J., Chen, A., Zhang, X. and Shu, D., 2012. Evidence for gill slits and a pharynx in Cambrian vetulicolians: implications for the early evolution of deuterostomes. *BMC Biology, 10*, pp.1–15.

The notochord

Annona, G., Holland, N.D. and D'Aniello, S., 2015. Evolution of the notochord. *EvoDevo, 6*, pp.1–13.

Chen, J.Y., Dzik, J., Edgecombe, G.D., Ramsköld, L. and Zhou, G.Q., 1995. A possible Early Cambrian chordate. *Nature, 377*(6551), pp.720–722.

Satoh, N., Tagawa, K. and Takahashi, H., 2012. How was the notochord born? *Evolution & Development, 14*(1), pp.56–75.

Shu, D.G., Luo, H.L., Conway Morris, S., Zhang, X.L., Hu, S.X., Chen, L., Han, J.I.A.N., Zhu, M., Li, Y. and Chen, L.Z., 1999. Lower Cambrian vertebrates from south China. *Nature, 402*(6757), pp.42–46.

Shu, D.G., Morris, S.C., Han, J., Zhang, Z.F., Yasui, K., Janvier, P., Chen, L.I.N.G., Zhang, X.L., Liu, J.N., Li, Y.O.N.G. and Liu, H.Q., 2003. Head and backbone of the Early Cambrian vertebrate *Haikouichthys*. *Nature, 421*(6922), pp.526–529.

Sui, Z., Zhao, Z. and Dong, B., 2021. Origin of the chordate notochord. *Diversity, 13*(10), p.462.

Tian, Q., Zhao, F., Zeng, H., Zhu, M. and Jiang, B., 2022. Ultrastructure reveals ancestral vertebrate pharyngeal skeleton in yunnanozoans. *Science, 377*(6602), pp.218–222.

Willey, A., 1891. On the development of the atrial chamber of Amphioxus. *Proceedings of the Royal Society of London, 48*(292–295), pp.80–89.

Witten, P.E. and Hall, B.K., 2021. The ancient, segmented, active and permanent notochord. In A. Pradel, J.S.S. Denton, and P. Janvier (eds.), *Ancient Fishes and Their Living Relatives: a Tribute to John G. Maisey* (pp.215–224). Verlag Dr. Friedrich Pfeil, Munich.

Zhang, X.G. and Hou, X.G., 2004. Evidence for a single median fin-fold and tail in the Lower Cambrian vertebrate, *Haikouichthys ercaicunensis*. *Journal of Evolutionary Biology, 17*(5), pp.1162–1166.

Visceral and somatic

Bichat, M.-F.-X., 1855. *Recherches Physiologiques sur la Vie et le Mort*. Fortin, Masson, Paris.

Nomaksteinsky, M., Kassabov, S., Chettouh, Z., Stoeklé, H.C., Bonnaud, L., Fortin, G., Kandel, E.R. and Brunet, J.F., 2013. Ancient origin of somatic and visceral neurons. *BMC Biology*, *11*, pp.1–11.

Romer, A.S., 1972. The vertebrate as a dual animal—somatic and visceral. *Evolutionary Biology*, *6*, pp.121–156.

## The Cambrian shore

### Gondwana: the no-shows at Chengjiang

Caron, J.B., Scheltema, A., Schander, C. and Rudkin, D., 2006. A soft-bodied mollusc with radula from the Middle Cambrian Burgess Shale. *Nature*, *442*(7099), pp.159–163.

Clausen, S., Hou, X.G., Bergström, J. and Franzén, C., 2010. The absence of echinoderms from the Lower Cambrian Chengjiang fauna of China: palaeoecological and palaeogeographical implications. *Palaeogeography, Palaeoclimatology, Palaeoecology*, *294*(3–4), pp.133–141.

Deline, B., Thompson, J.R., Smith, N.S., Zamora, S., Rahman, I.A., Sheffield, S.L., Ausich, W.I., Kammer, T.W. and Sumrall, C.D., 2020. Evolution and development at the origin of a phylum. *Current Biology*, *30*(9), pp.1672–1679.

Durham, J.W., 1974. Systematic position of *Eldonia ludwigi* Walcott. *Journal of Paleontology*, *48*, pp.751–755.

Mazurek, D. and Zatoń, M., 2011. Is *Nectocaris pteryx* a cephalopod? *Lethaia*, *44*(1), pp.2–4.

Shu, D.G., Morris, S.C., Han, J., Zhang, Z.F. and Liu, J.N., 2004. Ancestral echinoderms from the Chengjiang deposits of China. *Nature*, *430*(6998), pp.422–428.

Smith, M.R., 2013. Nectocaridid ecology, diversity, and affinity: early origin of a cephalopod-like body plan. *Paleobiology*, *39*(2), pp.297–321.

Smith, M.R., 2014. Ontogeny, morphology and taxonomy of the soft-bodied Cambrian 'mollusc' *Wiwaxia*. *Palaeontology*, *57*(1), pp.215–229.

Smith, M.R. and Caron, J.B., 2010. Primitive soft-bodied cephalopods from the Cambrian. *Nature*, *465*(7297), pp.469–472.

Topper, T.P., Guo, J., Clausen, S., Skovsted, C.B. and Zhang, Z., 2019. A stem group echinoderm from the basal Cambrian of China and the origins of Ambulacraria. *Nature Communications*, *10*(1), p.1366.

Zhang, Z., Smith, M.R. and Shu, D., 2015. New reconstruction of the *Wiwaxia* scleritome, with data from Chengjiang juveniles. *Scientific Reports*, *5*(1), p.14810.

### Gondwana: the finds at Jinyangkou, Balang, and Emu Bay

Fu, D., Tong, G., Dai, T., Liu, W., Yang, Y., Zhang, Y., Cui, L., Li, L., Yun, H., Wu, Y. and Sun, A., 2019. The Qingjiang biota—a Burgess Shale–type fossil Lagerstätte from the early Cambrian of South China. *Science*, *363*(6433), pp.1338–1342.

Hu, S., Luo, H., Hou, S. and Erdtmann, B.D., 2007. Eocrinoid echinoderms from the lower Cambrian Guanshan fauna in Wuding, Yunnan, China. *Chinese Science Bulletin*, *52*(5), pp.717–719.

Jago, J.B. and Cooper, B.J., 2011. The Emu Bay Shale lagerstätte: a history of investigations. *Australian Journal of Earth Sciences*, *58*(3), pp.235–241.

Lee, M.S., Jago, J.B., García-Bellido, D.C., Edgecombe, G.D., Gehling, J.G. and Paterson, J.R., 2011. Modern optics in exceptionally preserved eyes of Early Cambrian arthropods from Australia. *Nature*, *474*(7353), pp.631–634.

Paterson, J.R., García-Bellido, D.C., Jago, J.B., Gehling, J.G., Lee, M.S. and Edgecombe, G.D., 2016. The Emu Bay Shale Konservat-Lagerstätte: a view of Cambrian life from East Gondwana. *Journal of the Geological Society*, *173*(1), pp.1–11.

Smith, A.B., Zamora, S. and Álvaro, J.J., 2013. The oldest echinoderm faunas from Gondwana show that echinoderm body plan diversification was rapid. *Nature Communications*, *4*(1), p.1385.

Yuanlong, Z., Maoyan, Z.H.U., Babcock, L.E., Jinliang, Y., Parsley, R.L., Jin, P., Xinglian, Y. and Yue, W., 2005. Kaili Biota: a taphonomic window on diversification of metazoans from the basal Middle Cambrian: Guizhou, China. *Acta Geologica Sinica-English Edition*, *79*(6), pp.751–765.

Zamora, S. and Rahman, I.A., 2014. Deciphering the early evolution of echinoderms with Cambrian fossils. *Palaeontology*, *57*(6), pp.1105–1119.

Zamora, S., Rahman, I.A. and Smith, A.B., 2012. Plated Cambrian bilaterians reveal the earliest stages of echinoderm evolution. *PLoS One*, *7*(6), p.e38296.

Laurentia

Harvey, T.H. and Butterfield, N.J., 2011. Great Canadian Lagerstätten 2: macro-and microfossils of the Mount Cap formation (early and middle Cambrian, Northwest Territories). *Geoscience Canada*, *38*(4), pp.165–173.

Kimmig, J., Strotz, L.C., Kimmig, S.R., Egenhoff, S.O. and Lieberman, B.S., 2019. The Spence Shale Lagerstätte: an important window into Cambrian biodiversity. *Journal of the Geological Society*, *176*(4), pp.609–619.

Pates, S. and Daley, A.C., 2019. The Kinzers Formation (Pennsylvania, USA): the most diverse assemblage of Cambrian Stage 4 radiodonts. *Geological Magazine*, *156*(7), pp.1233–1246.

Siberia

Debrenne, F., Zhuravlev, A.Y. and Kruse, P.D., 2002. Class Archaeocyatha Bornemann, 1884. *Systema Porifera: A Guide to the Classification of Sponges* (pp.1539–1699). Springer, Boston, MA.

Kershaw, S., Munnecke, A. and Jarochowska, E., 2018. Understanding Palaeozoic stromatoporoid growth. *Earth-Science Reviews*, *187*, pp.53–76.

Rowland, S.M., 2001. Archaeocyaths—a history of phylogenetic interpretation. *Journal of Paleontology*, *75*(6), pp.1065–1078.

Stearn, C.W., Halim-Dihardja, M.K. and Nishida, D.K., 1987. An oil-producing stromatoporoid patch reef in the Famennian (Devonian) Wabamun Formation, Normandville Field, Alberta. *Palaios*, *2*, pp.560–570.

Vacelet, J., 2002. Recent 'Sphinctozoa', Order Verticillitida, Family Verticillitidae Steinmann, 1882. In J.N.A. Hooper and R.W.M. Van Soest (eds.) *Systema Porifera: a Guide to the Classification of Sponges* (pp.1097–1098). Kluwer AcadernicIPlenurn Publishers, New York.

Wörheide, G., 2008. A hypercalcified sponge with soft relatives: *Vaceletia* is a keratose demosponge. *Molecular Phylogenetics and Evolution*, *47*(1), pp.433–438.

Zhuravlev, A.Y., 1993. A functional morphological approach to the biology of the Archaeocyatha. *Neues Jahrbuch für Geologie und Paläontologie, Abhandlungen*, *190*, pp.315–327.

Zhuravlev, A.Y. and Wood, R.A., 1996. Anoxia as the cause of the mid-Early Cambrian (Botomian) extinction event. *Geology*, *24*(4), pp.311–314.

Baltica

Butterfield, N.J. and Harvey, T.H.P., 2012. Small carbonaceous fossils (SCFs): a new measure of early Paleozoic paleobiology. *Geology*, *40*(1), pp.71–74.

The worldwide Cambrian fauna

Holmes, J.D., García-Bellido, D.C. and Lee, M.S., 2018. Comparisons between Cambrian Lagerstätten assemblages using multivariate, parsimony and Bayesian methods. *Gondwana Research*, *55*, pp.30–41.

## The Yangtze Platform, Central China, about 535 Mya

Kuanchuanpu

Bottjer, D.J., Yin, Z., Zhao, F. and Zhu, M., 2020. Comparative taphonomy and phylogenetic signal of phosphatized Weng'an and Kuanchuanpu Biotas. *Precambrian Research*, *349*, p.105408.

Scalidophorans

Elder, H.Y. and Hunter, R.D., 1980. Burrowing of *Priapulus caudatus* (Vermes) and the significance of the direct peristaltic wave. *Journal of Zoology*, *191*(3), pp.333–351.

Hammond, R.A., 1980. The hydraulic system in a burrowing worm, *Priapulus caudatus*. *Journal of Zoology*, *192*(4), pp.489–499.

Liu, Y., Xiao, S., Shao, T., Broce, J. and Zhang, H., 2014. The oldest known priapulid-like scalidophoran animal and its implications for the early evolution of cycloneuralians and ecdysozoans. *Evolution & Development*, *16*(3), pp.155–165.

Neuhaus, B. and Higgins, R.P., 2002. Ultrastructure, biology, and phylogenetic relationships of Kinorhyncha. *Integrative and Comparative Biology*, *42*(3), pp.619–632.

Por, F.D. and Bromley, H.J., 1974. Morphology and anatomy of *Maccabeus tentaculatus* (Priapulida: Seticoronaria). *Journal of Zoology*, *173*(2), pp.173–197.

Shao, T., Tang, H., Liu, Y., Waloszek, D., Maas, A. and Zhang, H., 2018. Diversity of cnidarians and cycloneuralians in the Fortunian (early Cambrian) Kuanchuanpu Formation at Zhangjiagou, South China. *Journal of Paleontology*, *92*(2), pp.115–129.

Trott, T.J., 1998. Gustatory responses of *Priapulus caudatus* de Lamarck, 1816 (Priapulida, Priapulidae): feeding behavior and chemoreception by a living fossil. *Marine and Freshwater Behaviour and Physiology*, *31*(2), pp.251–257.

Wang, D., Vannier, J., Schumann, I., Wang, X., Yang, X.G., Komiya, T., Uesugi, K., Sun, J. and Han, J., 2019. Origin of ecdysis: fossil evidence from 535-million-year-old scalidophoran worms. *Proceedings of the Royal Society B*, *286*(1906), p.20190791.

Zhang, H., Xiao, S., Liu, Y., Yuan, X., Wan, B., Muscente, A.D., Shao, T., Gong, H. and Cao, G., 2015. Armored kinorhynch-like scalidophoran animals from the early Cambrian. *Scientific Reports*, *5*(1), p.16521.

Puzzles

Han, J., Kubota, S., Uchida, H.O., Stanley, G.D., Jr., Yao, X., Shu, D., Li, Y. and Yasui, K., 2010. Tiny sea anemone from the Lower Cambrian of China. *PLoS One*, *5*(10), p.e13276.

Han, J., Li, G., Wang, X., Yang, X., Guo, J., Sasaki, O. and Komiya, T., 2018. *Olivooides*-like tube aperture in early Cambrian carinachitids (Medusozoa, Cnidaria). *Journal of Paleontology*, *92*(1), pp.3–13.

Han, J., Morris, S.C., Ou, Q., Shu, D. and Huang, H., 2017. Meiofaunal deuterostomes from the basal Cambrian of Shaanxi (China). *Nature*, *542*(7640), pp.228–231.

Liu, Y., Carlisle, E., Zhang, H., Yang, B., Steiner, M., Shao, T., Duan, B., Marone, F., Xiao, S. and Donoghue, P.C., 2022. *Saccorhytus* is an early ecdysozoan and not the earliest deuterostome. *Nature*, *609*(7927), pp.541–546.

An odd egg

Dong, X.P., Vargas, K., Cunningham, J.A., Zhang, H., Liu, T., Chen, F., Liu, J., Bengtson, S. and Donoghue, P.C., 2016. Developmental biology of the early Cambrian cnidarian *Olivooides*. *Palaeontology*, *59*(3), pp.387–407.

Han, J., Li, G., Kubota, S., Ou, Q., Toshino, S., Wang, X., Yang, X., Uesugi, K., Masato, H., Sasaki, O., Kano, H., Sato, T. and Komiya, T., 2016. Internal microanatomy and zoological affinity of the early Cambrian *Olivooides*. *Acta Geologica Sinica-English Edition*, *90*(1), pp.38–65.

Liu, Y., Li, Y., Shao, T., Zhang, H., Wang, Q. and Qiao, J., 2014. *Quadrapyrgites* from the lower Cambrian of South China: growth pattern, post-embryonic development, and affinity. *Chinese Science Bulletin*, *59*, pp.4086–4095.

Wang, X., Han, J., Vannier, J., Ou, Q., Yang, X., Uesugi, K., Sasaki, O. and Komiya, T., 2017. Anatomy and affinities of a new 535-million-year-old medusozoan from the Kuanchuanpu Formation, South China. *Palaeontology*, *60*(6), pp.853–867.

**The Siberian Platform, Russia, about 540 Mya**

Dzik, J., 1994. Evolution of 'small shelly fossils' assemblages of the Early Paleozoic. *Acta Palaeontologica Polonica*, *39*(3), pp.247–313.

Matthews, S.T. and Missarzhevsky, V.V., 1975. Small shelly fossils of late Precambrian and early Cambrian age: a review of recent work. *Journal of the Geological Society*, *131*(3), pp.289–303.

Porter, S.M., 2004. Closing the phosphatization window: testing for the influence of taphonomic megabias on the pattern of small shelly fossil decline. *Palaios*, *19*(2), pp.178–183.

Swords and shields

Ahrendt, D., Krzywinski, S., Justo I Massot, E. and Krzywinski, J., 2019. Hybrid material designs by the example of additive manufacturing for novel customized stab protective clothing. *Light Weight Armour Group for Defense and Security*, pp.286–294.

Knoll, A.H., 2003. Biomineralization and evolutionary history. *Reviews in Mineralogy and Geochemistry*, *54*(1), pp.329–356.

Murdock, D.J., 2020. The 'biomineralization toolkit' and the origin of animal skeletons. *Biological Reviews*, *95*(5), pp.1372–1392.

Shi, Y., Yan, T., Lai, W., Lu, D. and Lin, C., 2023. Fabric preparation of stab-resistant functional clothing. In L.C. Jain (ed.), *Design Studies and Intelligence Engineering* (pp.206–211). IOS Press, New York, USA.

The mysterious small shellies

Bengtson, S., 2004. Early skeletal fossils. *The Paleontological Society Papers*, *10*, pp.67–78.

Butterfield, N.J. and Nicholas, C.J., 1996. Burgess Shale–type preservation of both non-mineralizing and 'shelly' Cambrian organisms from the Mackenzie Mountains, northwestern Canada. *Journal of Paleontology*, *70*(6), pp.893–899.

Kouchinsky, A.V., 2000. Mollusks, hyoliths, stenothecoids, and coeloscleritophorans. In R. Riding (ed.) *The Ecology of the Cambrian Radiation* (pp.326–349). Columbia University Press, New York, USA.

Kouchinsky, A., Bengtson, S., Landing, E., Steiner, M., Vendrasco, M. and Ziegler, K., 2017. Terreneuvian stratigraphy and faunas from the Anabar Uplift, Siberia. *Acta Palaeontologica Polonica*, *62*(2), pp.311–440.

Parkhaev, P.Y. and Demidenko, Y.E., 2010. Zooproblematica and Mollusca from the Lower Cambrian Meishucun section (Yunnan, China) and taxonomy and systematics of the Cambrian small shelly fossils of China. *Paleontological Journal*, *44*(8), pp.883–1161.

Rozanov, A.Y., 1992. The Cambrian radiation of shelly fossils. *Trends in Ecology & Evolution*, *7*(3), pp.84–87.

Steiner, M., Li, G., Qian, Y., Zhu, M. and Erdtmann, B.D., 2007. Neoproterozoic to early Cambrian small shelly fossil assemblages and a revised biostratigraphic correlation of the Yangtze Platform (China). *Palaeogeography, Palaeoclimatology, Palaeoecology*, *254*(1–2), pp.67–99.

Thomas, R.D., Runnegar, B. and Matt, K., 2020. *Pelagiella exigua*, an early Cambrian stem gastropod with chaetae: lophotrochozoan heritage and conchiferan novelty. *Palaeontology*, *63*(4), pp.601–627.

Thomas, R.D.K., Shearman, R.M. and Stewart, G.W., 2000. Evolutionary exploitation of design options by the first animals with hard skeletons. *Science*, *288*(5469), pp.1239–1242.

Topper, T.P., Guo, J., Clausen, S., Skovsted, C.B. and Zhang, Z., 2019. A stem group echinoderm from the basal Cambrian of China and the origins of Ambulacraria. *Nature Communications*, *10*(1), p.1366.

Wood, R.A., 2011. Paleoecology of the earliest skeletal metazoan communities: implications for early biomineralization. *Earth-Science Reviews*, *106*(1–2), pp.184–190.

Wood, R.A., Ivantsov, A.Y. and Zhuravlev, A.Y., 2017. First macrobiota biomineralization was environmentally triggered. *Proceedings of the Royal Society B: Biological Sciences*, *284*(1851), p.20170059.

Yang, B., Steiner, M., Zhu, M., Li, G., Liu, J. and Liu, P., 2016. Transitional Ediacaran–Cambrian small skeletal fossil assemblages from South China and Kazakhstan: implications for chronostratigraphy and metazoan evolution. *Precambrian Research*, *285*, pp.202–215.

The Halkieria animal

Bengtson, S., 1990. The solution to a jigsaw puzzle. *Nature*, *345*(6278), pp.765–766.

Morris, S.C. and Peel, J.S., 1990. Articulated halkieriids from the Lower Cambrian of north Greenland. *Nature, 345*(6278), pp.802–805.

Hammarlund, E.U., Smith, M.P., Rasmussen, J.A., Nielsen, A.T., Canfield, D.E. and Harper, D.A., 2019. The Sirius Passet Lagerstätte of North Greenland—a geochemical window on early Cambrian low-oxygen environments and ecosystems. *Geobiology, 17*(1), pp.12–26.

Harper, D.A., Hammarlund, E.U., Topper, T.P., Nielsen, A.T., Rasmussen, J.A., Park, T.Y.S. and Smith, M.P., 2019. The Sirius Passet Lagerstätte of North Greenland: a remote window on the Cambrian explosion. *Journal of the Geological Society, 176*(6), pp.1023–1037.

Vinther, J. and Nielsen, C., 2005. The early Cambrian *Halkieria* is a mollusc. *Zoologica scripta, 34*(1), pp.81–89.

The Microdictyon animal

Budd, G., 1993. A Cambrian gilled lobopod from Greenland. *Nature, 364*(6439), pp.709–711.

Hou, X., Siveter, D.J., Siveter, D.J., Aldridge, R.J., Cong, P., Gabbott, S.E., Ma, X., Purnell, M.A. and Williams, M., 2007. *The Cambrian fossils of Chengjiang China, the flowering of early animal life.* Blackwell, Oxford.

Liu, J. and Dunlop, J.A., 2014. Cambrian lobopodians: a review of recent progress in our understanding of their morphology and evolution. *Palaeogeography, Palaeoclimatology, Palaeoecology, 398*, pp.4–15.

Liu, J., Shu, D., Han, J., Zhang, Z. and Zhang, X., 2008. Origin, diversification, and relationships of Cambrian lobopods. *Gondwana Research, 14*(1–2), pp.277–283.

Maas, A., Mayer, G., Kristensen, R.M. and Waloszek, D., 2007. A Cambrian micro-lobopodian and the evolution of arthropod locomotion and reproduction. *Chinese Science Bulletin, 52*, pp.3385–3392.

Ou, Q., Shu, D. and Mayer, G., 2012. Cambrian lobopodians and extant onychophorans provide new insights into early cephalization in Panarthropoda. *Nature Communications, 3*(1), p.1261.

Young, F.J. and Vinther, J., 2017. Onychophoran-like myoanatomy of the Cambrian gilled lobopodian *Pambdelurion whittingtoni. Palaeontology, 60*(1), pp.27–54.

Anabarites

Kouchinsky, A., Bengtson, S., Feng, W., Kutygin, R. and Val'kov, A., 2009. The Lower Cambrian fossil anabaritids: affinities, occurrences and systematics. *Journal of Systematic Palaeontology, 7*(3), p.241.

Landing, E.D. and Kouchinsky, A., 2016. Correlation of the Cambrian evolutionary radiation: geochronology, evolutionary stasis of earliest Cambrian (Terreneuvian) small shelly fossil (SSF) taxa, and chronostratigraphic significance. *Geological Magazine, 153*(4), pp.750–756.

Zhang, Y.N., Liu, Y.H., Shao, T.Q., Wang, Q., Qin, J.C., Li, L.L. and Zhou, X.Y., 2022. The morphology and affinity of anabaritids revealed by their internal molds from the Cambrian Fortunian, southern Shaanxi, China. *Palaeoworld, 31*(4), pp.600–606.

Protohertzina

Aldridge, R.J. and Purnell, M.A., 1996. The conodont controversies. *Trends in Ecology & Evolution, 11*(11), pp.463–468.

Bengtson, S., 1983. The early history of the Conodonta. In A. Martinsson and S. Bengtson (eds.) *Taxonomy, Ecology and Identity of Conodonts: Proceedings of ECOS III, Lund, 1982* (pp.5–19). Universitetsforlaget, Oslo.

Briggs, D.E. and Caron, J.B., 2017. A large Cambrian chaetognath with supernumerary grasping spines. *Current Biology, 27*(16), pp.2536–2543.

Casenove, D., Goto, T. and Vannier, J., 2011. Relation between anatomy and lifestyles in recent and Early Cambrian chaetognaths. *Paleobiology, 37*(4), pp.563–576.

Demidenko, Y.E., 2006. New Cambrian lobopods and chaetognaths of the Siberian Platform. *Paleontological Journal, 40*, pp.234–243.

Szaniawski, H., 2002. New evidence for the protoconodont origin of chaetognaths. *Acta Palaeontologica Polonica, 47*(3), pp.405–419.

Telford, M.J., 2019. Evolution: arrow worms find their place on the tree of life. *Current Biology*, *29*(5), pp.R152–R154.

Vannier, J., Steiner, M., Renvoisé, E., Hu, S.X. and Casanova, J.P., 2007. Early Cambrian origin of modern food webs: evidence from predator arrow worms. *Proceedings of the Royal Society B: Biological Sciences*, *274*(1610), pp.627–633.

Cloudina

Cai, Y., Cortijo, I., Schiffbauer, J.D. and Hua, H., 2017. Taxonomy of the late Ediacaran index fossil *Cloudina* and a new similar taxon from South China. *Precambrian Research*, *298*, pp.146–156.

Han, J., Cai, Y., Schiffbauer, J.D., Hua, H., Wang, X., Yang, X., Uesugi, K., Komiya, T. and Sun, J.I.E., 2017. A *Cloudina*-like fossil with evidence of asexual reproduction from the lowest Cambrian, South China. *Geological Magazine*, *154*(6), pp.1294–1305.

Iten, H., van, Marques, A.C., Leme, J.D.M., Pacheco, M.L.F. and Simoes, M.G., 2014. Origin and early diversification of the phylum Cnidaria Verrill: major developments in the analysis of the taxon's Proterozoic–Cambrian history. *Palaeontology*, *57*(4), pp.677–690.

Moczydłowska, M., Westall, F. and Foucher, F., 2014. Microstructure and biogeochemistry of the organically preserved Ediacaran metazoan *Sabellidites*. *Journal of Paleontology*, *88*(2), pp.224–239.

Schiffbauer, J.D., Wong, C., Davis, C., Selly, T., Nelson, L.L. and Pruss, S.B., 2024. Reassessing the diversity, affinity, and construction of terminal Ediacaran tubiform fossils from the La Ciénega Formation, Sonora, Mexico. *Journal of Paleontology*, *98*(2), pp.266–282.

Shore, A. and Wood, R., 2021. Environmental and diagenetic controls on the morphology and calcification of the Ediacaran metazoan *Cloudina*. *Scientific Reports*, *11*(1), p.12341.

Shore, A.J., Wood, R.A., Butler, I.B., Zhuravlev, A.Y., McMahon, S., Curtis, A. and Bowyer, F.T., 2021. Ediacaran metazoan reveals lophotrochozoan affinity and deepens root of Cambrian Explosion. *Science Advances*, *7*(1), p.eabf2933.

## Chapter 8. The Ediacaran Fossils

### The ring pit

Ford, T.D., 1958. Precambrian fossils from Charnwood Forest, *Proceedings of the Yorkshire Geological Society*, *31*(3), pp.211–217.

### The red mine

Glaessner, M.F., 1958. New fossils from the base of the Cambrian in South Australia. *Transactions of the Royal Society of South Australia*, *81*(1), p.4.

Glaessner, M.F., 1959. Precambrian Coelenterata from Australia, Africa and England. *Nature*, *183*(4673), pp.1472–1473.

Sprigg, R.C., 1947. Early Cambrian (?) jellyfishes from the Flinders Ranges, South Australia. *Transactions of the Royal Society of South Australia*, *71*, p.212–224.

Sprigg, R.C., 1948. Jellyfish from the Basal Cambrian in South Australia. *Nature*, *161*(4093), pp.568–569.

Sprigg, R., 1988. On the 1946 discovery of the Precambrian Ediacabian fossil fauna in South Australia. *Earth Sciences History*, *7*(1), pp.46–51.

Turner, S. and Oldroyd, D., 2009. Reg Sprigg and the discovery of the Ediacara Fauna in South Australia: its approach to the high table. In D. Sepkowski and M. Ruse (eds.), *The Paleobiological Revolution* (pp.254–278). University of Chicago Press, Chicago, USA.

### Nama Fm, Namibia, about 543 Mya

Bags

Elliott, D.A., Trusler, P.W., Narbonne, G.M., Vickers-Rich, P., Morton, N., Hall, M., Hoffmann, K.H. and Schneider, G.I., 2016. *Ernietta* from the late Edicaran Nama Group, Namibia. *Journal of Paleontology*, *90*(6), pp.1017–1026.

Gibson, B.M., Rahman, I.A., Maloney, K.M., Racicot, R.A., Mocke, H., Laflamme, M. and Darroch, S.A., 2019. Gregarious suspension feeding in a modular Ediacaran organism. *Science Advances, 5*(6), p.eaaw0260.

Ivantsov, A.Y., Narbonne, G.M., Trusler, P.W., Greentree, C. and Vickers-Rich, P., 2016. Elucidating Ernietta: new insights from exceptional specimens in the Ediacaran of Namibia. *Lethaia, 49*(4), pp.540–554.

Narbonne, G.M., Saylor, B.Z. and Grotzinger, J.P., 1997. The youngest Ediacaran fossils from southern Africa. *Journal of Paleontology, 71*(6), pp.953–967.

Pflug, H.D., 1966. Neue Fossilreste aus den Nama-Schichten in Südwest-Afrika. *Paläontologische Zeitschrift, 40*, pp.14–25.

Pickford, M.H.L., 1995. Review of the Riphean, Vendian and early Cambrian palaeontology of the Otavi and Nama Groups, Namibia. *Communications of the Geological Survey of Namibia, 10*, pp.57–81.

Wood, R., Bowyer, F.T., Alexander, R., Yilales, M., Uahengo, C.I., Kaputuaza, K., Ndeunyema, J. and Curtis, A., 2023. New Ediacaran biota from the oldest Nama Group, Namibia (Tsaus Mountains), and re-definition of the Nama Assemblage. *Geological Magazine, 160*(9), pp.1673–1686.

**Flinders Ranges, South Australia, about 550 Mya**

Martin Glaessner

Glaessner, M.F., 1961. Pre-Cambrian animals. *Scientific American, 204*, pp.72–79.

Glaessner, M.F., 1966. Precambrian palaeontology. *Earth-Science Reviews, 1*(1), pp.29–50.

Glaessner, M.F., 1979. *Parvancorina*—an arthropod from the late Precambrian (Ediacarian) of South Australia. *Annalen des Naturhistorischen Museums in Wien, 83*, pp.83–90.

Hall, C.M., Droser, M.L., Gehling, J.G. and Dzaugis, M.E., 2015. Paleoecology of the enigmatic *Tribrachidium*: new data from the Ediacaran of South Australia. *Precambrian Research, 269*, pp.183–194.

Lin, J.P., Gon, S.M., III, Gehling, J.G., Babcock, L.E., Zhao, Y.L., Zhang, X.L., Hu, S.X., Yuan, J.L., Yu, M.Y. and Peng, J., 2006. A *Parvancorina*-like arthropod from the Cambrian of South China. *Historical Biology, 18*(1), pp.33–45.

McMenamin, M.A., 2003. *Spriggina* is a trilobitoid ecdysozoan. *Abstracts with Programs. Geological Society of America 35*(6), p.105.

Rahman, I.A., Darroch, S.A., Racicot, R.A. and Laflamme, M., 2015. Suspension feeding in the enigmatic Ediacaran organism *Tribrachidium* demonstrates complexity of Neoproterozoic ecosystems. *Science Advances, 1*(10), p.e1500800.

Pit-and-ridge creatures

Gehling, J.G., 1987. Earliest known echinoderm—a new Ediacaran fossil from the Pound subgroup of South Australia. *Alcheringa, 11*(4), pp.337–345.

Glaessner, M.F., 1979. *Parvancorina*—an arthropod from the late Precambrian (Ediacarian) of South Australia. *Annalen des Naturhistorischen Museums in Wien*, pp.83–90.

Hall, C.M., Droser, M.L., Gehling, J.G. and Dzaugis, M.E., 2015. Paleoecology of the enigmatic Tribrachidium: new data from the Ediacaran of South Australia. *Precambrian Research, 269*, pp.183–194.

Lin, J.P., Gon, S.M., III, Gehling, J.G., Babcock, L.E., Zhao, Y.L., Zhang, X.L., Hu, S.X., Yuan, J.L., Yu, M.Y. and Peng, J., 2006. A *Parvancorina*-like arthropod from the Cambrian of South China. *Historical Biology, 18*(1), pp.33–45.

Rahman, I.A., Darroch, S.A., Racicot, R.A. and Laflamme, M., 2015. Suspension feeding in the enigmatic Ediacaran organism Tribrachidium demonstrates complexity of Neoproterozoic ecosystems. *Science Advances, 1*(10), p.e1500800.

### White Sea coast, Russia, about 555 Mya

Dickinsonia

Bobrovskiy, I., Hope, J.M., Ivantsov, A., Nettersheim, B.J., Hallmann, C. and Brocks, J.J., 2018. Ancient steroids establish the Ediacaran fossil *Dickinsonia* as one of the earliest animals. *Science*, *361*(6408), pp.1246–1249.

Dzik, J. and Ivantsov, A.Y., 2002. Internal anatomy of a new Precambrian dickinsoniid dipleurozoan from northern Russia. *Neues Jahrbuch fur Geologie und Palaontologie Monatshefte*, (*7*), pp.385–396.

Gold, D.A., Runnegar, B., Gehling, J.G. and Jacobs, D.K., 2015. Ancestral state reconstruction of ontogeny supports a bilaterian affinity for *Dickinsonia*. *Evolution & Development*, *17*(6), pp.315–324.

Hoekzema, R.S., Brasier, M.D., Dunn, F.S. and Liu, A.G., 2017. Quantitative study of developmental biology confirms *Dickinsonia* as a metazoan. *Proceedings of the Royal Society B: Biological Sciences*, *284*(1862), p.20171348.

Ivantsov, A.Y., 2017. The most probable Eumetazoa among late Precambrian macrofossils. *Invertebrate Zoology*, *14*(2), pp.127–133.

Ivantsov, A.Y. and Zakrevskaya, M., 2023. Body plan of *Dickinsonia*, the oldest mobile animals. *Earth and Environmental Science Transactions of the Royal Society of Edinburgh*, *114*(1-2), pp.95–108.

Keretsa

Ivantsov, A.Y., 2017. The most probable Eumetazoa among late Precambrian macrofossils. *Invertebrate Zoology*, *14*(2), pp.127–133.

Kimberella

Bobrovskiy, I., Nagovitsyn, A., Hope, J.M., Luzhnaya, E. and Brocks, J.J., 2022. Guts, gut contents, and feeding strategies of Ediacaran animals. *Current Biology*, *32*(24), pp.5382–5389.

Fedonkin, M.A., Simonetta, A. and Ivantsov, A.Y., 2007. New data on *Kimberella*, the Vendian mollusc-like organism (White Sea region, Russia): palaeoecological and evolutionary implications. *Geological Society, London, Special Publications*, *286*(1), pp.157–179.

Fedonkin, M.A. and Waggoner, B.M., 1997. The Late Precambrian fossil *Kimberella* is a mollusc-like bilaterian organism. *Nature*, *388*(6645), pp.868–871.

Gehling, J.G., Runnegar, B.N. and Droser, M.L., 2014. Scratch traces of large Ediacara bilaterian animals. *Journal of Paleontology*, *88*(2), pp.284–298.

Ivantsov, A.Y., 2009. New reconstruction of *Kimberella*, problematic Vendian metazoan. *Paleontological Journal*, *43*, pp.601–611.

Ivantsov, A.Y., 2013. Trace fossils of Precambrian metazoans 'Vendobionta' and 'Mollusks'. *Stratigraphy and Geological Correlation*, *21*, pp.252–264.

Ivantsov, A., Nagovitsyn, A. and Zakrevskaya, M., 2019. Traces of locomotion of Ediacaran macroorganisms. *Geosciences*, *9*(9), p.395.

### Charnwood Forest, England, about 560 Mya

Boynton, H.E., 1999. New fossils in the Precambrian of Charnwood Forest, Leicestershire, England. *Mercian Geologist*, *14*(4), pp.197–200.

Boynton, H.E. and Ford, T.D., 1995. Ediacaran fossils from the Precambrian (Charnian Supergroup) of Charnwood Forest, Leicestershire, England. *Mercian Geologist*, *13*(4), pp.165–182.

Ford, T.D., 1999. The Precambrian fossils of Charnwood Forest. *Geology Today*, *15*(6), pp.230–234.

Howe, M.P., Evans, M., Carney, J.N. and Wilby, P.R., 2012. New perspectives on the globally important Ediacaran fossil discoveries in Charnwood Forest, UK: prequel to. *Proceedings of the Yorkshire Geological Society*, *59*(2), pp.137–144.

Kenchington, C.G., Harris, S.J., Vixseboxse, P.B., Pickup, C. and Wilby, P.R., 2018. The Ediacaran fossils of Charnwood Forest: shining new light on a major biological revolution. *Proceedings of the Geologists' Association*, *129*(3), pp.264–277.

Discs

Boyce, W.D. and Reynolds, K., 2008. The Ediacaran fossil Aspidella terranovica Billings, 1872 from St John's Convention Centre test pit CjAe-33. *Current Research*, *8*(1), pp.55–61.

Burzynski, G. and Narbonne, G.M., 2015. The discs of Avalon: relating discoid fossils to frondose organisms in the Ediacaran of Newfoundland, Canada. *Palaeogeography, Palaeoclimatology, Palaeoecology*, *434*, pp.34–45.

Gehling, J.G., Narbonne, G.M. and Anderson, M.M., 2000. The first named Ediacaran body fossil, *Aspidella terranovica*. *Palaeontology*, *43*(3), pp.427–456.

Ivantsov, A.Y., 2016. Reconstruction of *Charniodiscus yorgensis* (macrobiota from the Vendian of the White Sea). *Paleontological Journal*, *50*, pp.1–12.

**Avalon Peninsula, Newfoundland, about 565 Mya**

Anderson, M.M. and Misra, S.B., 1968. Fossils found in the Pre-Cambrian Conception Group of south-eastern Newfoundland. *Nature*, *220*(5168), pp.680–681.

Clapham, M.E., Narbonne, G.M. and Gehling, J.G., 2003. Paleoecology of the oldest known animal communities: Ediacaran assemblages at Mistaken Point, Newfoundland. *Paleobiology*, *29*(4), pp.527–544.

Hofmann, H.J., O'Brien, S.J. and King, A.F., 2008. Ediacaran biota on Bonavista Peninsula, Newfoundland, Canada. *Journal of Paleontology*, *82*(1), pp.1–36.

Liu, A.G. and Matthews, J.J., 2017. Great Canadian lagerstätten 6. Mistaken Point ecological reserve, Southeast Newfoundland. *Geoscience Canada*, *44*(2), pp.63–76.

Liu, A.G., Mcilroy, D., Matthews, J.J. and Brasier, M.D., 2014. Confirming the metazoan character of a 565 Ma trace-fossil assemblage from Mistaken Point, Newfoundland. *Palaios*, *29*(8), pp.420–430.

Misra, S.B., 1969. Late Precambrian (?) fossils from southeastern Newfoundland. *Geological Society of America Bulletin*, *80*(11), pp.2133–2140.

Wilby, P.R., Carney, J.N. and Howe, M.P., 2011. A rich Ediacaran assemblage from eastern Avalonia: evidence of early widespread diversity in the deep ocean. *Geology*, *39*(7), pp.655–658.

Fronds

Bamforth, E.L., Narbonne, G.M. and Anderson, M.M., 2008. Growth and ecology of a multi-branched Ediacaran rangeomorph from the Mistaken Point assemblage, Newfoundland. *Journal of Paleontology*, *82*(4), pp.763–777.

Butterfield, N.J., 2022. Constructional and functional anatomy of Ediacaran rangeomorphs. *Geological Magazine*, *159*(7), pp.1148–1159.

Dunn, F.S., Liu, A.G. and Donoghue, P.C., 2018. Ediacaran developmental biology. *Biological Reviews*, *93*(2), pp.914–932.

Dunn, F.S., Liu, A.G., Grazhdankin, D.V., Vixseboxse, P., Flannery-Sutherland, J., Green, E., Harris, S., Wilby, P.R. and Donoghue, P.C., 2021. The developmental biology of *Charnia* and the eumetazoan affinity of the Ediacaran rangeomorphs. *Science Advances*, *7*(30), p.eabe0291.

Dunn, F.S., Wilby, P.R., Kenchington, C.G., Grazhdankin, D.V., Donoghue, P.C. and Liu, A.G., 2019. Anatomy of the Ediacaran rangeomorph *Charnia masoni*. *Papers in Palaeontology*, *5*(1), pp.157–176.

Gehling, J.G. and Narbonne, G.M., 2007. Spindle-shaped Ediacara fossils from the Mistaken Point assemblage, Avalon zone, Newfoundland. *Canadian Journal of Earth Sciences*, *44*(3), pp.367–387.

Laflamme, M., Gehling, J.G. and Droser, M.L., 2018. Deconstructing an Ediacaran frond: three-dimensional preservation of *Arborea* from Ediacara, South Australia. *Journal of Paleontology*, *92*(3), pp.323–335.

Liu, A.G. and Dunn, F.S., 2020. Filamentous connections between Ediacaran fronds. *Current Biology*, *30*(7), pp.1322–1328.

Narbonne, G.M., 2004. Modular construction of early Ediacaran complex life forms. *Science*, *305*(5687), pp.1141–1144.

Bubbles or seaweeds?

Goldring, R., 1969. Criteria for recognizing Pre-Cambrian fossils. *Nature*, *223*(5210), pp.1076–1076. [Reply from Anderson and Misra follows.]

Seilacher's heresy

Seilacher, A., 1989. Vendozoa: organismic construction in the Proterozoic biosphere. *Lethaia*, *22*(3), pp.229–239.

Seilacher, A., 1992. Vendobionta and Psammocorallia: lost constructions of Precambrian evolution. *Journal of the Geological Society*, *149*(4), pp.607–613.

Jellyfish and sea pens

Brasier, J.A.M., 2007. *Charnia* and sea pens are poles apart. *Journal of the Geological Society of London*, *164*, pp.49–51.

What did the fronds eat?

Bernhard, J.M. and Bowser, S.S., 1992. Bacterial biofilms as a trophic resource for certain benthic foraminifera. *Marine Ecology Progress Series*, *83*, pp.263–272.

Bowser, S.S., Alexander, S.P., Stockton, W.L. and Delaca, T.E., 1992. Extracellular matrix augments mechanical properties of pseudopodia in the carnivorous foraminiferan *Astrammina rara*: role in prey capture. *Journal of Protozoology*, *39*(6), pp.724–732.

Sakaguchi, M., Suzaki, T., Khan, S.M.K. and Hausmann, K., 2002. Food capture by kinetocysts in the heliozoon *Raphidiophrys contractilis*. *European Journal of Protistology*, *37*(4), pp.453–458.

**Anhui Province, East China, about 600 Mya**

Wan, B., Yuan, X., Chen, Z., Guan, C., Pang, K., Tang, Q. and Xiao, S., 2016. Systematic description of putative animal fossils from the early Ediacaran Lantian Formation of South China. *Palaeontology*, *59*(4), pp.515–532.

Yang, C., Li, Y., Selby, D., Wan, B., Guan, C., Zhou,C. and Li, X.H., 2022. Implications for Ediacaran biological evolution from the ca. 602 Ma Lantian biota in China. *Geology*, *50*(5), pp.562–566.

Yuan, X., Chen, Z., Xiao, S., Wan, B., Guan, C., Wang, W., Zhou, C. and Hua, H., 2013. The Lantian biota: a new window onto the origin and early evolution of multicellular organisms. *Chinese Science Bulletin*, *58*, pp.701–707.

**The hiatus**

The worldwide PreCambrian fauna

Narbonne, G.M., Laflamme, M., Trusler, P.W., Dalrymple, R.W. and Greentree, C., 2014. Deep-water Ediacaran fossils from northwestern Canada: taphonomy, ecology, and evolution. *Journal of Paleontology*, *88*(2), pp.207–223.

Soldatenko, Y., El Albani, A., Ruzina, M., Fontaine, C., Nesterovsky, V., Paquette, J.L., Meunier, A. and Ovtcharova, M., 2019. Precise U-pb age constrains on the Ediacaran biota in Podolia, east European platform, Ukraine. *Scientific Reports*, *9*(1), p.1675.

Xiao, S., Chen, Z., Pang, K., Zhou, C. and Yuan, X., 2021. The Shibantan lagerstätte: Insights into the Proterozoic–Phanerozoic transition. *Journal of the Geological Society*, *178*(1), pp.jgs2020–2135.

The Ediacaran Period

Cloud, P.E. and Glaessner, M.F., 1982. The Ediacarian Period and System: Metazoa inherit the Earth. *Science 217*, 783–792.

Knoll, A., Walter, M., Narbonne, G. and Christie-Blick, N., 2006. The Ediacaran Period: a new addition to the geologic time scale. *Lethaia*, *39*(1), pp.13–30.

The end of the Ediacarans

Bottjer, D.J., Hagadorn, J.W. and Dornbos, S.Q., 2000. The Cambrian substrate revolution. *GSA Today*, *10*(9), pp.1–7.

Conway Morris, S., 1993. Ediacaran-like fossils in Cambrian Burgess Shale-type faunas of North America. *Palaeontology*, *36*(3), pp.593–635.

Cuthill, J.F.H., 2022. Ediacaran survivors in the Cambrian: suspicions, denials and a smoking gun. *Geological Magazine*, *159*(7), pp.1210–1219.

Darroch, S.A., Cribb, A.T., Buatois, L.A., Germs, G.J., Kenchington, C.G., Smith, E.F., Mocke, H., O'Neil, G.R., Schiffbauer, J.D., Maloney, K.M. and Racicot, R.A., 2021. The trace fossil record of the Nama Group, Namibia: exploring the terminal Ediacaran roots of the Cambrian explosion. *Earth-Science Reviews*, *212*, p.103435.

Darroch, S.A., Smith, E.F., Nelson, L.L., Craffey, M., Schiffbauer, J.D. and Laflamme, M., 2023. Causes and consequences of end-Ediacaran extinction: an update. *Cambridge Prisms: Extinction*, *1*, p.e15.

Darroch, S.A., Sperling, E.A., Boag, T.H., Racicot, R.A., Mason, S.J., Morgan, A.S., Tweedt, S., Myrow, P., Johnston, D.T., Erwin, D.H. and Laflamme, M., 2015. Biotic replacement and mass extinction of the Ediacara biota. *Proceedings of the Royal Society B: Biological Sciences*, *282*(1814), p.20151003.

Hagadorn, J.W., Fedo, C.M. and Waggoner, B.M., 2000. Early Cambrian Ediacaran-type fossils from California. *Journal of Paleontology*, *74*(4), pp.731–740.

Jensen, S., Gehling, J.G. and Droser, M.L., 1998. Ediacara-type fossils in Cambrian sediments. *Nature*, *393*(6685), pp.567–569.

Retallack, G.J., 2015. Reassessment of the Silurian problematicum Rutgersella as another post-Ediacaran vendobiont. *Alcheringa: An Australasian Journal of Palaeontology*, *39*(4), pp.573–588.

Seilacher, A., Buatois, L.A. and Mángano, M.G., 2005. Trace fossils in the Ediacaran–Cambrian transition: behavioral diversification, ecological turnover and environmental shift. *Palaeogeography, Palaeoclimatology, Palaeoecology*, *227*(4), pp.323–356.

Seilacher, A. and Pflüger, F., 1994. From biomats to benthic agriculture: a biohistoric revolution. In W.E. Krumbein, D.M. Peterson, and L.J. Stal (eds.), *Biostabilization of Sediments* (pp.97–105). Bibliotheks-und-Informationssystem der Carl von Ossietzky Universität, Oderburg.

The cause of the hiatus

Darroch, S.A., Smith, E.F., Nelson, L.L., Craffey, M., Schiffbauer, J.D. and Laflamme, M., 2023. Causes and consequences of end-Ediacaran extinction: An update. *Cambridge Prisms: Extinction*, *1*, p.e15.

Darroch, S.A., Sperling, E.A., Boag, T.H., Racicot, R.A., Mason, S.J., Morgan, A.S., Tweedt, S., Myrow, P., Johnston, D.T., Erwin, D.H. and Laflamme, M., 2015. Biotic replacement and mass extinction of the Ediacara biota. *Proceedings of the Royal Society B: Biological Sciences*, *282*(1814), p.20151003.

Hagadorn, J.W., Fedo, C.M. and Waggoner, B.M., 2000. Early Cambrian Ediacaran-type fossils from California. *Journal of Paleontology*, *74*(4), pp.731–740.

The missing sponges

Antcliffe, J.B., Callow, R.H. and Brasier, M.D., 2014. Giving the early fossil record of sponges a squeeze. *Biological Reviews*, *89*(4), pp.972–1004.

Botting, J.P. and Muir, L.A., 2018. Early sponge evolution: a review and phylogenetic framework. *Palaeoworld*, *27*(1), pp.1–29.

Brain, C.B., Prave, A.R., Herd, D.A., Allison, S.G., Hoffmann, K.H., Fallick, A.E., Botha, A., Sturrock, C., Young, I. and Condon, D.J., 2012. The first animals: ca. 760-million-year-old sponge-like fossils from Namibia. *South African Journal of Science*, *108*(1), pp.1–8.

Li, C.W., Chen, J.Y. and Hua, T.E., 1998. Precambrian sponges with cellular structures. *Science*, *279*(5352), pp.879–882.

Wang, X., Hu, S., Gan, L., Wiens, M. and Müller, W.E., 2010. Sponges (Porifera) as living metazoan witnesses from the Neoproterozoic: biomineralization and the concept of their evolutionary success. *Terra Nova*, *22*(1), pp.1–11.

Wang, X., Liu, A.G., Chen, Z., Wu, C., Liu, Y., Wan, B., Pang, K., Zhou, C., Yuan, X. and Xiao, S., 2024. A late-Ediacaran crown-group sponge animal. *Nature*, *630*(8018), pp.905–911.

The conulariids

Brood, K., 1995. Morphology, structure, and systematics of the conulariids. *GFF*, *117*(3), pp.121–137.

Iten, H., van, de Moraes Leme, J., Simoes, M.G., Marques, A.C. and Collins, A.G., 2006. Reassessment of the phylogenetic position of conulariids (? Ediacaran-Triassic) within the subphylum medusozoa (phylum cnidaria). *Journal of Systematic Palaeontology*, *4*(2), pp.109–118.

Ivantsov, A.Y. and Fedonkin, M.A., 2002. Conulariid-like fossil from the Vendian of Russia: a metazoan clade across the Proterozoic/Palaeozoic boundary. *Palaeontology*, *45*(6), pp.1219–1229.

Ivantsov, A.Y., Vickers-Rich, P., Zakrevskaya, M.A. and Hall, M., 2019. Conical thecae of Precambrian macroorganisms. *Paleontological Journal*, *53*, pp.1134–1146.

Leme, J.M., de, Simões, M.G., Rodrigues, S.C., van Iten, H. and Marques, A.C. 2008. Major developments in conulariid research: problems of interpretation and future perspectives. *Ameghiniana*, *45*, pp.407–420.

Lucas, S.G., 2012. The extinction of the conulariids. *Geosciences*, *2*(1), pp.1–10.

Pacheco, M.L.F., Galante, D., Rodrigues, F., Leme, J.M. de, Bidola, P., Hagadorn, W., Stockmar, M., Herzen, J., Rudnitzki, I.D., Pfeiffer F. and Marques, A.C., 2015. Insights into the skeletonization, lifestyle, and affinity of the unusual Ediacaran fossil *Corumbella*. *PLoS One*, *10*(3), p.e0114219.

The missing polyps

Buss, L.W. and Seilacher, A., 1994. The Phylum Vendobionta: a sister group of the Eumetazoa? *Paleobiology*, *20*(1), pp.1–4.

Dunn, F.S., Kenchington, C.G., Parry, L.A., Clark, J.W., Kendall, R.S. and Wilby, P.R., 2022. A crown-group cnidarian from the Ediacaran of Charnwood Forest, UK. *Nature Ecology & Evolution*, *6*(8), pp.1095–1104.

Liu, A.G., Matthews, J.J., Menon, L.R., McIlroy, D. and Brasier, M.D., 2014.*Haootia quadriformis* n. gen., n. sp., interpreted as a muscular cnidarian impression from the Late Ediacaran period (approx. 560 Ma). *Proceedings of the Royal Society B: Biological Sciences*, *281*(1793), p.20141202.

Liu, A.G., Matthews, J.J., Menon, L.R., McIlroy, D. and Brasier, M.D., 2015. The arrangement of possible muscle fibres in the Ediacaran taxon *Haootia quadriformis*. *Proceedings of the Royal Society B: Biological Sciences*, *282*(1803), p.20142949.

Miranda, L.S., Collins, A.G. and Marques, A.C., 2015. Is *Haootia quadriformis* related to extant Staurozoa (Cnidaria)? Evidence from the muscular system reconsidered. *Proceedings of the Royal Society B: Biological Sciences*, *282*(1803), p.20142396.

## Chapter 9. Sea Changes

Li, Z.X., Bogdanova, S., Collins, A.S., Davidson, A., De Waele, B., Ernst, R.E., Fitzsimons, I.C.W., Fuck, R.A., Gladkochub, D.P., Jacobs, J. and Karlstrom, K.E., 2008. Assembly, configuration, and break-up history of Rodinia: a synthesis. *Precambrian Research*, *160*(1–2), pp.179–210.

Mukherjee, I., Large, R.R., Corkrey, R. and Danyushevsky, L.V., 2018. The Boring Billion, a slingshot for complex life on Earth. *Scientific Reports*, *8*(1), p.4432.

### Ice ages and impacts

Kaufman, A.J., Knoll, A.H. and Narbonne, G.M., 1997. Isotopes, ice ages, and terminal Proterozoic earth history. *Proceedings of the National Academy of Sciences*, *94*(13), pp.6600–6605.

Snowball Earth?

Chakraborty, N., 2022. Snowball Earth: a critical review on Neoproterozoic glaciation. *Journal of Geointerface*, *1*, pp.50–56.

Donnadieu, Y., Ramstein, G., Fluteau, F., Besse, J. and Meert, J., 2002. Is high obliquity a plausible cause for Neoproterozoic glaciations? *Geophysical Research Letters*, *29*(23), pp.42-1–42-4.

Fairchild, I.J. and Kennedy, M.J., 2007. Neoproterozoic glaciation in the Earth System. *Journal of the Geological Society, 164*(5), pp.895–921.

Hoffman, P.F., Abbot, D.S., Ashkenazy, Y., Benn, D.I., Brocks, J.J., Cohen, P.A., Cox, G.M., Creveling, J.R., Donnadieu, Y., Erwin, D.H. and Fairchild, I.J., 2017. Snowball Earth climate dynamics and Cryogenian geology-geobiology. *Science Advances, 3*(11), p.e1600983.

Hoffman, P.F., Kaufman, A.J., Halverson, G.P. and Schrag, D.P., 1998. A Neoproterozoic snowball Earth. *Science, 281*(5381), pp.1342–1346.

Hyde, W.T., Crowley, T.J., Baum, S.K. and Peltier, W.R., 2000. Neoproterozoic 'snowball Earth' simulations with a coupled climate/ice-sheet model. *Nature, 405*(6785), pp.425–429.

Kirschivink, J.L., 1992. Late Proterozoic low-latitude global glaciation: the snowball Earth. *The Proterozoic Biosphere, 52*, pp.51–52.

Levrard, B. and Laskar, J., 2003. Climate friction and the Earth's obliquity. *Geophysical Journal International, 154*(3), pp.970–990.

Micheels, A. and Montenari, M., 2008. A snowball Earth versus a slushball Earth: results from Neoproterozoic climate modeling sensitivity experiments. *Geosphere, 4*(2), pp.401–410.

Williams, D.M., Kasting, J.F. and Frakes, L.A., 1998. Low-latitude glaciation and rapid changes in the Earth's obliquity explained by obliquity–oblateness feedback. *Nature, 396*(6710), pp.453–455.

The Acraman asteroid

Gostin, V.A., Haines, P.W., Jenkins, R.J., Compston, W. and Williams, I.S., 1986. Impact ejecta horizon within late Precambrian shales, Adelaide geosyncline, South Australia. *Science, 233*(4760), pp.198–200.

Williams, G.E., 1986. The Acraman impact structure: source of ejecta in late Precambrian shales, South Australia. Science *233*(4760), pp.200–203.

Williams, G.E. and Wallace, M., 2003. The Acraman asteroid impact, South Australia: magnitude and implications for the late Vendian environment. *Journal of the Geological Society, 160*(4), pp.545–554.

## Light carbon

Kump, L.R. and Arthur, M.A., 1999. Interpreting carbon-isotope excursions: carbonates and organic matter. *Chemical Geology, 161*(1–3), pp.181–198.

Miyazaki, Y., Planavsky, N.J., Bolton, E.W. and Reinhard, C.T., 2018. Making sense of massive carbon isotope excursions with an inverse carbon cycle model. *Journal of Geophysical Research: Biogeosciences,*
*123*(8), pp.2485–2496.

Oehlert, A.M. and Swart, P.K., 2014. Interpreting carbonate and organic carbon isotope covariance in the sedimentary record. *Nature Communications, 5*(1), p.4672.

Sial, A.N., Gaucher, C., Ferreira, V.P., Pereira, N.S., Cezario, W.S., Chiglino, L. and Lima, H.M., 2015. Isotope and elemental chemostratigraphy. In Mu. Ramkumar (ed.) *Chemostratigraphy* (pp.23–64). Elsevier, Amsterdam.

The Shuram excursion

Gong, Z., Wei, G.Y., Fakhraee, M., Alcott, L.J., Jiang, L., Zhao, M. and Planavsky, N.J., 2023. Revisiting marine redox conditions during the Ediacaran Shuram carbon isotope excursion. *Geobiology, 21*(4), pp.407–420.

Kimura, H., Matsumoto, R., Kakuwa, Y., Hamdi, B. and Zibaseresht, H., 1997. The Vendian-Cambrian δ13C record, North Iran: evidence for overturning of the ocean before the Cambrian explosion. *Earth and Planetary Science Letters, 147*(1–4), pp.E1–E7.

Le Guerroué, E., Allen, P.A. and Cozzi, A., 2006. Chemostratigraphic and sedimentological framework of the largest negative carbon isotopic excursion in Earth history: the Neoproterozoic Shuram Formation (Nafun Group, Oman). *Precambrian Research, 146*(1–2), pp.68–92.

Le Guerroué, E., Allen, P.A., Cozzi, A., Etienne, J.L. and Fanning, M., 2006. 50 Myr recovery from the largest negative δ13C excursion in the Ediacaran ocean. *Terra Nova, 18*(2), pp.147–153.

# General Index

*For the benefit of digital users, indexed terms that span two pages (e.g., 52–53) may, on occasion, appear on only one of those pages.*

Figures are indicated by an italic *f*.